ST. MARY'S SCHOOL
340 LEACOCK DRIVE
BARRIE, ON
L4N 6K1

ADDISON WESLEY

Math
Makes Sense

8

Author Team

Trevor Brown

Antonietta Lenjosek

Cathy Heideman

Jim Mennie

Georgia Konis-Chatzis

A. J. Keene

Margaret Sinclair

Craig Featherstone

Jason Johnston

Bryn Keyes

Elizabeth Wood

Elizabeth Milne

Don Jones

Sharon Jeroski

PEARSON
Addison
Wesley

Publisher
Claire Burnett

Publishing Team
Lesley Haynes
Enid Haley
Alison Rieger
Tricia Carmichael
Ioana Gagea
Lynne Gulliver
Stephanie Cox
Kaari Turk
Judy Wilson

Elementary Math Team Leader
Anne-Marie Scullion

Product Manager
Diane Wyman

Editorial Contributors
John Burnett
Janine Leblanc
Christina Yu
Claire Sauvé

Photo Research
Karen Hunter
Christina Beamish

Design
Word & Image Design Studio Inc.

Copyright © 2006 Pearson Education Canada Inc.

ISBN 0-321-23578-9

Printed and bound in Canada.

1 2 3 4 5 -- TCP -- 09 08 07 06 05

The information and activities presented in this book have been carefully edited and reviewed. However, the publisher shall not be liable for any damages resulting, in whole or in part, from the reader's use of this material.

Brand names that appear in photographs of products in this textbook are intended to provide students with a sense of the real-world applications of mathematics and are in no way intended to endorse specific products.

The publisher wishes to thank the staff and students of Hollycrest Middle School for their assistance with photography.

Statistics Canada information is used with the permission of the Minister of Industry, as Minister responsible for Statistics Canada. Information on the availability of the wide range of data from Statistics Canada can be obtained from Statistics Canada's Regional Offices, its World Wide Web site at http://www.statcan.ca, and its toll-free access number 1-800-263-1136.

Program Consultants and Advisers

Program Consultants

Craig Featherstone
Maggie Martin Connell
Trevor Brown

Assessment Consultant
Sharon Jeroski

Elementary Mathematics Adviser
John A. Van de Walle

Program Advisers

Pearson Education thanks its Program Advisers, who helped shape the vision for *Addison Wesley Mathematics Makes Sense* through discussions and reviews of prototype materials and manuscript.

Anthony Azzopardi
Sandra Ball
Victoria Barlow
Lorraine Baron
Bob Belcher
Judy Blake
Steve Cairns
Christina Chambers
Daryl M. J. Chichak
Lynda Colgan
Marg Craig
Elizabeth Fothergill
Jennifer Gardner

Florence Glanfield
Linden Gray
Pamela Hagen
Dennis Hamaguchi
Angie Harding
Andrea Helmer
Peggy Hill
Auriana Kowalchuk
Gordon Li
Werner Liedtke
Jodi Mackie
Lois Marchand
Becky Matthews

Cathy Molinski
Cynthia Pratt Nicolson
Bill Nimigon
Stephen Parks
Eileen Phillips
Carole Saundry
Evelyn Sawicki
Leyton Schnellert
Shannon Sharp
Michelle Skene
Lynn Strangway
Laura Weatherhead
Mignonne Wood

Program Reviewers

Field Testers

Pearson Education would like to thank the teachers and students who field-tested *Addison Wesley Math Makes Sense* prior to publication. Their feedback and constructive recommendations have helped us to develop a quality mathematics program.

Aboriginal Content Reviewers

Early Childhood and School Services Division
Department of Education, Culture, and Employment
Government of Northwest Territories:

Steven Daniel, Coordinator, Mathematics, Science, and Secondary Education
Liz Fowler, Coordinator, Culture Based Education
Margaret Erasmus, Coordinator, Aboriginal Languages

Grade 8 Reviewers

Lorraine Baron
School District #23 (Kelowna), BC

Judy Blaney
Simcoe County District School Board, ON

Melanie Boultbee
Toronto District School Board, ON

Cathy Chaput
Wellington Catholic District School Board, ON

Kyla Cleator
Edmonton Public School Board, AB

Kathryn Day
Toronto District School Board, ON

Sharyl L. De Mille
Toronto District School Board, ON

Thomas Falkenberg
North Vancouver School District, BC

Laurie Grandin
Halton District School Board, ON

Simon Houzer
Toronto District School Board, ON

Bruce Merz
School District #23 (Central Okanagan), BC

Susan Mitchell
Red Deer Public School Board, AB

Clarissa Salinas Moldawa
Toronto District School Board, ON

Mark Moorhouse
Lakehead District School Board, ON

Stephen Parks
Independent Consultant, NB

Mary Anna Pokerznik
Edmonton Public School Board, AB

Ioannis (John) Poulimenos
Brant Haldimand Norfolk Catholic District School Board, ON

Barbara Seaton
Thames Valley District School Board, ON

Ann Marie Slak
Dufferin-Peel Catholic District School Board, ON

Wendy Swonnell
Independent Consultant, BC

Rich Tamblyn
Upper Canada District School Board, ON

Karyne Todd
Toronto District School Board, ON

James Tremblay
Durham Catholic District School Board, ON

Karl Walters
York Region District School Board, ON

Sharon S. You
Peel District School Board, ON

Table of Contents

Geometry

UNIT 8

Square Roots and Pythagoras

Strand
- Geometry and Spatial Sense
- Measurement

UNIT 9

Integers

Strand
- Number Sense and Numeration
- Geometry and Spatial Sense

Welcome to
Addison Wesley Math Makes Sense 8

Math helps you understand your world.

This book will help you improve your problem-solving skills and show you how you can use your math now, and in your future career.

The opening pages of **each unit** are designed to help you prepare for success.

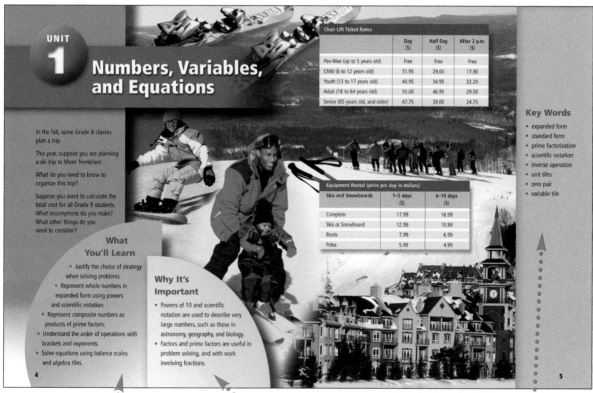

Find out **What You'll Learn** and **Why It's Important**. Check the list of **Key Words**.

Review some of the math concepts you've already met.

Study the **Example**.

Then try the **Check** questions to review your skills.

Skills You'll Need

Rounding Measurements

1 cm = 10 mm
and, 1.6 cm = 16 mm
When a measurement in centimetres has 1 decimal place,
the measurement is given to the nearest millimetre.

1 cm
1.6 cm

1 m = 100 cm
1.3 m = 130 cm
1.37 m = 137 cm ← When a measurement in metres has 2 decimal places, the measurement is given to the nearest centimetre.

1 m = 1000 mm
1.4 m = 1400 mm
1.43 m = 1430 mm
1.437 m = 1437 mm ← When a measurement in metres has 3 decimal places, the measurement is given to the nearest millimetre.

Example

Write each measurement to the nearest millimetre.
a) 25.2 mm   b) 3.58 cm

Solution

a) 25.2 mm
Since there are 2 tenths, round down.
25.2 mm is 25 mm to the nearest millimetre.

b) 3.58 cm
To round to the nearest millimetre, round to 1 decimal place.
Since there are 8 hundredths, round up.
3.58 cm is 3.6 cm to the nearest millimetre.

✓ Check

1. a) Round to the nearest metre.
   i) 4.38 m   ii) 57.298 m   iii) 158.5 cm

   b) Round to the nearest millimetre and nearest centimetre.
   i) 47.2 mm   ii) 47.235 cm   iii) 1.0579 m

6.1   **Investigating Circles**

Focus   Measure radius and diameter, and discover their relationship.

Which attribute do these objects share?

Explore

Work with a partner.
You will need circular objects, a compass, ruler, and scissors.
➤ Use a compass. Draw a large circle.
  Use a ruler.
  Draw a line segment that joins two points on the circle.
  Measure the line segment. Label the line segment with its length.
  Draw and measure other segments that join two points
  on the circle.
  Find the longest segment in the circle.
  Repeat the activity for other circles.

➤ Trace a circular object.
  Find a way to locate the centre of the circle.
  Measure the distance from the centre to the circle.
  Measure the distance across the circle, through its centre.
  Record the measurements in a table.
  Repeat the activity with other circular objects.
  What pattern do you see in your results?

Reflect & Share

Compare your results with those of another pair of classmates.
Where is the longest segment in any circle?
What relationship did you find between the distance across a circle
through its centre, and the distance from the centre to the circle?

In each Lesson:

Explore an idea or problem. You may use materials.

Reflect & Share your results with other students.

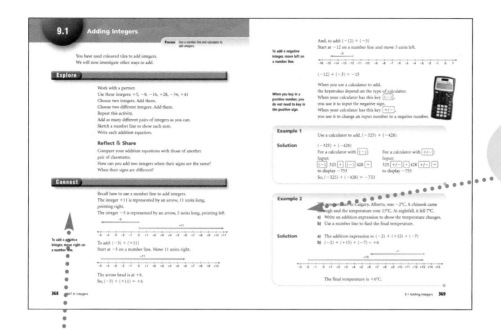

Examples show you how to use the ideas.

Connect summarizes the math.

Stay sharp with **Number Strategies**, **Mental Math**, and **Calculator Skills**.

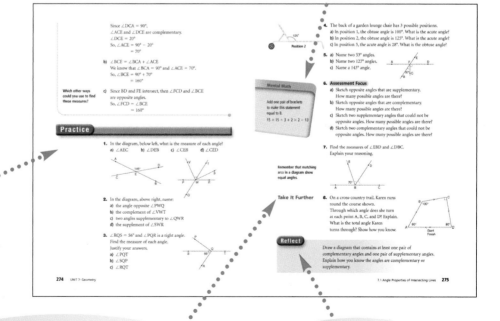

Practice questions reinforce the math.

Take It Further questions offer enrichment and extension.

Reflect on the big ideas of the lesson.

Use the **Mid-Unit Review** to refresh
your memory of key concepts.

Reading and Writing in Math helps you
understand how reading and writing about math
differs from other language skills you use. It may
present problems for you to solve.

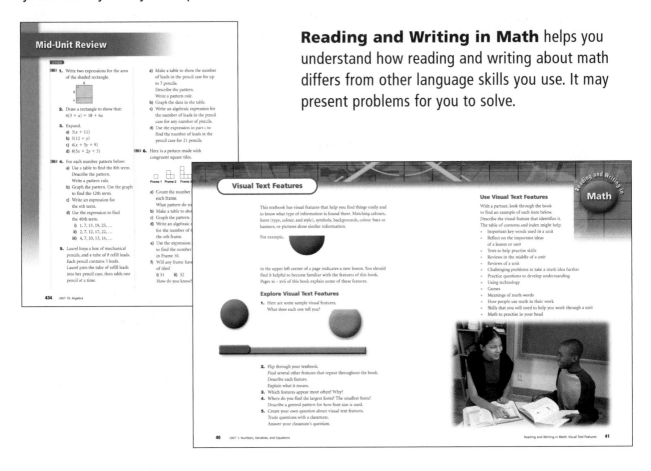

**What Do I Need to
Know?** summarizes
key ideas from the unit.

**What Should I Be
Able to Do?** allows
you to find out if you are
ready to move on. The
on-line tutorial **etext**
provides additional
support.

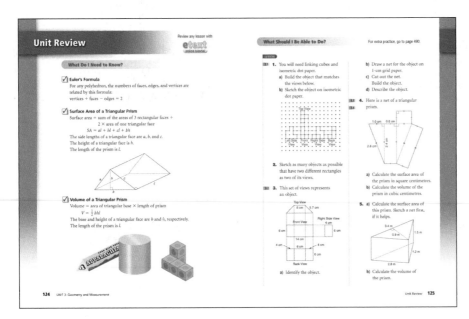

The **Practice Test** models the kind of test your teacher might give.

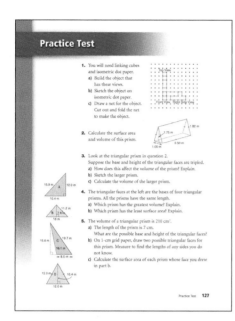

The **Unit Problem** presents problems to solve, or a project to do, using the math of the unit.

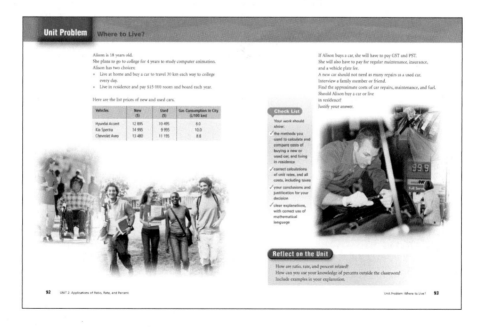

Keep your skills sharp with **Cumulative Review** and **Extra Practice.**

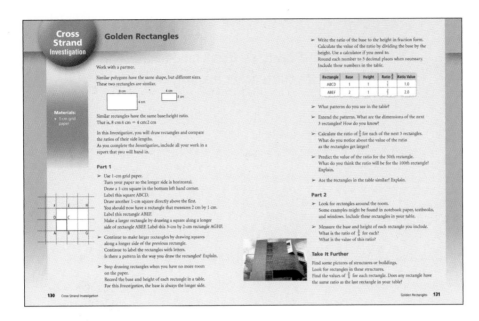

Explore some interesting math when you do the **Cross Strand Investigations**.

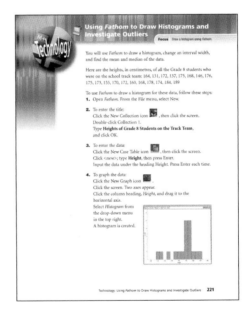

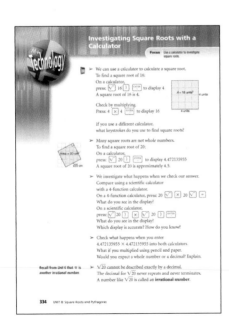

Icons remind you to use **technology**. Follow the step-by-step instructions for using a computer or calculator to do math.

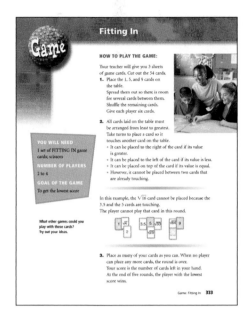

Play a **Game** with your classmates or at home to reinforce your skills.

The World of Work describes how people use mathematics in their careers.

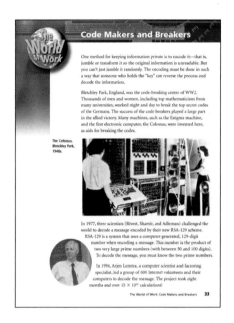

The **Illustrated Glossary** is a dictionary of important math words.

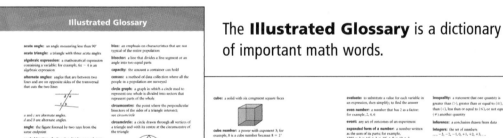

Cross Strand Investigation

Digital Roots

Work with a partner.

The **digital root** of a number is the result of adding the digits of the number until a single-digit number is reached.

For example, the digital root of 27 is: $2 + 7 = 9$
To find the digital root of 168:
Add the digits: $1 + 6 + 8 = 15$
Since 15 is not a single-digit number, add the digits: $1 + 5 = 6$
Since 6 is a single-digit number, the digital root of 168 is 6.

A digital root can also be found for the product
of a multiplication fact.
For the multiplication fact, 8×4:
$8 \times 4 = 32$
Add the digits in the product: $3 + 2 = 5$
Since 5 is a single-digit number, the digital root of 8×4 is 5.

You will explore the digital roots of the products in a multiplication table, then display the patterns you find.
As you complete the *Investigation*, include all your work in a report that you will hand in.

Materials:
- multiplication chart

Part 1

➤ Use a blank 12×12 multiplication chart.
Find each product.
Find the digital root of each product.
Record each digital root in the chart.
For example, for the product $4 \times 4 = 16$,
the digital root is $1 + 6 = 7$.

×	1	2	3	4	...
1					
2					
3					
4				7	
⋮					

➤ Describe the patterns in the completed chart.
Did you need to calculate the digital root of each product?
Explain.
Did you use patterns to help you complete the table? Explain.

➤ Look down each column. What does each column represent?

Part 2

➤ Use a compass to draw 12 circles.
Use a protractor to mark 9 equally spaced points on each circle.
Label these points in order, clockwise, from 1 to 9.
Use the first circle.
Look at the first two digital roots in the 1st column of your chart.
Find these numbers on the circle.
Use a ruler to join these numbers with a line segment.
Continue to draw line segments to join points that match the digital roots in the 1st column.
What figure have you drawn?

➤ Repeat this activity for each remaining column.
Label each circle with the number at the top of the column.

➤ Which circles have the same figure?
Which circle has a unique figure?
What is unique about the figure?
Why do some columns have the same pattern of digital roots?
Explain.

Take It Further

➤ Investigate if similar patterns occur in each case:
  • Digital roots of larger 2-digit numbers, such as 85 to 99
  • Digital roots of 3-digit numbers, such as 255 to 269
Write a report on what you find out.

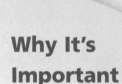

Numbers, Variables, and Equations

In the fall, some Grade 8 classes plan a trip.

This year, suppose you are planning a ski trip to Mont Tremblant.

What do you need to know to organize this trip?

Suppose you want to calculate the total cost for all Grade 8 students. What assumptions do you make? What other things do you need to consider?

What You'll Learn

• Justify the choice of strategy when solving problems.
• Represent whole numbers in expanded form using powers and scientific notation.
• Represent composite numbers as products of prime factors.
• Understand the order of operations with brackets and exponents.
• Solve equations using balance scales and algebra tiles.

Why It's Important

• Powers of 10 and scientific notation are used to describe very large numbers, such as those in astronomy, geography, and biology.
• Factors and prime factors are useful in problem solving, and with work involving fractions.

Chair Lift Ticket Rates	Day ($)	Half Day ($)	After 2 p.m. ($)
Pee-Wee (up to 5 years old)	Free	Free	Free
Child (6 to 12 years old)	31.95	29.00	17.90
Youth (13 to 17 years old)	40.95	34.95	22.20
Adult (18 to 64 years old)	55.00	46.95	29.00
Senior (65 years old, and older)	47.75	39.00	24.75

Equipment Rental (price per day in dollars)		
Skis and Snowboards	**1–5 days ($)**	**6–10 days ($)**
Complete	17.99	16.99
Skis or Snowboard	12.99	10.99
Boots	7.99	6.99
Poles	5.99	4.99

Key Words

- expanded form
- standard form
- prime factorization
- scientific notation
- inverse operation
- unit tiles
- zero pair
- variable tile

Skills You'll Need

Understanding Exponents

The exponent 3 indicates how many 5s are multiplied.

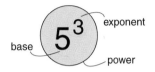

$5^3 = 5 \times 5 \times 5$ **expanded form**

$\quad = 125$ **standard form**

5^3 is a power of 5.

We say: 5 to the power of 3, or 5 cubed

Example 1

Write 81 as a power in more than one way.

Solution

Since $81 = 9 \times 9$,

then $81 = 9^2$

Since $9 = 3 \times 3$,

then $81 = 3 \times 3 \times 3 \times 3$, or 3^4

✓ Check

1. Write in exponent form.

a) $4 \times 4 \times 4$ b) $2 \times 2 \times 2 \times 2 \times 2 \times 2 \times 2$

c) 7×7 d) $12 \times 12 \times 12 \times 12 \times 12$

2. Write in expanded form, then in standard form.

a) 5^4 b) 11^2 c) 2^8 d) 12^3

3. Write as a power. Do this more than one way, if possible.

a) 27   b) 36   c) 64   d) 4

e) 125   f) 8   g) 343   h) 625

Understanding Powers of 10

This place-value chart shows some of the powers of 10 written in words and exponent form.

One hundred million	Ten million	One million	One hundred thousand	Ten thousand	One thousand	One hundred	Ten
10^8	10^7	10^6	10^5	10^4	10^3	10^2	10^1

Example 2

Express each number in standard form and as a power of 10.
a) one hundred thousand
b) one hundred
c) one thousand
d) one hundred million

Solution

Use the place-value chart.
a) One hundred thousand is 100 000, which is 10^5.
b) One hundred is 100, which is 10^2.
c) One thousand is 1000, which is 10^3.
d) One hundred million is 100 000 000, which is 10^8.

✓ Check

4. a) Express each number as a power of 10.
   i) 10 000
   ii) 10 000 000
   iii) $10 \times 10 \times 10$
   iv) $10 \times 10 \times 10 \times 10 \times 10 \times 10 \times 10 \times 10$
   b) What patterns do you see in part a?
   How can you use these patterns to write a number as a power of 10?

5. Write in standard form.
   a) 10^4
   b) 10^6
   c) 10^{10}
   d) 10^{12}

6. A student says that $10^6 - 10^4 = 10^2$.
   Is the student correct? Explain.

7. Add.
   a) $10^2 + 10^3$
   b) $10^3 + 10^4$
   c) $10^4 + 10^5$

Solving Equations

To solve an equation means to find the value of the variable that makes the equation true.
Two ways to solve an equation are:
- by systematic trial
- by inspection

Example 3

Solve by systematic trial. $3x + 4 = 28$

Solution

$3x + 4 = 28$
Choose a value for x and substitute.

Try: $x = 5$	$3(5) + 4 = 15 + 4 = 19$	too small
Try: $x = 10$	$3(10) + 4 = 30 + 4 = 34$	too large, but not by much
Try: $x = 8$	$3(8) + 4 = 24 + 4 = 28$	correct

The solution is $x = 8$.

Example 4

Solve by inspection. $100 - 7x = 58$

Solution

$100 - 7x = 58$
Think: 100 subtract 7 times a number is 58.
We know: $100 - 42 = 58$; so, 7 times a number is 42.
We know: $7 \times 6 = 42$; so, the number is 6.
The solution is $x = 6$.

✓ Check

8. Solve each equation.

a) $10x - 9 = 81$ b) $37 + 20x = 117$ c) $88 = 4 + 7x$

d) $9 = 19x - 124$ e) $x + 29 = 46$ f) $x - 85 = 17$

g) $\frac{x}{5} = 9$ h) $\frac{3x}{10} = 6$ i) $\frac{3x}{10} + 3 = 6$

Numbers in the Media

We use numbers to describe situations, to compare quantities, to make decisions, and to support points of view.

In the media, the numbers indicate that Canada is a good place to live.

How can we use the data in the table below to justify Canada's high ranking as a place to live?

Canada still ranked in the top 5 places to live — United Nations

Statistics indicate that most Canadians will finish high school

Life expectancy reaches a new high — 79.3 years

Literacy rate approaches 100%

Country	Estimated Annual Income ($)	Adult Literacy Rate (% of population)	Infant Mortality (per 1000 live births)	Number of Physicians (per 100 000 people)	Estimated Population (mid-2004)
Bangladesh	2 035	41.1	51	23	141 340 476
Brazil	10 879	86.4	30	206	184 101 109
Canada	36 299	99.9	5	187	31 964 434
China	5 435	90.9	31	164	1 298 847 624
Ethiopia	1 008	41.5	114	3	67 851 281
Mexico	12 967	90.5	24	156	104 959 594
New Zealand	26 481	99.9	6	219	3 993 817
Norway	42 340	99.9	4	367	4 574 560

Explore

Work with a partner.
Use the data in the table above.
Make up 4 problems for your partner to solve.
Try to write a problem that can be solved using each one of these methods:

- mental math
- pencil and paper
- estimation
- calculator

Reflect & Share

Compare problems and solutions with your partner.
Justify your choice of method for solving each problem.

When calculating with rounded numbers,
an estimated answer is appropriate.
Use mental math to solve a problem with numbers that are easy to
handle, and when the solving requires few steps.
Use pencil and paper, or a calculator, when the problem requires more
complex calculations.

Example

In North America, a new computer is purchased every two seconds.

Find how many new computers are purchased:
a) in two minutes   b) in one month

Solution

a) Use mental math.
   There are 120 s in 2 min.
   One computer is purchased every 2 s;
   so, the number purchased in 120 s is:
   $120 \div 2 = 60$
   60 computers are purchased in 2 min.

b) Use pencil and paper, or a calculator.
   Use 30 for the number of days in one month.
   There are 24 h in 1 day.
   Number of hours in one month: 30×24 h $= 720$ h
   Number of minutes in one month: 720×60 min $= 43\,200$ min
   Number of seconds in one month: $43\,200 \times 60$ s $= 2\,592\,000$ s
   Number of computers purchased in one month:
   $2\,592\,000 \div 2 = 1\,296\,000$
   In North America, about 1 300 000 computers
   are purchased in one month.
   We can write this as 1.3 million.

There are 60 min in 1 h.
There are 60 s in 1 min.

"Evaluate" means find the answer.

1. Use mental math to evaluate.
Describe the strategies you used.
   a) 88 + 56   **b)** 118 + 296   **c)** 2958 − 1998   **d)** 25 × 25 × 4

2. State whether you will use mental math,
paper and pencil, or a calculator; then evaluate.
   a) 29 + 41   **b)** 19 × 21   **c)** 7872 ÷ 1000   **d)** 7850 − 6975
   e) 777 + 333   **f)** 9876 ÷ 3   **g)** 25% of 500   **h)** 80% of 150

3. Explain whether you would need an exact answer or an estimate.
   a) You want to know the number of books read this year by
   Grade 8 students in Canada.
   b) You want to know the amount of money needed to pay for your
   Grade 8 graduation clothes.
   c) You want to know how much money you collected for a charity
   walk-a-thon.

4. **Assessment Focus** The table shows the U.S. and worldwide box
office earnings for the top five movies of all time, as of 2004.

Movie	Year	U.S. Box Office Revenue ($ millions)	World Box Office Revenue ($ millions)
Titanic	1997	600.8	1835.4
Lord of the Rings— Return of the King	2003	377.0	1129.2
Harry Potter and the Sorcerer's Stone	2001	317.6	975.8
Star Wars Episode 1— The Phantom Menace	1999	431.1	925.5
Lord of the Rings— The Two Towers	2002	341.7	924.7

   a) What were the total earnings worldwide of the top five movies?
   b) How much more money did *Titanic* earn in the world than each
   of the other movies?
   c) Name two movies that together earned about as much as *Titanic*.
   d) Which movies more than tripled their U.S. box office revenues?
   e) Is revenue a fair way to rank the top five movies? Explain.
   f) Pose a problem that could be solved using these data.
   Solve the problem you posed.

Mount Everest
8800 m above
sea level

sea level

Marianas Trench
10 900 m below
sea level

5. The tallest mountain in the world is Mount Everest.
Its summit is about 8800 m above sea level.
The lowest point in the ocean is the Marianas Trench.
Its deepest point is about 10 900 m below sea level.

    a) What is the distance between the highest point on land and
the lowest point in the ocean?

    b) About how many ladders would you need to bridge the
distance in part a?
State your assumptions and show your work.

6. Here are some statistics about water use.

- 100 L for 1 bath
- 600 L to produce 1 kg of wheat
- 2000 L to produce 1 kg of rice
- 40 000 L to produce 1 kg of beef

Answer each question.
Explain which calculation method you used.

    a) How much water would be needed for 25 baths?

    b) How much water would be needed to produce
3.75 kg of wheat?

    c) Suppose you bought 2.45 kg of beef.
How much water was used to produce it?

    d) Twenty-nine thousand litres of water would produce how
much rice? How much beef? How much wheat?

7. In the year 2003, Statistics Canada reported that workers in the
mining industry were paid a higher wage than those in other
similar jobs.

Sector	Weekly Income ($)
Mining	1176
Forestry	840
Manufacturing	832
Construction	817

Make up a problem using the data above.
Trade your problem with a classmate.
Solve your classmate's problem.

8. According to one source, the average Canadian uses between 300 L and 350 L of water per day.
   a) Does this sound reasonable? Explain.
   b) What would 300 L of water look like? What would 300 L fill?

9. Canadians were asked to choose their most popular participation sport. Here are the top five results in 1998 for people aged 15 and older.

Sport	Females		Males	
	Number	%	Number	%
Golf	476 000	3.9	1 325 000	11.1
Ice Hockey	65 000	0.5	1 435 000	12.0
Baseball	386 000	3.1	953 000	8.0
Swimming	688 000	5.6	432 000	3.6
Basketball	237 000	1.9	550 000	4.6

The percents in the table have been rounded to the nearest tenth.

In 1998, there were about 12 323 000 females aged 15 and older, and about 11 937 000 males aged 15 and older.
   a) For which sport is the number of females about 3 times the number for basketball?
   b) For which sport is the number of males about $\frac{2}{3}$ the number for ice hockey?
   c) In 2004, Canada's population aged 15 and older was about 25 849 000. How many people do you think chose basketball? What assumptions did you make?
   d) Write a problem that could be solved using these data. Solve your problem.

Take It Further

10. Describe how you could add the numbers from 1 to 50 quickly, without using a calculator.

Mental Math

Use all these numbers with addition and subtraction.

5, 7, 14, 21, 28, 30, 35, 40

Make two different expressions equal to 100.

Reflect

Give an example of a problem that requires an exact answer.
Give an example of a problem that requires estimation.
Solve each problem.
Show your solutions.

Focus | Represent composite numbers as products of prime factors.

In earlier grades, you broke numbers apart by finding their factors. We will now further investigate these factors.

Explore

Work on your own.
Use a copy of this factor square.

4	×	15	×	2	×	9	×	40	×
×	18	×	2	×	2	×	5	×	3
2	×	3	×	24	×	36	×	20	×
×	16	×	10	×	30	×	10	×	12
2	×	60	×	6	×	4	×	2	×
×	11	×	50	×	28	×	5	×	4
2	×	15	×	20	×	3	×	24	×
×	8	×	2	×	5	×	2	×	6
2	×	10	×	3	×	7	×	18	×
×	3	×	5	×	60	×	24	×	5

Draw a loop around factors of 120.
Look for 2 factors, 3 factors, and so on, up to as many factors as possible.
One set of factors is looped for you: 2 × 4 × 15
List all the factors you find.
Have you found all possible factors of 120?
Organize your results to justify your answer to this question.

Reflect & Share

Compare your results with those of a classmate.
Identify the factorization of 120 that has the most numbers.
What is special about these factors?
Use powers to write the product of these factors a simpler way.

A composite number has more than two factors. A prime number has exactly two factors: itself and 1.

All numbers, except 1, can be written as the product of two different factors.

The number 36 can be written as: $1 \times 36, 2 \times 18, 3 \times 12, 4 \times 9, 6 \times 6$

36 can also be written as the product of three factors:

for example, $2 \times 2 \times 9, 2 \times 3 \times 6$

And, as the product of four factors: $2 \times 2 \times 3 \times 3$

Each of these four factors is a prime number.

The prime factors of 36 are 2 and 3.

So, we say that $2 \times 2 \times 3 \times 3$ is a product of prime factors, or a **prime factorization** of 36.

Any composite number can be written as a product of prime factors. For example, to write 144 as a product of prime factors, begin with 12×12.

Continue to factor 12×12 until all the factors are prime numbers.

$$12 \times 12 = (4 \times 3) \times (4 \times 3)$$
$$= (2 \times 2 \times 3) \times (2 \times 2 \times 3)$$
$$= 2 \times 2 \times 2 \times 2 \times 3 \times 3$$

The prime factors of 144 are 2 and 3.

Use powers to write the prime factorization a simpler way:

$$144 = 2^4 \times 3^2$$

Another method to find the prime factors is to divide by prime numbers.

Example 1

Find the prime factorization of 200.

Solution

Divide 200 by its least prime factor, which is 2.

2 $\mid$ 200	Keep dividing by 2 until the quotient is an odd number.
2 $\mid$ 100	
2 $\mid$ 50	
5 $\mid$ 25	Neither 2 nor 3 is a factor of 25.
5 $\mid$ 5	So, divide by the next prime factor, 5.
1	Continue to divide by prime factors until the quotient is 1.

All the divisors are the prime factors of 200.

The prime factors of 200 are 2 and 5.

$$200 = 2 \times 2 \times 2 \times 5 \times 5$$
$$= 2^3 \times 5^2$$

We can use prime factors to find common factors and common multiples.

Example 2

a) Find all the common factors of 18 and 24.

b) Find the first 3 common multiples of 18 and 24.

Solution

a) Write each number as a product of its prime factors. Identify common factors.

$$18 = 2 \times 3 \times 3 \qquad 24 = 2 \times 2 \times 2 \times 3$$

common factors

The greatest common factor of 18 and 24 is 6.

The common factors are 2, 3; and $2 \times 3 = 6$.

Note that 2 and 3 are prime common factors, and 6 is a composite common factor.

b) Use the prime factors from part a.

The first common multiple is the product of all the prime factors, but the common factors are only included once.

common factors

The first common multiple is called the lowest common multiple.

$$18 = 2 \times 3 \times 3 \qquad\qquad 24 = 2 \times 2 \times 2 \times 3$$

remaining factors

The first common multiple is the product of the common factors and the remaining factors.

That is: $2 \times 3 \times 3 \times 2 \times 2 = 6 \times 12 = 72$

common factors   remaining factors

The second common multiple is: $72 \times 2 = 144$

The third common multiple is: $72 \times 3 = 216$

Since 1 is a factor of all numbers, when we list factors or common factors we do not need to include 1.

1. Write each product as a number in standard form.
   a) $2^2 \times 3^2$
   b) $7^2 \times 2^3$
   c) $5^2 \times 3^3$
   d) $2^2 \times 3^2 \times 5$
   e) $2^7 \times 3$
   f) 7×3^4
   g) $3^2 \times 7^2$
   h) $10^2 \times 7$

2. List the prime factors of each number.
   a) 21
   b) 14
   c) 100
   d) 125
   e) 19
   f) 50
   g) 77
   h) 96

3. Write each number as a product of prime factors. Use exponents where possible.
   a) 48
   b) 63
   c) 400
   d) 16
   e) 120
   f) 55
   g) 36
   h) 88

4. Use the prime factors from questions 2 and 3. Find all the common factors of each pair of numbers.
   a) 55, 88
   b) 48, 120
   c) 96, 63
   d) 125, 400

5. Use the prime factors from questions 2 and 3. Find the first 3 common multiples of each pair of numbers.
   a) 16, 21
   b) 36, 96
   c) 77, 88
   d) 36, 63

6. A number has 2, 3, and 5 as factors.
   a) Which is the least possible number?
   b) Find two more numbers with these factors.

7. According to a student, the least number that has 2, 3, 4, and 5 as factors can be found by multiplying: $2 \times 3 \times 4 \times 5$
   Is the student correct? Explain.

8. Can you find the greatest number with factors 11, 23, and 37? Explain.

9. a) Find the least number with the factors 14, 27, and 38.
   b) Write the prime factorization of this number.

10. a) Write a four-digit number that is divisible by 5 and 7. Explain how you did it.
    b) Write the prime factorization of this number.

11. A number has 21 and 77 as factors.
   a) Which is the least possible number?
   b) Which other numbers are factors of the number in part a? Explain your reasoning.

12. **Assessment Focus** The prime factorization of a number is $2^2 \times 5^2 \times 7$.
   a) What is the number?
   b) Use the prime factors to list all the factors of that number. Show your work.

13. "The total number of prime factors for a perfect square is always an even number."
   a) Do you agree or disagree with this statement? Justify your answer with at least three different perfect squares.
   b) Use prime factors to show that 3025 is a perfect square.

14. Is $4^2 \times 4^2$ a prime factorization of 256? Explain.

Take It Further

15. Can you find a number less than 150 that is divisible by 4 different prime numbers? Justify your answer.

16. On graduation day, 100 Grade 8 students lined up outside the school.
As they entered the school, they passed their lockers.
The first student opened all the locker doors.
The second student closed every second locker door.
The third student changed the position of every third locker door. If the door was open, the student closed it.
If the door was closed, the student opened it.
The fourth student changed the position of every fourth locker door. This pattern continues.
Which doors are open after 100 students have entered the school? Explain how you know.

Reflect

What makes the factorization of a number a *prime* factorization?
Use examples in your explanation.

Focus Represent whole numbers in expanded form and scientific notation.

In mid-July, 2004, the population of Canada was estimated to be 31 964 434.
We can write this number in different ways.

In expanded form:

$31\ 964\ 434 = 30\ 000\ 000 + 1\ 000\ 000 + 900\ 000 + 60\ 000 + 4000 + 400 + 30 + 4$

Using powers of 10 in this expanded form:

First, write each number as a product of a whole number and a power of 10:

$$31\ 964\ 434 = (3 \times 10\ 000\ 000) + (1 \times 1\ 000\ 000) + (9 \times 100\ 000) +$$
$$(6 \times 10\ 000) + (4 \times 1000) + (4 \times 100) + (3 \times 10) + 4$$

Then, write each power of 10 in exponent form:

$$31\ 964\ 434 = 3 \times 10^7 + 1 \times 10^6 + 9 \times 10^5 + 6 \times 10^4 + 4 \times 10^3 + 4 \times 10^2 + 3 \times 10^1 + 4$$

Explore

Work with a partner.
Use a calculator if you need to.
Here are some headlines relating to the Internet in 2004.

Experts predict e-commerce will top $300 000 000 000

1 000 000 000 people will be connected to the Internet by 2005

In 2000, Canadian adolescents surfed an average of 197 hours

Connected schools in Canada reach 15 035

For each headline:
* Write each number in expanded form using powers of 10.
* Write each number as a product of two factors,
  where one factor is a power of 10.

Reflect & Share

Compare your results with those of another pair of classmates.
Talk about the different ways you wrote factors for each product.
How many ways could you do this for each number?
Was each expression correct? How could you check?

In *Lesson 1.1*, you learned that the U.S. box office revenue for the movie *Titanic* was $600 800 000.
We can write large numbers like this in scientific notation.

A number expressed in **scientific notation** is written as a product of two factors:
- One factor is a number greater than or equal to 1, and less than 10.
- The other factor is a power of 10.

For example,

$$
\begin{aligned}
41 &= 4.1 \times 10^1 \\
410 &= 4.10 \times 10^2 \\
4105 &= 4.105 \times 10^3 \\
41\ 057 &= 4.1057 \times 10^4 \\
410\ 578 &= 4.105\ 78 \times 10^5
\end{aligned}
$$

These numbers are written in scientific notation.

The number of digits in the standard form of the number **is** 1 more than the exponent in scientific notation.

The exponent in the power of 10 indicates the number of places to the position of the decimal point when the number is written in standard form.

$4.105\ 78 \times 10^5 = 410\ 578.$

5 places

Example

1 barrel of oil is 159 L.

Norman Wells, Northwest Territories, is the fourth largest producing oil field in Canada. Its estimated reserves are 140 million barrels of oil. Write this number in scientific notation.

Solution

Write 140 million as 140 000 000.
For the factor greater than 1 and less than 10, mark a decimal point after the first digit:
1.40 000 000, or 1.4 ◄——— Delete the zeros that indicate place value at the right of the number.

There are 9 digits in 140 000 000.
The exponent in the power of 10 is 1 less than 9, which is 8.
$140\ 000\ 000 = 1.4 \times 10^8$

The *Example* illustrates that, when a large number is written in scientific notation, it is easier to read.

Calculators display scientific notation in different ways.

Key in: 1230000 $\boxed{\times}$ 1230000 $\boxed{\genfrac{}{}{0pt}{}{\text{ENTER}}{=}}$ to see how your calculator

displays 1.5129×10^{12}.

Practice

1. Write each number in expanded form using powers of 10. How could you check your answer?
   a) 834 000
   b) 98 977 183
   c) 7 000 010
   d) 23 232

2. Which is greater? Explain.
   a) $4 \times 10^3 + 6 \times 10^2 + 6 \times 10^1 + 7$ or   4327
   b) $2 \times 10^4 + 4 \times 10^3 + 2 \times 10^2 + 4 \times 10^1$ or   2432
   c) $7 \times 10^7 + 7 \times 10^3$ or   777 777

3. For each power of 10, write the exponent that makes each statement true. How do you know?
   a) $7000 = 7 \times 10^?$
   b) $400\ 000 = 4 \times 10^?$
   c) $2\ 890\ 000 = 2.89 \times 10^?$
   d) $20\ 000 = 2 \times 10^?$
   e) $704 = 7.04 \times 10^?$
   f) $71 = 7.1 \times 10^?$

4. Write each number in scientific notation. Check with a calculator.
   a) 1 532 000
   b) 31 000
   c) 4 600 000 000
   d) 150
   e) 6 000 100
   f) 147 032

5. Order the numbers in each set from least to greatest.
   a) 1.6×10^3, 1616, 6.1×10^2, 616
   b) 2.453×10^6, 248 555, 2.4531×10^6, 2 453 101

6. Look at a number in scientific notation. Why is the exponent in the power of 10 one less than the number of digits in the number? Use a place-value chart or another way to show why this is true.

7. The diameter of Earth
is 12 756 km.
The diameter of Uranus is
approximately four times as
great as Earth's diameter.
Write the approximate diameter of
Uranus in scientific notation.
Justify your answer.

8. Here are some facts about the central nervous system.
Write the numbers that would complete the table.

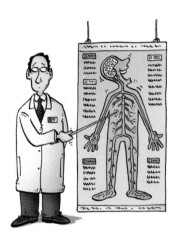

		Standard Form	Scientific Notation
a)	Number of neurons in the central nervous system		1×10^{11}
b)	Bits of information the brain is able to record in one day	8 600 000	
c)	Approximate number of nerve cells in the body		3×10^{10}
d)	Average number of nerve endings per square centimetre in a human hand	208	

9. **Assessment Focus**

a) Is 12.756×10^3 in scientific notation?
If your answer is yes, explain how you know.
If your answer is no, how would you write the number
in scientific notation?

b) Why do we use scientific notation?

The mean of five numbers is the sum of the numbers divided by 5.

10. The mean of five numbers is 7.5×10^4.
Four of the numbers are: 50 000, 100 000, 75 000, 80 000
What is the 5th number? How do you know?

11. Different colours of light have different frequencies.
The frequency of red light is 4.3×10^{14} Hz.
The frequency of violet light is 7.5×10^{14} Hz.

The frequency of 1 Hz (one Hertz) means 1 wave per second.

a) Which frequency is greater and by how much?

b) Are numbers written in scientific notation easier to compare
than numbers in standard form? Explain.

12. The table lists Canadian provinces and territories.
It shows the year each joined Confederation, and the
approximate 2004 population.

	Date of Entry	Population
New Brunswick	1867	751 400
Nova Scotia	1867	937 000
Ontario	1867	12 392 700
Quebec	1867	7 542 800
Manitoba	1870	1 170 300
Northwest Territories	1870	42 800
British Columbia	1871	4 196 400
Prince Edward Island	1873	137 900
Yukon	1898	31 200
Alberta	1905	3 201 900
Saskatchewan	1905	995 400
Newfoundland and Labrador	1949	517 000
Nunavut	1999	29 600

Premier Paul Okalik in Nunavut in 1999

a) Write each population in scientific notation.
b) Write the sum of the populations of the 4 original provinces
in scientific notation.
c) Which 3 provinces or territories together have a population
approximately equal to that of Nova Scotia?
d) For parts b and c, is it easier to add the numbers in standard
form or in scientific notation? Explain.

Take It Further

13. Approximately how many times has your heart beaten in your
lifetime? Give your answer in as many different forms as
possible. Explain how you solved this problem.

14. Use only the operations of multiplication or division.
Write 0.004 32 in terms of a number between 1 and 10,
and a power of 10.

Reflect

How can you tell if a number is written in scientific
notation? Give examples.

Mid-Unit Review

LESSON

1.1 **1.** Here is a list of the top 10 all-time point scorers in NHL history.

Player	Goals	Assists	Total Points
Gretzky	894	1963	2857
Messier	694	1193	1887
Howe	801	1049	1850
Francis	549	1249	1798
Dionne	731	1040	1771
Yzerman	678	1043	1721
Lemieux	683	1018	1701
Esposito	717	873	1590
Bourque	410	1169	1579
Coffey	396	1135	1531

a) Estimate the total number of goals scored by the top ten players. Explain how you estimated.

b) Which players have about twice as many assists as goals? Explain.

c) Write your own problem about the data. Solve your problem and describe how you solved it.

1.2 **2.** Write each number as a product of prime factors. Use exponents where possible.

a) 444 b) 162

c) 102 d) 1225

3. Find the first 2 common multiples of each pair of numbers.

a) 15, 27 b) 16, 28

c) 18, 32 d) 20, 36

4. Find the common factors of each pair of numbers.

a) 100, 120 b) 56, 80

c) 72, 27 d) 48, 92

5. a) Why is 2^3 a prime factorization of a number?

b) Why is 4^3 not a prime factorization of a number?

6. Write each number in standard form. Then, order the numbers from least to greatest.

$2^4 \times 5^2$, $5^2 \times 11$,

$3^2 \times 7 \times 11$,

$2^5 \times 7 \times 13^2$

7. a) Is it possible that 3 consecutive whole numbers can be prime factors? Explain.

b) Is it possible that 2 consecutive whole numbers can be prime factors? Explain.

1.3 **8.** Write each number in expanded form using powers of 10.

a) 806 087 137 b) 20 020 220

9. Write each number in scientific notation.

a) 5 600 000 b) 773 291

c) 9 200 000 000 d) 62

10. Which numbers are in scientific notation? Explain how you know.

a) 66.8×10^5 b) 4.163×10^4

c) 73×10^7 d) 2×10^8

Explore

Work on your own.

Two students evaluated this expression: $(3 + 5.4) + 5.2 \times 10^2$

Jinni did this:

$(3 + 5.4) + 5.2 \times 10^2$
$= 8.4 + 5.2 \times 10^2$
$= 13.6 \times 10^2$
$= 13.6 \times 100$
$= 1360$

Todd did this:

$(3 + 5.4) + 5.2 \times 10^2$
$= 8.4 + 5.2 \times 10^2$
$= 8.4 + 52^2$
$= 8.4 + 2704$
$= 2712.4$

Which student is correct?
Or, are the evaluations of both students incorrect?
If so, how was each evaluation incorrect?
How would you evaluate the expression?
Show your work.

Reflect & Share

Compare your answer to this problem with that of another student.
Who is correct: Jinni, Todd, your classmate, or you?
Explain how you know.

Connect

To avoid getting different answers when we evaluate an expression, mathematicians have agreed on the order in which operations should be performed:
• Do the operations in brackets.
• Do the exponents.
• Multiply and divide, in order, from left to right.
• Add and subtract, in order, from left to right.
So, when this order is followed, everyone should get the same answer.

Example 1

Evaluate.

$15.5^2 - 2.4 \times (3.1 + 4.7)^2$

Solution

Use a calculator.
Do the operation in brackets first.

$$15.5^2 - 2.4 \times (3.1 + 4.7)^2$$
$$= 15.5^2 - 2.4 \times (7.8)^2 \qquad \text{Deal with the exponents.}$$
$$= 240.25 - 2.4 \times 60.84 \qquad \text{Then multiply.}$$
$$= 240.25 - 146.016 \qquad \text{Then subtract.}$$
$$= 94.234$$

If you have a scientific calculator, you can input the expression in *Example 1* directly.

To evaluate $15.5^2 - 2.4 \times (3.1 + 4.7)^2$, key in:

15.5 $\boxed{x^2}$ $\boxed{-}$ 2.4 $\boxed{\times}$ $\boxed{(}$ 3.1 $\boxed{+}$ 4.7 $\boxed{)}$ $\boxed{x^2}$ $\boxed{\substack{\text{ENTER} \\ =}}$ to display 94.234

We also use the order of operations when we evaluate an algebraic expression by substituting for the variable.

Example 2

The power, in watts (W), supplied to a circuit by a 9-V battery is expressed as $9c - 0.5c^2$, where c is the current in amperes (A). Find the power supplied when the current is 8 A.

Solution

Substitute $c = 8$.

$$9c - 0.5c^2 = 9(8) - 0.5(8)^2$$

Deal with the exponent first.

$$9(8) - 0.5(8)^2 = 9 \times 8 - 0.5 \times 64 \quad \text{Then multiply.}$$
$$= 72 - 32$$
$$= 40$$

When the current is 8 A, the power is 40 W.

1. Evaluate.

a) $7 \times 12 - 48$ **b)** $15 + 3 + 12 - 6$ **c)** $(5 + 6) \times 11$

d) $(34 + 46) - 5 \times 11$ **e)** $89 - (76 + 13)$ **f)** $144 \div (36 \times 2)$

2. Evaluate.

a) 3.2×10^4 **b)** $66.15 \div 10.5^2$

c) $18.3 - (7.2 - 3.5)^2$ **d)** $(22.3 + 1.1)^2 - (22.3 - 1.1)^2$

e) $10.8 + 6.3^2 - 1.2 \times 2.1$ **f)** $20.8 \div 1.3 \times (14.8 + 17.2)$

Calculator Skills

Copy the decimals below. Replace each comma with $\times$ or $\div$ to make an expression equal to 1.

0.4, 0.5, 0.2, 0.25, 0.4, 0.1

3. Cody bought 3 DVDs at $24.99 each and 2 compact discs at $14.99 each.

Write an expression to show how much he spent before taxes.

4. Forty metres of fencing are available to enclose a rectangular pen.

When the length of the pen is l metres, the area of the pen, in square metres, is expressed as $20l - l^2$.

What is the area of the pen for each value of l?

a) 4 m **b)** 10 m **c)** 13 m

5. When a 3-m springboard diver leaves the diving board, her height above the water depends on the time since she left the board. When the time is t seconds, the diver's height above the water, in metres, is expressed as $3 + 8.8t - 4.9t^2$.

Find the height of the diver after each time.

a) 0.5 s **b)** 1 s **c)** 1.5 s

6. Copy each statement.

Insert brackets to make each statement true.

a) $10 + 2 \times 3^2 - 2 = 106$ **b)** $10 + 2 \times 3^2 - 2 = 24$

c) $10 + 2 \times 3^2 - 2 = 84$ **d)** $10 + 2 \times 3^2 - 2 = 254$

7. Copy each statement.

Insert brackets to make each statement true.

a) $20 \div 2 + 2 \times 2^2 + 6 = 26$ **b)** $20 \div 2 + 2 \times 2^2 + 6 = 30$

c) $20 \div 2 + 2 \times 2^2 + 6 = 8$ **d)** $20 \div 2 + 2 \times 2^2 + 6 = 120$

8. Evaluate each expression.
Then write the expressions in order from greatest to least.
   a) $7^2, 2^7, 4^5, 5^4$
   b) $3^2, 2^3, (3 - 2)^2, (3 + 2)^2$
   c) $(7.5 + 1)^2, (10.5 - 1)^2, 61.5 + 2^2, 103.5 - 1^2$
   d) $(2.2 + 8)^2, (8 - 2.2)^2, 8 + 2^2, 8 - 2.2^2$

9. Skylar wants to join the local gym.
The cost in dollars for a membership can be expressed as:
$100 + 39.99m$
where 100 is the initiation fee in dollars,
39.99 is the monthly fee in dollars, and
m is the number of months for which a person signs up.
How much will it cost Skylar to join the gym for 24 months?

10. **Assessment Focus** Use the numbers 2, 4, 6, 8, and any
operations or brackets to make an expression
that equals each number. Show your work.
   a) 24   b) 40   c) 60   d) 80

11. Use four 4s and any operations or brackets to make each
whole number from 1 to 10.

12. A student said: "The sum of the squares of two numbers is equal
to the square of the sum of the numbers."
Do you agree with this statement? Justify your answer.

Take It Further

13. Write your birth date in this form: year/month/date
Use the digits in this number in the order they are written
to make an equation.
For example, if your birth date is April 15, 1992, write it as
92/04/15, then use the numbers 9 2 0 4 1 5:
$9 \times (2 \times 0) = (4 + 1) - 5$

14. Use each digit from 1 to 9 once, and any operations, to write an
expression with answer 144.

Reflect

Why is it necessary to follow the order of operations?
Include examples in your explanation.

Using a Model to Solve Equations

Focus Solve equations by using a model and by inverse operations.

Sometimes, systematic trial and inspection are not suitable ways to solve an equation. We shall develop other methods to solve an equation.

The balance point of the scales is called the fulcrum.

When scales are balanced, the mass in one pan is equal to the mass in the other pan.

We can write an equation to describe the masses in grams.

$20 = 10 + 5 + 5$

Explore

Work with a partner.
Use a two-pan balance if it is available.
Otherwise, draw diagrams.

Here are some balance scales.
Some masses are known. Other masses are unknown.

Balance A   **Balance B**

➤ The scales are balanced.
  For each two-pan balance:
  • Write an equation to represent the masses.
  • Find the value of the unknown mass.

➤ Make up your own two-pan balance problem.
  Make sure the scales are balanced and one mass is unknown.
  Solve your problem.

Reflect & Share

Trade problems with another pair of classmates.
Compare strategies for finding the value of the unknown mass.

➤ Here is another two-pan balance problem.
Mass A is an unknown mass.

We can remove the same mass from each pan, or add the same
mass to each pan; and the scales will still be balanced.
We need to find out which mass balances Mass A.
If we remove 7 g from the left pan,
then Mass A is alone in that pan.
To remove 7 g from the right pan,
we replace 10 g with 3 g and 7 g; then remove 7 g.

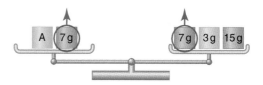

We are left with Mass A in the left pan balancing 18 g in the right
pan. So, Mass A is 18 g.

We can write this problem as an equation.
Then, we solve the equation algebraically.
Let x grams represent the unknown mass.
Then: $x + 7 \quad = \quad 10 + 15$

Left pan   balances   right pan

$$x + 7 \quad = \quad 25$$

Subtract 7 from each side of the equation.

$$x + 7 - 7 = 25 - 7$$
$$x = 18$$

In the solution of the equation $x + 7 = 10 + 15$,
we used the **inverse operation**.
That is, to isolate x, we *subtracted* 7 from +7 to get $+ 7 - 7 = 0$
on the left side.

We can verify the solution to this problem in two ways.

- Replace Mass A with 18 g.
  Then, in the left pan: 18 g + 7 g = 25 g
  And, in the right pan: 10 g + 15 g = 25 g
  Since the masses are equal, the solution is correct.

- Substitute $x = 18$ in the equation $x + 7 = 10 + 15$.
  Left side: $x + 7 = 18 + 7$ Right side $= 10 + 15$
          $= 25$ $= 25$

Since the left side of the equation equals the right side, the solution is correct.

Example

Solve each equation. Verify the solution.

a) $10 + x = 5 + 8$ **b)** $x - 6 = 10 - 3$

Solution

a) $10 + x = 5 + 8$

Add the numbers on the right side.

$10 + x = 13$

Subtract 10 from each side to isolate x.

$10 + x - 10 = 13 - 10$

$x = 3$

To verify $x = 3$ is correct, substitute $x = 3$ in $10 + x = 5 + 8$.

Left side $= 10 + x$ Right side $= 5 + 8$

$= 10 + 3$ $= 13$

$= 13$

Since the left side equals the right side, $x = 3$ is correct.

b) $x - 6 = 10 - 3$

Subtract the numbers on the right side.

$x - 6 = 7$

Add 6 to each side to isolate x.

$x - 6 + 6 = 7 + 6$

$x = 13$

To verify $x = 13$ is correct, substitute $x = 13$ in $x - 6 = 10 - 3$.

Left side $= x - 6$ Right side $= 10 - 3$

$= 13 - 6$ $= 7$

$= 7$

Since the left side equals the right side, $x = 13$ is correct.

1. Find the value of the unknown mass on each two-pan balance.

a)

| 20 g | A | | | 50 g |

b)

| 75 g | | | B | 20 g |

c)

| 15 g | C | | | 30 g | 50 g |

d)

| 20 g | 20 g | 25 g | | | D | 15 g |

2. Solve each equation.

a) $x + 3 = 5$ b) $x + 5 = 10$ c) $x + 10 = 17$
d) $x - 3 = 5$ e) $x - 5 = 10$ f) $x - 10 = 17$

3. Solve each equation. Verify the solution.

a) $3 + x = 5 + 9$ b) $x - 3 = 11 - 8$
c) $4 + 7 = x - 8$ d) $21 - 13 = 7 + x$

Number Strategies

Find a square number that is the sum of 2 two-digit square numbers.

How many numbers can you find?

4. Five more than a number is 24.
Let x represent the number.
Then an equation is $5 + x = 24$.
Solve the equation. What is the number?

5. **Assessment Focus** The masses for a two-pan balance are multiples of 5 g.
a) Sketch a two-pan balance to represent this equation:
$x + 35 = 60$
How many different balances can you sketch?
b) Solve the equation. Verify the solution.
Show your work.

Take It Further **6.** a) Sketch a two-pan balance to represent the equation:
$5x + 10 = 105$
b) Solve the equation. Verify the solution.

Reflect

Write an equation, then solve it using a two-pan balance.
Write an equation you cannot solve using a two-pan balance.
Solve the equation algebraically.

Code Makers and Breakers

One method for keeping information private is to encode it—that is, jumble or transform it so the original information is unreadable. But you can't just jumble it randomly. The encoding must be done in such a way that someone who holds the "key" can reverse the process and decode the information.

Bletchley Park, England, was the code-breaking centre of WW2. Thousands of men and women, including top mathematicians from many universities, worked night and day to break the top secret codes of the Germans. The success of the code breakers played a large part in the allied victory. Many machines, such as the Enigma machine, and the first electronic computer, the Colossus, were invented here, as aids for breaking the codes.

The Colossus, Bletchley Park, 1940s.

In 1977, three scientists (Rivest, Shamir, and Adleman) challenged the world to decode a message encoded by their new RSA-129 scheme. RSA-129 is a system that uses a computer-generated, 129-digit number when encoding a message. This number is the product of two very large prime numbers (with between 50 and 100 digits). To decode the message, you must know the two prime numbers.

In 1994, Arjen Lenstra, a computer scientist and factoring specialist, led a group of 600 Internet volunteers and their computers to decode the message. The project took eight months and over 15×10^{16} calculations!

Focus | Use algebra tiles and inverse operations to solve equations.

Recall the integer tiles you have used.

This tile represents $+1$, or 1.   This tile represents -1.

These two tiles are called **unit tiles**.

One red unit tile and one yellow unit tile combine to model 0.

$\left.\begin{array}{l} \blacksquare \ {-1} \\ \square \ {+1} \end{array}\right\}$ These two unit tiles form **a zero pair**.

We also use tiles to represent variables.
This tile represents x.

We call it an x-tile, or a **variable tile**.
Unit tiles and variable tiles are collectively called algebra tiles.

Explore

Work with a partner.
You will need algebra tiles.

➤ Use algebra tiles to represent this equation: $x + 5 = 9$
   Use the tiles to solve the equation.
   Sketch the tiles you used.
➤ Repeat the process for the equation $x - 5 = 9$.
➤ Write your own equation to solve using algebra tiles.
   Solve your equation.

Reflect & Share

Share your strategies for solving an equation using algebra tiles with another pair of classmates.
Write the algebraic solutions for your equations.
How did you use zero pairs in your solutions?

We extend the idea of balanced scales to model an equation with algebra tiles.

We draw a vertical line in the centre of the page: it represents the balance point of the scales and the equal sign in the equation.

We arrange tiles on each side of the line to represent an equation. Whatever we do to one side of the equation, we also do to the other side.
To solve the equation $x - 3 = 10$:

On the left side, put algebra tiles to represent $x - 3$.

On the right side, put algebra tiles to represent 10.

To isolate the x-tile, add 3 yellow unit tiles to make zero pairs. Remove zero pairs.

Add 3 yellow unit tiles to this side, too.

The tiles above show the solution $x = 13$.

Here is the algebraic solution.

$$x - 3 = 10$$

Use the inverse operation.

Remember that: $-3 + 3 = 0$

$$x - 3 + 3 = 10 + 3$$

Add 3 to each side to isolate x.

$$x = 13$$

We can use algebra tiles to solve an equation when the x-term is greater than $1x$.

Example 1

a) Use algebra tiles to solve this equation: $2x + 1 = 9$

b) Solve the equation in part a algebraically.

c) Verify the solution.

Solution

a) $2x + 1 = 9$

Add 1 red unit tile to each side.
Remove zero pairs.

Arrange the tiles on each side into 2 equal groups.

Compare groups.
One x-tile equals 4 unit tiles.
So, $x = 4$

b) $2x + 1 = 9$ Subtract 1 from each side.

$2x + 1 - 1 = 9 - 1$

$2x = 8$

$\dfrac{2x}{2} = \dfrac{8}{2}$ Divide each side by 2.

$x = 4$

c) To verify the solution, substitute $x = 4$ into $2x + 1 = 9$.

Left side $= 2x + 1$ Right side $= 9$
$$= 2(4) + 1$$
$$= 8 + 1$$
$$= 9$$

Since the left side equals the right side, $x = 4$ is correct.

Example 2

a) Use algebra tiles to solve this equation: $2 = 3x - 4$
b) Solve the equation in part a algebraically.
c) Verify the solution.

Solution

a) $2 = 3x - 4$

Add 4 yellow unit tiles to each side.
Remove zero pairs.

Arrange the tiles on each side into equal groups.

2 yellow unit tiles equal one x-tile.
So, $x = 2$

b)

$$2 = 3x - 4 \qquad \text{Add 4 to each side.}$$
$$2 + 4 = 3x - 4 + 4$$
$$6 = 3x$$
$$\frac{6}{3} = \frac{3x}{3} \qquad \text{Divide each side by 3.}$$
$$2 = x$$

c) To verify the solution, substitute $x = 2$ into $2 = 3x - 4$.

Left side $= 2$ Right side $= 3(2) - 4$
$$= 6 - 4$$
$$= 2$$

Since the left side equals the right side, $x = 2$ is correct.

Practice

1. Use algebra tiles to solve each equation.
   a) $x + 4 = 8$ **b)** $3 + x = 10$ **c)** $12 = x + 2$
   d) $x - 4 = 8$ **e)** $10 = x - 3$ **f)** $12 = x - 2$

2. Solve each equation in question 1 algebraically.

3. Five more than a number is 11.
   Let x represent the number.
   Then, an equation is $5 + x = 11$.
   Solve the equation.
   What is the number?

4. Four less than a number is 13.
   Let x represent the number.
   Then, an equation is $x - 4 = 13$.
   Solve the equation.
   What is the number?

5. **a)** Use algebra tiles to solve each equation.
   b) Solve each equation algebraically.
   c) Verify each solution.
   - **i)** $2x + 7 = 13$
   - **ii)** $11 = 3x - 1$
   - **iii)** $4x + 13 = 17$
   - **iv)** $9 = 5x - 6$

6. Five times a number is 30.
Let x represent the number.
Then, an equation is $5x = 30$.
Solve the equation. What is the number?

7. The perimeter of a regular octagon is 104 cm.
Let x centimetres represent the length of one side.
Then, an equation to represent the perimeter is $8x = 104$.
Solve the equation.
What is the side length of the octagon?

8. Seven more than three times a number is 28.
Let x represent the number.
Then, an equation is $7 + 3x = 28$.
Solve the equation. What is the number?

9. Assessment Focus Look at the problems
in questions 3, 4, 6, 7, and 8.
Make up a problem similar to these.
Write an equation for your problem.
Solve the equation using algebra tiles, and algebraically.
Verify the solution. Show your work.

Take It Further

10. Solve each equation algebraically.
a) $3x + 12 = 6$ **b)** $5 - 2x = 11$ **c)** $-2 = 3x + 10$

11. Write an equation for each number.
Solve the equation to find each number.
Verify the solution.
a) Five more than two times a number is 1.
What is the number?
b) Five less than two times a number is -1.
What is the number?

Number Strategies

Find the least number that obeys these clues:

The number is divisible by 5 and by 11.

When the number is divided by 9, the remainder is 2.

Reflect

Write an equation you can solve with algebra tiles
but not with a two-pan balance.
Explain why the equation cannot be solved
with a two-pan balance.
Solve the equation. Verify the solution.

Visual Text Features

This textbook has visual features that help you find things easily and to know what type of information is found there. Matching colours, fonts (type, colour, and style), symbols, backgrounds, colour bars or banners, or pictures show similar information.

For example,

in the upper left corner of a page indicates a new lesson. You should find it helpful to become familiar with the features of this book. Pages xi – xvii of this book explain some of these features.

Explore Visual Text Features

1. Here are some sample visual features.
   What does each one tell you?

2. Flip through your textbook.
   Find several other features that repeat throughout the book.
   Describe each feature.
   Explain what it means.
3. Which features appear most often? Why?
4. Where do you find the largest fonts? The smallest fonts?
   Describe a general pattern for how font size is used.
5. Create your own question about visual text features.
   Trade questions with a classmate.
   Answer your classmate's question.

Use Visual Text Features

With a partner, look through the book
to find an example of each item below.
Describe the visual feature that identifies it.
The table of contents and index might help.

- Important key words used in a unit
- Reflect on the important ideas
  of a lesson or unit
- Tests to help practise skills
- Reviews in the middle of a unit
- Reviews of a unit
- Challenging problems to take a math idea further
- Practice questions to develop understanding
- Using technology
- Games
- Meanings of math words
- How people use math in their work
- Skills that you will need to help you work through a unit
- Math to practise in your head

Unit Review

Review any lesson with

online tutorial

What Do I Need to Know?

✓ **Prime Factorization**
Use repeated division to find prime factors.
Write the prime factors as powers when possible.
For example, $630 = 2 \times 3^2 \times 5 \times 7$

✓ **Expanded Form with Powers of 10**
Use place value to write a number as the sum of its parts.
For example, $4356 = 4 \times 10^3 + 3 \times 10^2 + 5 \times 10 + 6$

✓ **Scientific Notation**
It is used to write a large number as the product of a number greater than or equal to 1 and less than 10, and a power of 10.
For example, $7\ 500\ 000 = 7.5 \times 10^6$

✓ **Order of Operations**
To evaluate an expression with different operations, follow this order:
- Do the operations in brackets.
- Do the exponents.
- Multiply and divide, in order, from left to right.
- Add and subtract, in order, from left to right.

What Should I Be Able to Do?

For extra practice, go to page 488.

LESSON

1.1

1. Here are some data about voluntary organizations in 2003:
   - Canadians took out 139 000 000 memberships in these organizations.
   - Nineteen million Canadians contributed more than 2 000 000 000 h of voluntary time.
   - These organizations had $112 000 000 000 in revenues.

   Write a problem about these data. Solve your problem. Justify the strategy you used.

1.2 **2.** Write each number as a product of prime factors.

   a) 64   **b)** 42   **c)** 60   **d)** 30

3. Write each product in standard form.

   a) $2^4 \times 3$ **b)** $5^2 \times 2^2$

4. A number has 11 and 23 as factors.

   a) Which is the least number this could be?

   b) Which is the greatest number with these factors that can be displayed on your calculator?

5. Which expression shows the prime factorization of 600? Explain.

   a) $1 \times 2^3 \times 3 \times 5^2$

   b) $2^2 \times 3 \times 5^2$

   c) $2^3 \times 5^2 \times 3$

   d) $2 \times 2 \times 2 \times 75$

6. For each pair of numbers below:

   i) Find all the common factors.

   ii) Find the first 3 common multiples.

   a) 15, 35   **b)** 20, 100
   c) 25, 75   **d)** 30, 36

7. **a)** Find two pairs of numbers for which the lowest common multiple is the product of the numbers.

   b) Find two pairs of numbers for which the lowest common multiple is less than the product of the numbers.

c) How can you tell if the lowest common multiple of two numbers is less than or equal to the product of the numbers?

1.3 **8.** Here are the areas of the ten largest lakes in the world, to the nearest 100 km^2.

Lake	Area (km^2)
Huron, North America	5.96×10^4
Aral Sea, Asia	3.07×10^4
Caspian Sea, Asia-Europe	371 000
Great Bear, North America	31 300
Superior, North America	82 100
Baikal, Asia	3.15×10^4
Victoria, Africa	69 500
Malawi, Africa	28 900
Michigan, North America	57 800
Tanganyika, Africa	32 900

   a) Order the lakes from greatest area to least area. Explain how you did this.

   b) Which two lakes together have an area approximately equal to that of Lake Superior? How do you know?

9. Write each number in expanded form using powers of 10.

   a) 9 337 000   **b)** 977 183
   c) 106 040 055   **d)** 73 532

10. Write each number in scientific notation.

   a) 1 500 000   **b)** 42 000
   c) 600 000 000   **d)** 27

11. Write each number in standard form.
a) 6×10^3 b) 8.43×10^6
c) 7.2×10^5 d) 3.28×10^8

1.4 **12.** Evaluate.
a) $83 - 6 \times 11$
b) $15 + (3 + 12) \times 6$
c) $(20 - 9)^2 - 3 \times 2$
d) $1.3 + 4.1^2 - 15$

13. A rectangular lot has a river along one side.

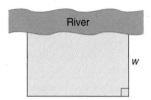

The fencing for the other 3 sides has a total length of 30 m. When the width of the lot is w metres, the area of the lot, in square metres, is expressed as $30w - 2w^2$. What is the area of the lot for each value of w?
a) 5 m b) 9 m c) 12 m

1.5 **14.** Solve each equation.
a) $x + 2 = 7$ b) $x - 3 = 5$
c) $13 = 4 + x$ d) $7 = x - 2$

15. Jan collects foreign stamps. Her friend gives her 8 stamps. Jan then has 21 stamps. How many stamps did Jan have to start with?
Let x represent the number of stamps.

Then, an equation is $8 + x = 21$.
Solve the equation.
Answer the question.

16. Solve each equation. Verify the solution.
a) $3 + 11 = 5 + x$
b) $x - 3 = 11 - 8$
c) $16 - 9 = x + 4$
d) $x - 7 = 8 - 5$

1.6 **17.** Solve each equation. Verify the solution.
a) $6 + 3x = 17 - 2$
b) $9 + 12 = 2x - 1$
c) $5x - 3 = 9 - 2$
d) $14 - 3 = 4x + 7$

18. One paperback book costs $7. How many books can be bought for $133?
Let x represent the number of books.
Then, an equation is $7x = 133$.
Solve the equation.
Answer the question.

19. Jaya has 26 hockey cards. Jaya has 1 fewer than 3 times the number her brother, Kumar, has. How many cards does Kumar have?
Let x represent the number of cards Kumar has.
Then, an equation is $3x - 1 = 26$.
Solve the equation.
Answer the question.

Practice Test

1. Canada is rich in natural resources.
   We have abundant water, yet we must conserve this resource.
   The average Canadian toilet uses 20 L of water per flush.
   a) About how many litres of water are flushed in one household
      per day? Justify your answer.
   b) Suppose you installed a low-consumption toilet that used only
      6 L of water per flush.
      How much water could you save in one day?

2. How is expanded form with powers of 10 like scientific notation?
   How are they different? Give examples to support your answer.

3. a) Write a 4-digit number that is divisible by 11 and 17.
   b) Write the prime factorization of this number.

4. A cell phone company has a monthly service charge of $25 plus
   10¢ for each minute or part of a minute of air time.
   a) What is the cost if a person uses 100 min of air time in
      one month?
   b) In each expression below, m represents the number of minutes
      of air time.
      Which expression represents the monthly cost in dollars?
      Explain how you know.
      i) $10m + 0.25$ ii) $(25 + 0.10)m$
      iii) $25 + 0.10m$ iv) $0.10 + 25m$
   c) Find the cost if a person uses 175 min of air time in 3 months.

5. Solve each equation. Verify the solution.
   a) $5 + x = 20 - 3$
   b) $11 = 2x - 9$
   c) $6 + 3x = 40 - 1$

6. Three less than 5 times a number is 62.
   What is the number?
   Show your work.

Part 1

1. The top ten North American ski resorts are listed, with their elevations.

Whistler/Blackcomb, B.C.: 2.284×10^3 m

Big Sky, Montana: 3.399×10^3 m

Jackson Hole, Wyoming: 3185 m

Kicking Horse, British Columbia: 2450 m

Steamboat, Colorado: 3221 m

Aspen Highlands, Colorado: 3559 m

Telluride, Colorado: 3.737×10^3 m

Heavenly, California: 3068 m

Vail, Colorado: 3527 m

Sun Valley, Idaho: 2789 m

Order the resorts from greatest elevation to least elevation.

Suppose you are in charge of planning a four-day Grade 8 ski trip.

2. You decide to ski in Mont Tremblant, Quebec.
Ninety-six students will go on the trip.
They are organized in groups based on the following rules:
- Groups must have an even number of students.
- All groups must have the same number of students.
- Groups cannot have fewer than 4 or more than 10 students.

a) What is the minimum number of students that could be in each group?

b) What is the maximum number of students that could be in each group?

3. When you are skiing, you must be aware of extreme temperatures. As you ascend the mountain, the temperature drops. An expression for calculating the temperature at a given elevation is: $c - e \div 150$; where
c is the temperature in degrees Celsius at sea level, and
e is the elevation in metres.

a) Suppose the temperature at sea level is 10°C.
What is the temperature at an elevation of 1050 m?

b) Mont Tremblant has an elevation of 900 m.
Suppose the temperature at sea level is 0°C.
What is the temperature at the peak?

4. Two local bus companies, Company A and Company B,
offer packages for school trips.
The rate for each company is given by these expressions:
Company A: $300 + $55n
Company B: $100 + $75n
where n is the number of people on a bus.
Suppose 96 students and 8 adults go on the trip.
Which company should you choose? Justify your answer.

Check List

Your work should
show:

✓ all calculations in
detail

✓ your understanding
of algebraic
expressions

✓ the strategies you
used to solve the
problems

✓ a clear explanation
of the thinking
behind your
solutions

Part 2

Plan the ski trip for your class.
Decide how many adults will go with you.
There needs to be 1 adult for every 12 students.
The hotel accommodation in Mont Tremblant is
$66 per person per night.
Use the information on these pages and on page 5.
Show all your work. Include costs for transportation,
accommodation, and recreation.
State any assumptions
you made.

Reflect on the Unit

Write about the different ways to represent a number,
and how variables are used in expressions and equations.

Applications of Ratio, Rate, and Percent

The First Nations University of Canada is in Saskatchewan.

- In 2003, the ratio of full-time students to part-time students was about 5:1. There were about 120 part-time students. About how many full-time students were there?

- The fees for a Canadian student are 50% of the fees for a foreign student. A foreign student paid about $7780 in 2003. How much did a Canadian student pay that year?

What You'll Learn

- Solve proportions.
- Use scales to calculate distance.
- Investigate similar figures.
- Solve problems involving rates, percents, and ratios.
- Solve problems involving discount, sales tax, commission, and simple interest.

Why It's Important

- You use ratios to change recipes to make a larger or smaller pot of soup.
- You use rates to estimate how long it will take to walk a certain distance.
- You use percents to calculate sales tax.

Key Words

- equivalent ratios
- unit rate
- proportion
- similar figures
- discount
- sales tax
- commission
- simple interest
- principal
- amount

Skills You'll Need

What Is a Ratio?

Recall that a ratio is a comparison of two quantities measured in the same unit.

Here are some data for the 2002–2003 hockey season.

Northeast Division	Games Played	Wins	Losses	Ties	Overtime Losses
Ottawa	82	52	21	8	1
Toronto	82	44	28	7	3
Boston	82	36	31	11	4
Montreal	82	30	35	8	9
Buffalo	82	27	37	10	8

For the Toronto Maple Leafs:
The ratio of wins to games played was 44:82. This is a part-to-whole ratio.
The ratio of wins to losses was 44:28. This is a part-to-part ratio.

Example 1

Write each ratio in simplest form.

a) 44:82   **b)** 44:28

Solution

To write a ratio in simplest form, divide the terms by their greatest common factor.

a)   44:82   Divide each term by 2.
$= (44 \div 2):(82 \div 2)$
$=$ 22:41

b)   44:28   Divide each term by 4.
$= (44 \div 4):(28 \div 4)$
$=$ 11:7

In *Example 1*, the ratios 44:82 and 22:41 are **equivalent**,
and the ratios 44:28 and 11:7 are equivalent.

1. Use the table on page 50. For each team below, find:
    i) the ratio of wins to losses
    ii) the ratio of games played to ties
    Write each ratio in simplest form.
    a) Ottawa   **b)** Boston   **c)** Montreal   **d)** Buffalo

2. Identify any pairs of equivalent ratios in question 1.

What Is a Rate?

A rate is a comparison of two quantities measured in different units.

In 5 min, Ryan typed 425 words.
So, in $\frac{5}{5}$, or 1 min, Ryan typed: $\frac{425 \text{ words}}{5} = 85$ words
Ryan's typing rate is 85 words/min.
This is a **unit rate** because it represents the words typed in *one* minute.

Example 2

Meghan drove 300 km in 4 h.
a) What was her average speed?
b) Draw a graph to illustrate Meghan's journey.
    Explain how the average speed is represented in the graph.

Solution

a) In 4 h, Meghan drove 300 km.
    So, in 1 h, Meghan drove: $\frac{300 \text{ km}}{4} = 75$ km
    Meghan's average speed was 75 km/h.

b) On a grid, label the horizontal axis *Time*
    and the vertical axis *Distance*.
    Mark a point to represent 300 km in 4 h.
    Draw a line through this point and
    the origin.
    From the graph, the distance
    travelled after 1 h is 75 km.
    So, the average speed is 75 km/h.

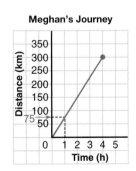

Meghan's Journey

3. Write each pair of quantities as a unit rate.
   a) 225 heartbeats in 3 min   **b)** $18.70 for 2 tickets
   c) $6.75 for 3 golf balls   **d)** $78.00 for 8 h work

4. Jose travelled 480 km in 6 h.
   a) Draw a graph to illustrate Jose's journey. Find his average speed.
   b) At this average speed, how long will Jose take to travel 700 km?

Relating Fractions, Decimals, and Percents

This hundredths chart represents one whole, or 1.
The shaded part of this chart can be described three ways.

As a fraction: $\frac{24}{100}$
As a percent: 24%
As a decimal: 0.24

Each way above can be written
in any of the other two ways.

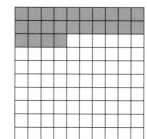

Recall that 1 whole is 100%.

Example 3

   a) Write 5% as a decimal.   **b)** Write 1.65 as a percent.   **c)** Write $\frac{3}{8}$ as a decimal.

Solution

   a) $5\% = \frac{5}{100}$
   $= 0.05$

   b) $1.65 = 1\frac{65}{100}$
   $= \frac{165}{100}$
   $= 165\%$

   c) $\frac{3}{8} = 3 \div 8$
   $= 0.375$

✓ **Check**

5. a) Write each fraction as a decimal and as a percent.
   i) $\frac{3}{10}$ **ii)** $\frac{4}{5}$ **iii)** $\frac{21}{20}$ **iv)** $\frac{3}{100}$
   b) Write each percent as a decimal and as a fraction.
   i) 25%   **ii)** 34%   **iii)** 250%   **iv)** 2%
   c) Write each decimal as a percent and as a fraction.
   i) 0.15   **ii)** 0.07   **iii)** 0.4   **iv)** 1.15

Using Proportions to Solve Ratio Problems

Focus Set up a proportion to solve a ratio problem.

We use ratios when we change a recipe.

Explore

Work on your own.
Here is a recipe for apple pie.

500 mL flour

200 mL margarine

500 g sliced apples

125 g sugar

Each pie serves 6 people.
Frank has only 350 g of sliced apples.
How much of each other ingredient does
Frank need to make the pie?
How many people will Frank's pie serve? Explain.

Reflect & Share

Compare your answers with those of a classmate.
What strategies did you use to solve the problems?

Connect

In a recycle drive last week at Cue Elementary School, Mr. Bozyk's
Grade 8 class collected bottles and recycled some of them.
The ratio of bottles recycled to bottles collected was 3:4.
This week, his class collected 24 bottles.
Mr. Bozyk told the students that the ratio of bottles recycled to
bottles collected is the same as the ratio the preceding week.
How can the students find how many bottles are recycled this week?

Last week's ratio of bottles recycled to bottles collected is 3:4.
This week's ratio of bottles recycled to bottles collected is ?:24.
These two ratios are equal.
Let r represent the number of bottles recycled.
Then, $r:24 = 3:4$
A statement that two ratios are equal is a **proportion**.

We can write each ratio in the proportion in fraction form:
$\frac{r}{24} = \frac{3}{4}$

To find the value of r, solve the proportion.

To isolate r, multiply each side of the proportion by 24.

$$24 \times \frac{r}{24} = \frac{3}{4} \times 24$$
$$\frac{24r}{24} = \frac{72}{4}$$
$$r = 18$$

$\frac{72}{4}$ means $72 \div 4$.

18 bottles were recycled.

Example

These suitcases have the same length to width ratio.

Calculate the width of Suitcase B.

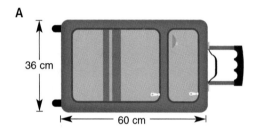

A

36 cm

60 cm

B

50 cm

Solution

Write the ratio with the variable as the first term of the ratio. This makes it easier to solve the proportion.

Let the width of Suitcase B be w centimetres.

For Suitcase B, width to length is w:50.

For Suitcase A, width to length is 36:60.

Write this ratio in simplest form.

Divide each term by 12.

$$36{:}60 = (36 \div 12){:}(60 \div 12)$$
$$= 3{:}5$$

The ratios for Suitcase A and Suitcase B are equal.

Write each ratio in fraction form.

Then write a proportion.

$$\frac{w}{50} = \frac{3}{5}$$

Multiply each side by 50.

$$50 \times \frac{w}{50} = \frac{3}{5} \times 50$$
$$\frac{50w}{50} = \frac{150}{5}$$
$$w = 30$$

Suitcase B is 30 cm wide.

Practice

1. Find each missing term.
   a) $\frac{t}{18} = \frac{6}{3}$ b) $\frac{v}{60} = \frac{3}{10}$ c) $\frac{x}{15} = \frac{2}{3}$
   d) $a{:}7 = 30{:}70$ e) $b{:}45 = 5{:}15$ f) $l{:}8 = 15{:}6$

2. In the NHL, the ratio of shots taken to goals scored by an all-star player is 9:2. The player has a 50-goal season.
   How many shots did he take?

3. Fatima plants 5 tree seedlings for every 3 that Shamar plants. Shamar plants 6 trees in 1 min.
   a) How many trees does Fatima plant in 1 min?
   b) Did you write a proportion to solve this problem?
      If so, how else could you have solved it?

4. An ad said that 4 out of 5 dentists recommend a certain chewing gum for their patients. Suppose 185 dentists were interviewed. Find the number of dentists who recommend this gum.

5. A bike is in fourth gear.
   When the pedals turn 3 times, the rear wheel turns 7 times.
   When the pedals turn twice, how many times does the rear wheel turn?

6. At the annual Grade 8 ski trip, for every 2 students who skied, 3 snowboarded. Ninety-six students snowboarded.
   How many students skied?

7. **Assessment Focus** In rectangle MNPQ, the ratio of the length of MN to the length of MP is 4:5.
   a) Does this ratio tell you how long MN is? Explain.
   b) What if MN is 12 cm long. How long is MP?
      Use a diagram to illustrate your answer.
   Show your work.

Find 3 prime numbers that have a sum of 91 and a product of 14 573.

8. Last week, Marcia played goalkeeper for her hockey team. She stopped 20 out of 30 shots on goal. This week, Marcia faced 36 shots. She stopped shots on goal in the same ratio as the previous week. How many shots did Marcia stop?

9. At the movie theatre, 65 student tickets were sold for one performance.
   a) The ratio of adult tickets sold to student tickets sold was 3:5. How many adult tickets were sold?
   b) The ratio of adult tickets sold to child tickets sold was 3:2. How many child tickets were sold?
   c) One adult ticket costs $13. One student ticket costs $7.50. One child ticket costs $4.50. How much money was made for that performance?

10. Forty-five students take piano lessons. The ratio of the numbers of students who take piano lessons to violin lessons is 15:8. The ratio of the numbers of students who take violin lessons to clarinet lessons is 8:9.
   a) How many students take violin lessons?
   b) How many students take clarinet lessons?

Take It Further

11. Suppose you want to find the height of a flagpole. You know your height. You can measure the length of the shadow of the flagpole. Your friend can measure your shadow. How can you use ratios to find the height of the flagpole? Explain. Include a sketch in your explanation.

Reflect

Can you use a proportion to solve any ratio problem? If your answer is yes, which ratio problem *would* you not use a proportion to solve? If your answer is no, which ratio problem *could* you not use a proportion to solve?

Explore

Work in a group.

You will need 1-cm grid paper and a metre stick, measuring tape, or trundle wheel.

Measure the perimeter of the classroom.

Make a scale drawing of the classroom floor.

What scale did you use?

Use your drawing to find the length of a diagonal of the classroom floor.

Reflect & Share

Compare your drawings with those of another group of classmates.

Justify your choice of scale to your classmates.

Compare your results for the length of a diagonal.

If the results are different, try to find out why.

Connect

A desktop is a rectangle with dimensions 1.06 m by 51 cm.

To make a scale drawing, we need to fit the rectangle on a piece of paper.

The paper measures 28 cm by 21.5 cm.

We need a margin around the drawing; so, we will use no more than 24 cm by 18 cm.

To choose a scale, find the ratios of the corresponding dimensions.

Multiply by 100 to convert metres to centimetres.

length of paper:length of desk
= 24 cm:1.06 m
= 24 cm:106 cm
= 24:106
$= \frac{24}{24} : \frac{106}{24}$
= 1:4.416

The desk is more than 4 times as long as the paper.

width of paper:width of desk
= 18 cm:51 cm
= 18:51
$= \frac{18}{18} : \frac{51}{18}$
= 1:2.83

The desk is almost 3 times as wide as the paper.

So, we will choose a scale of 1:5, or 1 cm represents 5 cm.

The desktop has dimensions 106 cm by 51 cm.

So, the scale drawing has dimensions:

$\frac{106 \text{ cm}}{5}$ by $\frac{51 \text{ cm}}{5}$, or 21.2 cm by 10.2 cm

To illustrate the scale drawing in this textbook,

we will use a smaller scale of 1:10, or 1 cm represents 10 cm.

Then, the scale drawing has dimensions:

$\frac{106 \text{ cm}}{10}$ by $\frac{51 \text{ cm}}{10}$, or 10.6 cm by 5.1 cm.

The scale drawing is labelled with the dimension of the desktop.

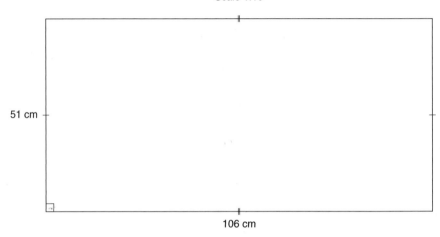

Scale 1:10

51 cm

106 cm

The straight-line distance is sometimes referred to "as the crow flies."

In a similar way:

We can use the scale on a map to calculate the distance between two towns on the map.

On a map of Ontario, the straight-line distance between Windsor and Toronto is 3.6 cm.

The scale on the map is written as the ratio of the map distance to the actual distance, 1:10 000 000.

This means that 1 cm on the map represents 10 000 000 cm actual distance.

So, the actual distance between Windsor and Toronto, as the crow flies, is:

3.6 cm × 10 000 000

There are 100 cm in 1 m.

= 36 000 000 cm   Divide by 100.

There are 1000 m in 1 km.

= 360 000 m   Divide by 1000.

= 360 km

The straight-line distance from Windsor to Toronto is 360 km.

We can write, then solve, a proportion to calculate a distance on a scale drawing.

Example

Here is a scale drawing
of a leaf-cutting ant.
Calculate the actual
length of the ant.

Scale 3:1

Solution

The scale is 3:1.
That is, length on drawing
to actual length is 3:1.
Or, actual length to length on drawing is 1:3.
Let the actual length be *l* centimetres.
The length on the drawing is measured as 4.9 cm.
So, actual length to length on drawing is *l*:4.9.
The proportion is

$$l:4.9 \;=\; 1:3$$

Write these ratios in fraction form.

$$\frac{l}{4.9} \;=\; \frac{1}{3}$$

Multiply each side by 4.9.

$$4.9 \times \frac{l}{4.9} \;=\; \frac{1}{3} \times 4.9$$
$$\frac{4.9l}{4.9} \;=\; \frac{4.9}{3}$$
$$l \;\doteq\; 1.63$$

The actual length of the ant is about 1.6 cm.

In the *Example*, the first term of the scale is greater than the
second term.
This indicates that the scale drawing is an enlargement.

Practice

1. Measure, then use the scale to find each actual measurement.

   a) Scale 1:90   **b)** Scale 65:1

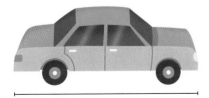

2. A square field has edge length 275 m.
Choose a scale. Make a scale drawing of the field.
Justify the scale you used.

3. Measure a fingernail.
Choose a scale. Make a scale drawing of your fingernail.
Justify the scale you used.

4. The Eiffel Tower in Paris, France, is 321 m tall,
including the television antenna.
Measure the height of the tower in this picture.
What is the scale for this picture?

Number Strategies

How many different ways
can 24 bottles of water
be arranged in a
rectangular array?

What if the number of
bottles is doubled. Which
rectangular arrays are
possible now?

5. a) The scale on a map of British Columbia is 1:5 000 000.
The map distance between Kelowna and Salmon Arm
is 2.1 cm.
What is the actual distance between these towns?
b) The scale on a map of Japan is 1:500 000.
The actual distance between Tokyo and Yokohama is 30 km.
What is the map distance between these cities?

6. A blueprint for a new house has a scale of 1:50.
Here are the dimensions of two rooms.
Calculate the dimensions of each room on the blueprint.
a) 4.8 m by 6.4 m   **b)** 3.1 m by 4.2 m

7. The Horseshoe Falls at Niagara are 53 m high. A person takes a
photo of the Falls. What if the photograph just fits on this page.
What is the scale of the photo?

8. A ladybug is 4 mm long.
A scale drawing of the bug is 5.6 cm long.
What is the scale of the drawing?

9. Assessment Focus Make a scale drawing of the floor of a
room in your home. Include any furniture on the floor.
Justify your choice of scale. Show your work.

Reflect

Describe two types of scale drawing.
Explain how the scale describes the actual measurements of
an object and its measurements on the drawing.

Exploring Similar Figures

Focus Investigate relationships among side lengths, angles, perimeter, and area.

Software, such as *The Geometer's Sketchpad*, uses transformations called *dilatations* to construct similar figures.

1. Open *The Geometer's Sketchpad*.
    From the **File** menu, choose **New Sketch**.

2. From the **Graph** menu, choose **Show Grid**.
    The screen has grid lines and two numbered axes.
    Click on each axis and the two red dots.
    The axes and the dots are highlighted.
    From the **Display** menu, choose **Hide Objects**.
    The axes and the dots disappear.
    The screen appears as a piece of grid paper.

3. From the **Graph** menu, choose **Snap Points**.

4. From the **Edit** menu, click **Preferences**.
    The Preferences dialog box appears.
    Change the first two Precision settings to units.
    Change the third Precision setting to hundredths.
    Click **OK**.

Constructing Similar Rectangles

5. To draw a rectangle:
    From the **Toolbox**, choose ⟋ .
    Click and drag to construct a rectangle with base 12 units and height 8 units.

6. To label the sides of the rectangle:
    From the **Toolbox**, choose ↖ .
    Click each segment of the rectangle to select it.
    Do not select the vertices.
    From the **Display** menu, choose **Show Labels**.
    Move the cursor to a label. The ↖ changes to 🖝 .
    To edit the label, double-click the label.
    Label the sides a, b, c, and d as shown.

7. To construct a smaller similar rectangle:

From the **Toolbox**, choose ![cursor icon].

Double-click the lower left vertex.

Draw a box around the rectangle to select it.

From the **Transform** menu, choose **Dilate**.

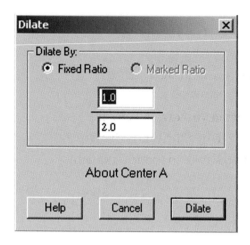

Dilate

Dilate By:
- ⦿ Fixed Ratio   ○ Marked Ratio

  1.0

  2.0

About Center A

Help   Cancel   Dilate

The Dilate dialog box should appear as shown. If not, change the settings to match those shown. The Fixed Ratio 1.0:2.0 is the ratio of corresponding sides of the new rectangle and the original rectangle. The new rectangle is a dilatation image of the original rectangle.

Click **Dilate**. The image rectangle is shown.

Click anywhere on the screen.

Click each side of the image rectangle to select it.

From the **Display** menu, choose **Show Labels**.

The sides are labelled a', b', c', d'.

8. To construct another smaller, similar rectangle:

From the **Toolbox**, choose ![cursor icon].

Draw a box around the smaller rectangle to select it.

From the **Transform** menu, choose **Dilate**.

The dialog box is unchanged.

Click **Dilate**.

Click anywhere on the screen.

Click each side of the image rectangle to select it.

From the **Display** menu, choose **Show Labels**.

The sides are labelled a'', b'', c'', d''.

There are now 3 similar rectangles: the original rectangle, the first image, and the second image.

Exploring Side Lengths of Similar Rectangles

9. To measure the base and height of each rectangle:

From the Toolbox, choose ![cursor icon].

Click anywhere on the screen.

Click the base (side b) and height (side c) for all three rectangles.

From the **Measure** menu, choose **Length**.

How do the bases and heights compare?

Click and drag a vertex of the original rectangle.

Observe the changing measures. Return to the original rectangle.

10. To calculate the base-to-height ratio of similar rectangles:
From the **Measure** menu, choose **Calculate**.
The *Sketchpad* calculator appears.
Calculate the base-to-height ratio of the original rectangle.
Select $\boxed{b = 12\ cm}$, then $\boxed{\div}$, then $\boxed{c = 8\ cm}$. Press $\boxed{\qquad OK \qquad}$.
Record your findings.
Repeat this step to calculate the base-to-height ratio of each smaller rectangle.
What is true about the base-to-height ratios of similar rectangles?
What happens to the ratios when you drag a vertex?
Does the size of the rectangle change the ratio? Explain.
Does the type of quadrilateral change the ratio? Explain.

11. To compare the base-to-base and height-to-height ratios of two similar rectangles:
From the **Measure** menu, choose **Calculate**.
Calculate the base-to-base ratio for the original rectangle (b) and its image rectangle (b′).
Calculate the height-to-height ratio of the original rectangle (c) and its image rectangle (c′). Record your findings.
Compare b′:b″ with c′:c″. Compare b:b″ with c:c″.
What do you notice?

Comparing Corresponding Angles in Similar Figures

12. Click and drag a vertex of the original rectangle to get 3 similar quadrilaterals. To measure an angle:
Click on 3 vertices in order.
From the **Measure** menu, choose **Angle**.
The measure of the angle described by the 3 vertices is shown.
Measure all the angles of each quadrilateral.
What do you notice about the measures of angles in similar figures?

Exploring Perimeters of Similar Rectangles

13. A figure must have an *interior* to measure its perimeter or area.
To construct an interior:
From the **Toolbox**, choose ![arrow tool].
Click anywhere on the screen.

Click the vertices of the original rectangle in order.
Do not click the segments.
From the **Construct** menu, choose **Quadrilateral Interior**.

14. To measure the perimeter:
From the **Toolbox**, choose ▲ .
From the **Measure** menu, choose **Perimeter**.
The vertices of the original rectangle now appear labelled A, B, C, D.

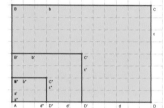

15. Repeat *Steps 13* and *14* for each of the smaller rectangles.

16. To compare the perimeters of two similar rectangles:
From the **Measure** menu, choose **Calculate**.
Calculate the perimeter-to-perimeter ratio for the original rectangle and its first image.

Compare the perimeter-to-perimeter ratio for the first and second image rectangles, and for the original rectangle and the second image. What do you notice?

17. For each rectangle, what is the ratio of perimeter:base?
What do you notice?

Exploring Areas of Similar Rectangles

18. To measure the area:
Repeat *Step 13* to construct quadrilateral interior.
From the **Toolbox**, choose ▲ .
From the **Measure** menu, choose **Area**.

19. Repeat *Step 18* for each image rectangle.

20. For the original rectangle and the first image rectangle, compare the ratio of the areas with the ratio of the bases.
Write each ratio in simplest form. What do you notice?

21. Repeat *Step 20* for the original rectangle and the second image rectangle.

Reflect

Use your observations to list some properties of similar rectangles.
Are these properties true for other polygons?
Use *The Geometer's Sketchpad* to find out.

Explore

Work on your own.
Use a calculator if it helps.

36 tea bags for $1.49   144 tea bags for $5.59

Which box of tea bags is the better buy?
What do you need to consider before you decide?

Reflect & Share

Compare your results with those of a classmate.
Did both of you choose the same box? If so, justify your choice.
If not, can both of you be correct? Explain.

Connect

Great Start cereal can
be purchased in three
different sizes
and prices.

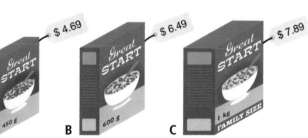

We want to find the least expensive cereal.
The smallest box costs the least, but that does not mean
it is the least expensive.
We find the unit cost for each cereal.
The cost of 1 g is very small, so we calculate the cost of
100 g for each cereal.

Your World

In a supermarket, the tag on the shelf with the bar code of an item often shows the cost of 1 g of the item.

With this information, you can compare the costs of different-size packages of the item.

Box A has mass 450 g and costs $4.69.

450 g is 4.5 × 100 g; so, the cost of 100 g of Box A is:
$\frac{\$4.69}{4.5} \doteq \$1.042$

Box B has mass 600 g and costs $6.49.

The cost of 100 g of Box B is: $\frac{\$6.49}{6} \doteq \$1.082$

Box C has mass 1000 g and costs $7.89.

The cost of 100 g of Box C is: $\frac{\$7.89}{10} = \$0.789$

Each unit cost can be written as a unit rate.

The least unit rate is $0.789/100 g, for Box C.

The greatest unit rate is $1.082/100 g, for Box B.

If you are buying cereal for a family, then the best buy is Box C.

If you do not eat much cereal, the best buy is probably Box A.

If you buy Box C because it has the least unit price, you might have to throw some of it away if it gets stale.

Example

Mariah is looking for a part-time job for 15 h a week.
She has been offered three positions.

Day Camp Counsellor	Cashier	Library Assistant
$7.50 per hour	$25.00 for 3 h	$44.00 for 5 h

a) Which job should Mariah accept?
b) How much will Mariah earn in a week?

Solution

a) Calculate the unit rate for each job.
   The unit rate is the hourly rate of pay.
   For day camp counsellor, the unit rate is $7.50/h.
   For cashier, the unit rate is: $\frac{\$25.00}{3 \text{ h}} \doteq \$8.33/h$
   For library assistant, the unit rate is: $\frac{\$44.00}{5 \text{ h}} = \$8.80/h$
   The library assistant job pays the most.
   So, if Mariah likes books, she should take the job in the library.

b) Mariah works 15 h a week, at a rate of $8.80/h.
   She will earn: 15 × $8.80 = $132.00
   Mariah will earn $132.00 a week as a part-time library assistant.

66   UNIT 2: Applications of Ratio, Rate, and Percent

1. Write a unit rate for each statement.
   a) $399 earned in 3 weeks
   b) 680 km travelled in 8 h
   c) 12 bottles of juice for $3.49
   d) 3 cans of soup for $0.99

2. Which is the better buy?
   a) 5 grapefruit for $1.99 or 8 grapefruit for $2.99
   b) 500 g of yogurt for $3.49 or 125 g for $0.79
   c) 100 mL of toothpaste for $1.79 or
      150 mL of toothpaste for $2.19
   d) 2 L of orange juice for $4.49 or 1 L for $2.89

$ 3.29

3. How much does 1 kg of butter cost at the rate shown in the picture at the left?

4. Mr. Gomez travelled 525 km in 6 h.
   a) What was the mean distance travelled in 1 h?
   b) How is the mean distance related to average speed?
   c) At this rate, how long will it take Mr. Gomez to travel 700 km?

5. a) Which is the greatest average speed?
      i) 60 km in 3 h   ii) 68 km in 4 h   iii) 70 km in 5 h
   b) Draw a graph to illustrate your answers in part a.

6. In the first 9 basketball games of the season,
   Lashonda scored 114 points.
   a) On average, how many points does Lashonda score per game?
   b) At this rate, how many points will Lashonda have after
      24 games?

7. Lakelse Lake, B.C., had the most snow for one day in Canada,
   which was 118.1 cm. Assume the snow fell at a constant rate.
   How much snow fell in 1 h?

8. Who has the greatest average typing speed?
   a) Mei-Lin types 350 words in 6 min.
   b) Nishant types 250 words in 5 min.
   c) Adam types 300 words in 5.5 min.

9. A 2.5-kg bag of grass seed covers an area of 1200 m^2.
   How much seed is needed to cover a square park with
   side length 500 m?

10. **Assessment Focus** The food we eat provides energy in calories. When we exercise, we burn calories.

These tables illustrate these data for different foods and different exercises.

These data are for a female with a mass of 56 kg. The data vary for men and women, and for different masses.

Food	Energy Provided (Calories)	Activity	Calories Burned per Hour
Medium apple	60	Skipping	492
Slice of white bread	70	Swimming	570
Medium peach	50	Cycling	216
Vanilla fudge ice cream	290	Aerobics	480
Chocolate iced doughnut	204	Walking	270

Mental Math

How many numbers less than 150 are divisible by 4 and by 6?

Explain how you know.

a) How long would a person have to:
   i) cycle to burn the calories in an apple?
   ii) walk to burn the calories in two slices of bread?
b) Suppose a person ate vanilla fudge ice cream and a chocolate iced doughnut.
   i) Which activity would burn the calories quickest? How long would it take?
   ii) Which 2 activities would burn the calories in about 2 h?
c) Use the data in the tables to write your own problem. Solve your problem. Show your work.

Take It Further

11. Population density is a rate. It compares the number of people in a population with the area of the land where they live. Population density is measured in number of people per square kilometre.
   a) Find the population density for each country.

   These data are for 2002.

   i) Canada: 30 007 094 people in 9 984 670 km^2
   ii) China: 1 279 557 000 people in 9 562 000 km^2
   iii) Japan: 127 538 000 people in 377 727 km^2
   b) How do the population densities in part a compare?

Reflect

What is a unit rate?
Describe the types of problems you can solve using unit rates.

Mid-Unit Review

LESSON

2.1　**1.** Find each missing term.

　　a) $x:10 = 60:12$　　b) $y:21 = 9:7$

　　c) $\frac{z}{15} = \frac{20}{60}$　　d) $\frac{a}{21} = \frac{4}{3}$

2. At the local high school, the ratio of boys to girls is 3 to 4.
There are 1200 boys in the school.
How many girls are there?

3. The numbers of bass and pike in a lake are estimated to be in the ratio of 5:3. There are approximately 500 fish in the lake.

　　a) How many bass are there?

　　b) How many pike are there?

2.2　**4.** A boat is 26.5 m long.
A picture of the boat is drawn to the scale of 1:200.
What is the length of the boat in the picture?

5. For each animal, its actual measurement is given. What scale would you use to draw each animal to fill a page of your notebook?
Show your work.

　　a) The bee hummingbird is the smallest bird on Earth.
It measures 5.7 cm from beak tip to tail tip.

　　b) The blue whale is the largest animal on Earth.
Its length can be 33 m.

6. Use the scales you chose in question 5.
Draw a line segment to represent each length in question 5.

7. An amoeba is an organism with one cell. A picture of an amoeba is drawn to a scale of 250:1. The amoeba in the picture is 9 cm wide. How wide is the actual amoeba?

2.3　**8.** Find the unit cost of each item.

　　a) 4 L of milk for $4.29

　　b) 2.4 kg of beef for $10.72

　　c) 454 g of margarine for $1.99

9. Which is the better buy?
Justify each answer.

　　a) 6.2 L of gas for $5.39 or
8.5 L of gas for $7.31

　　b) 5 bagels for $3.00 or
12 bagels for $5.99

　　c) 2 kg of potatoes for $1.38 or
5 kg of potatoes for $2.79

10. Hisan can type 86 words in 2 min.
At this rate, how many words can Hisan type in each time?

　　a) 1 min　b) 3 min　c) 12.5 min

2.2
2.3　**11.** On a map, the distance between two cities is 5.6 cm. The scale on the map is 1:3 000 000.
A cyclist travels this distance in 4 h. What is the cyclist's average speed?

Focus Calculate percents from less than 1% to greater than 100%.

Explore

Work with a partner.
Use a calculator if you need it.

There are 225 students in an elementary school.

- Eighteen students have no siblings.
  Estimate what fraction of the students in the school have no siblings.
  Write the fraction of students in the school who have no siblings.
  What percent of students in the school have no siblings?
- One-third of the students have two siblings.
  Estimate what percent of the students have two siblings.
  What fraction of the students do not have two siblings?
  What percent of the school is this?

Reflect & Share

What do you notice about the percents for one-third and two-thirds?
How are these different from other percents you have worked with?

Connect

When the whole is 1.0, you know that:

$$100\% = 1.0$$
$$10\% = 0.10$$
$$1\% = 0.01$$

We can extend this pattern to write percents less than 1% as decimals:

$$0.1\% = 0.001$$
$$\text{and} \quad 0.5\% = 0.005$$

Some fractions have percents that are repeating decimals.
We can use a calculator to show that: $\frac{1}{3} = 0.3333333\ldots$
We write this decimal as $0.\overline{3}$.

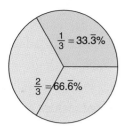

$\frac{1}{3} = 33.\overline{3}\%$

$\frac{2}{3} = 66.\overline{6}\%$

To write $\frac{1}{3}$ as a percent, we write $0.\overline{3}$ as $0.33\overline{3}$.

Then, $0.33\overline{3} = \frac{33.\overline{3}}{100}$

$= 33.\overline{3}\%$

And $\frac{2}{3} = 0.6666\ldots$

$= 0.66\overline{6}$

$= \frac{66.\overline{6}}{100}$

$= 66.\overline{6}\%$

Example 1

According to Statistics Canada, in 2001, the population of Ontario was about 11 410 000. About 0.35% of the population were aboriginal people living on reserves.

a) About how many aboriginal people lived on reserves?

b) Estimate to check the answer.

c) Illustrate the answer to part a with a diagram.

Solution

a) Find 0.35% of 11 410 000.

First write 0.35% as a decimal.

$0.35\% = \frac{0.35}{100}$

$= 0.0035$

Use a calculator.

Then, 0.35% of 11 410 000 $= 0.0035 \times 11\ 410\ 000$

$= 39\ 935$

In 2001, about 40 000 people lived on reserves in Ontario.

b) 0.35% is approximately 0.33%.

0.33% is approximately $\frac{1}{3}\%$, or $\frac{1}{3}$ of 1%.

1% of 11 410 000 is: $0.01 \times 11\ 410\ 000 = 114\ 100$

114 100 is about 100 000.

$\frac{1}{3}$ of 100 000 is about 33 000.

This estimate is close to the calculated answer.

c) To illustrate 0.35%, first show 1% on a number line.

Then, 0.35% is about $\frac{1}{3}$ of 1%.

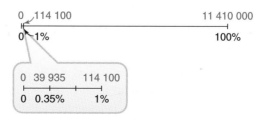

Example 2

According to Statistics Canada, the number of children in Canada between 10 and 14 years of age in 2001 was about 180% of their number in 1951.
In 1951, there were about 1 131 000 children in this age group.
a) In 2001, about how many children in this age group were there in Canada?
b) Estimate to check the answer.
c) Illustrate the answer to part a with a diagram.

Solution

a) Find 180% of 1 131 000.
First, write 180% as a decimal.
$$180\% = \frac{180}{100}$$
$$= 1.80$$
Then, 180% of 1 131 000 $= 1.80 \times 1\ 131\ 000$
$$= 2\ 035\ 800$$

In 2001, there were about 2 million children between the ages of 10 and 14 in Canada.

b) 180% is close to 200%.
1 131 000 is close to 1 000 000.
200% of 1 000 000 is: $2 \times 1\ 000\ 000 = 2\ 000\ 000$
This estimate is close to the calculated answer.

c)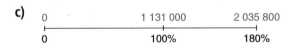

Practice

1. Write each percent as a decimal.
Draw a diagram to illustrate each percent.
a) 120% b) 250% c) 475% d) 0.3% e) 0.53% f) 0.75%

2. a) Write each fraction as a percent.
i) $\frac{1}{3}$ ii) $\frac{2}{3}$ iii) $\frac{3}{3}$ iv) $\frac{4}{3}$ v) $\frac{5}{3}$ vi) $\frac{6}{3}$
b) What patterns do you see in your answers in part a?
c) Use these patterns to write each fraction as a percent.
i) $\frac{7}{3}$ ii) $\frac{8}{3}$ iii) $\frac{9}{3}$

3. a) Find the percent of each number. Draw a diagram to illustrate.

　i) 200% of 360　**ii)** 20% of 360　**iii)** 2% of 360　**iv)** 0.2% of 360

b) What patterns do you see in your answers in part a?

c) Use the patterns in part a to find each percent.
Explain your work.

　i) 2000% of 360　　**ii)** 0.02% of 360

4. Six hundred eighteen runners registered for the marathon.
Of these runners, 0.8% completed the race in under 2 h 15 min.

a) How many runners completed the race in this time?

b) Estimate to check your answer.

5. At the local theatre, 120 people attended the production of
Romeo and Juliet on Friday.
The attendance on Saturday was 140% of the attendance on Friday.

a) How many people went to the theatre on Saturday?

b) Estimate to check your answer.

6. Fifty-six students signed up for the school play.
About 34% of these students were boys.
At the auditions, only 31 girls attended.
What percent of the girls who signed up for the play attended
the auditions?

7. (**Assessment Focus**) During the 1888 Gold Rush, a British
Columbia town had a population of 2000.
By 1910, the town had become a ghost town.
The population was 0.75% of its population in 1888.

a) Estimate the population in 1910. Justify your estimate.

b) Calculate the population in 1910.

c) Find the decrease in population from 1888 to 1910.
Show your work.

Reflect

How do you find a percent of a number in each case?

- The percent is less than 1%.
- The percent is greater than 100%.

Use an example to explain each case.
Include diagrams.

2.5 Solving Percent Problems

Focus Find the whole, when given a percent, and find percent increase and decrease.

Explore

Work with a partner.
Tasha conducted a survey of the students in her school.

➤ From the results, Tasha calculated that 60% of the students go to school by bus.
  Liam knows that 438 students go to school by bus.
  How can Liam use these data to find how many students are in the school?

➤ Tasha also found that 50% more students go by bus than walk or drive.
  About how many students walk or drive to school?
Sketch number lines to illustrate your work.

Reflect & Share

Compare your results with those of another pair of students.
Discuss the strategies you used to solve the problems.

Connect

Grady is 13 years old and 155 cm tall. His height at this age is about 90% of his height when he stops growing.

To estimate Grady's height when he stops growing:
90% of Grady's height is 155 cm.
So, 1% of his height is: $\frac{155 \text{ cm}}{90}$
And, 100% of his height is: $\frac{155 \text{ cm}}{90} \times 100 \doteq 172.2$ cm
So, when Grady stops growing, his height will be about 172 cm.

When we know a percent of the whole, we divide to find 1%, then multiply by 100 to find 100%, which is the whole.

Another type of problem involving percents is to find the percent increase or decrease. This is illustrated in *Example 1*.

Example 1

a) The price of a carton of milk in the school cafeteria increased from 95¢ to $1.25.
What was the increase as a percent?

b) At the same time, to encourage students to eat healthy food, the price of green salad decreased from $2.50 to $1.95.
What was this decrease as a percent?

Solution

a) The increase was: $\$1.25 - 95¢ = 125¢ - 95¢$
$$= 30¢$$

To find the percent increase, write the increase as a fraction of the original price: $\dfrac{30¢}{95¢}$

Use a calculator.

To write this fraction as a percent: $\dfrac{30}{95} \doteq 0.32$
$$= \dfrac{32}{100}$$
$$= 32\%$$

The price of a carton of milk increased by about 32%.

```
        0¢                      95¢   $1.25
        ├──────────────────────┼─────┤
        0%                     100%  132%
```

b) The decrease is: $\$2.50 - \$1.95 = 55¢$
To find the percent decrease, write the decrease as a fraction of the original price: $\dfrac{55¢}{\$2.50} = \dfrac{55}{250}$

Use a calculator.

To write this fraction as a percent: $\dfrac{55}{250} = 0.22$
$$= \dfrac{22}{100}$$
$$= 22\%$$

The price of a green salad decreased by 22%.

```
                             ├── 55¢ ──┤
        $0                 $1.95     $2.50
        ├────────────────────┼─────────┤
        0%                            100%
                            ├── 22% ──┤
```

Example 2

In 2004, the cost to join a gym was $169.00.
In 2005, the cost was 7% more.
How much did it cost to join in 2005?

Solution

Here are two methods to find the cost in 2005.

Method 1

The cost in 2004 was $169.00.
This is 100%.
The cost in 2005 was 7% more.
7% of $169.00
$= 0.07 \times \$169.00$
$= \$11.83$
So, the cost in 2005 is:
$\$169.00 + \$11.83 = \$180.83$
It cost $180.83 to join the gym in 2005.

Method 2

The cost in 2004 was $169.00.
This is 100%.
The cost in 2005 was 7% more.
This is 100% + 7%,
or 107% of $169.00.
107% of $169.00
$= 1.07 \times \$169.00$
$= \$180.83$

Practice

1. Find the number in each case.
   Illustrate each answer with a number line.
   a) 25% of a number is 5.   b) 75% of a number is 18.
   c) 4% of a number is 32.   d) 120% of a number is 48.

2. Find the amount in each case.
   a) 15% is 125 g.   b) 9% is 45 cm.   c) 0.8% is 12 g.

3. Write each increase as a percent.
   Illustrate each answer with a number line.
   a) The price of a house increased from $210 000 to $225 000.
   b) The elastic stretched from 10 cm to 13 cm.

4. The volume of the gas in a container was 1500 cm³.
   The gas was heated until its volume was 20% greater.
   What was the new volume of the gas?

One hectare (1 ha) is a unit of area, which is equal to 10 000 m².

5. Write each decrease as a percent.
   Illustrate each answer with a number line.
   a) The price of gasoline decreased from 79.9¢/L to 75.9¢/L.
   b) The area of rain forest in Jamaica decreased from 128 800 ha in 1981 to 122 000 ha in 1990.

6. There were about 193 000 miners in Canada in 1986.
   By 2001, the number of miners was 12% less.
   How many miners were there in 2001?

7. In a shipment of MP3 players, 2% are defective. There are 5 defective MP3 players. How many MP3 players are there in this shipment?

8. Scott delivers 14 newspapers in 10 min. This is 28% of his round.
   a) How many more papers does Scott have to deliver?
   b) Scott continues to deliver papers at the same rate. How long will it take him to deliver all his papers?

9. On average, a girl reaches 90% of her final height by the time she is 11 years old, and 98% of her final height when she is 17 years old.
   a) Anna is 11 years old. She is 150 cm tall. Estimate her height when she is 20 years old.
   b) Raji is 17 years old. She is 176 cm tall. Estimate her height when she is 30 years old.

10. **Assessment Focus** On average, a boy reaches 90% of his final height by the time he is 13 years old, and 98% of his final height by the time he is 18 years old. Use these data or the data in question 9 to estimate your height when you are 21 years old. Explain any assumptions you made. Show your work.

Take It Further

11. Here are some data about countries in North America.

Country	Population in 2004	Area (km²)
Canada	32 507 874	9 984 670
U.S.	293 027 571	9 629 091
Mexico	104 959 594	1 972 550

   a) How much space does each person have in each country?
   b) By what percent does:
      i) the population of Mexico exceed that of Canada?
      ii) the area of Canada exceed that of Mexico?
   c) Make up, then solve your own problem about these data.

Reflect

What is the difference between a percent increase and a percent decrease? Include examples in your explanation.

Explore

Work with a partner.

Choice for Tennis

$ 129.99 20% off!

Live for Sport

$ 109.99 10% off!

$ 99.99

Essential Equipment

Estimate which racquet is the least expensive.
Then calculate the sale price of each racquet to check your estimate.

Reflect & Share

Compare your answers with those of another pair of classmates.
If you used the same method, think of a different way to find each sale price.

Connect

When an item is on sale for 20% off, we say that
there is a **discount** of 20%.
A discount of 20% on an item for sale means that you pay:
100% − 20% = 80% of the regular price

A video game is marked with a discount of 30%.
Its regular price is $27.99.
The sale price of the video game is: 100% − 30%, or 70% of $27.99
70% of $27.99 = 0.7 × $27.99
= $19.59

This is the price before sales tax is added.
The provincial sales tax (PST) is 8%.
The goods and services tax (GST) is 7%.
So, the total sales tax is: 8% + 7% = 15%

So, the price you pay is: $19.59 + 15% of $19.59
We can calculate this directly as: 115% of $19.59 = 1.15 × $19.59
= $22.53

Many salespeople earn commission.
A **commission** is a percent of the money received from the sale of the item.

Example

A real estate agent sells a home for $219 000.
She earns 2.5% commission.
How much does the agent earn on this sale?

Solution

The real estate agent earns 2.5% of $219 000.
This is: 0.025 × $219 000 = $5475
The real estate agent earns $5475.

Practice

Use a calculator when you need to.
1. For each item below:
   a) Estimate the PST and GST.
   b) Calculate the PST and GST.
   c) Calculate the cost, including taxes.
     i)

$ 25.99

     ii)

$ 152.45

2. For each item below:

  a) Estimate the discount.

  b) Calculate the discount.

  c) Calculate the sale price before taxes.

  d) Calculate the sale price, including taxes.

  i)

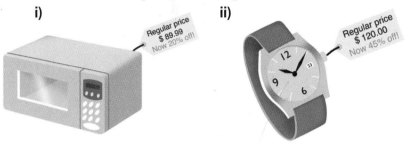

Regular price
$ 89.99
Now 20% off!

  ii)

Regular price
$ 120.00
Now 45% off!

3. In a sale, a blow dryer is marked down from $18.98 to $11.39.

  a) What is the percent decrease?

  b) Calculate the sale price, including taxes.

4. A car dealership offers two choices of discount on a car with a sticker price of $25 000.

  Choice A: $4000 rebate

  Choice B: 20% discount

  Which is the better deal for the customer?

  Justify your answer.

5. A video store offers these choices.

  Choice A: 30% off each video with regular price $25.00

  Choice B: Buy two videos for $40.00.

  Which is the better deal for the customer?

  Justify your answer.

6. A salesperson earns commission on a *sliding scale*: the more he sells, the greater his rate of commission. Jack earns 3% commission on sales up to and including $75 000 per month, and 5% commission on sales over $75 000.

  How much commission does Jack earn on sales of $90 000?

7. A new house was purchased for $304 000. After 3 years, its market value had increased by 28%. What is the new market value of the house?

Number Strategies

List the prime factors of each number.

72, 192, 210, 1890

8. A clothing store reduces the price of an item by 20% if it is unsold after 3 weeks.

It reduces the price a further 30% if the item is still unsold after 6 weeks.

And, after 7 weeks if the item is unsold, a further 10% reduction is made.

   a) Estimate the price, after 8 weeks, of a coat that had a regular price of $289.50.

   b) Calculate the sale price, including the sales taxes.

9. **Assessment Focus** Two stores have the same item for the same price.

   • Store A offers successive discounts of 5% one week, 10% the next week, and 10% the third week.

   • Store B offers one discount of 25% the third week.

   Which offer results in the greater discount?

   Justify your answer. Show your work.

10. During a 20% off sale, the sale price of a radio was $35.96. What was the regular price of the radio?

11. The price of a snowsuit is reduced by $28.38. This is a discount of 33%.

   a) What is the regular price of the snowsuit?

   b) What is the sale price of the snowsuit, including taxes?

Take It Further

12. For a promotion, a store offers to pay the sales taxes on any item you buy. You are actually paying taxes but they are calculated on a lower price.

   Suppose you buy an item for $100.

   The sales taxes are 15%.

   a) What is the true sale price of the item?

   b) How much tax are you really paying?

Reflect

Explain how percents are used in the world around you.
Include examples in your explanation.

2.7 Simple Interest

When you buy a Canada Savings Bond (CSB), you are lending the Canadian government money. The government pays you for borrowing your money. It pays a percent of what you invested. In 2004, if you invested money for 1 year, the government paid you 1.55% of the amount you invested.

Explore

Work on your own.
You will need a calculator.
Suppose you bought a $500 CSB in 2004.
Estimate how much the government will pay you for 1 year.
Calculate how much the government will pay you for 1 year.
What if you bought a $1000 CSB? A $1500 CSB?
How much does the government pay in each case?

Reflect & Share

Compare your answers and solutions with those of a classmate.
What patterns do you see in your answers?

Connect

Interest is the "rent" you pay for borrowing money,
or the "rent" paid to you when you save money.
You may pay interest when you buy something and pay for it later.

Mae bought a computer for college. It cost $1250.
Mae did not have to pay for 1 year.
The store charged her interest at a rate of 11.5% for that year.
The interest Mae paid at the end of the year was:
11.5% of $1250 = 0.115 × $1250
$$= \$143.75$$

So, Mae paid a total of: $1250 + $143.75 = $1393.75

When the interest is calculated at the end of the period for which the money is borrowed or invested, the interest is called **simple interest**.

Example 1

Joe borrowed $7500 from his mother to buy a car.
His mother charged Joe 3.5% simple interest each year for 2 years.
Joe will pay back the money in equal monthly payments over 2 years.
a) What simple interest does Joe pay?
b) What is each monthly payment?

Solution

a) Joe pays 3.5% each year for 2 years.
Each year, Joe pays 3.5% of $7500 = 0.035 × $7500
$$= \$262.50$$
So, for 2 years, the simple interest Joe pays is:
2 × $262.50 = $525.00

b) The total Joe pays is: $7500 + $525.00 = $8025
There are 12 months in 1 year, so there are 24 months in 2 years.
Joe makes 24 monthly payments.
Each monthly payment is: $8025 ÷ 24 = $334.38

In *Example 1*, we can calculate the simple interest directly.
The simple interest for 1 year is:
Money borrowed or invested × interest rate as a decimal
The simple interest for 2 years is double this:
Money borrowed or invested × interest rate as a decimal × 2

We write this formula for calculating simple interest: $I = Prt$
I is the simple interest.
P is the money borrowed, invested, or deposited;
it is called the **principal**.
r is the annual interest rate as a decimal.
t is the time in years.

Simple interest may be charged or paid for a time less than 1 year.

Example 2

Hannah borrows $1500 for 6 months.
The annual interest rate is 7%.
a) How much simple interest does Hannah pay?
b) How much does Hannah pay back?

Solution

a) Use the formula: $I = Prt$

The principal, P, is $1500.

Remember that r is the interest rate as a decimal.

The annual interest rate, r, is 7%, or 0.07.

Since time, t, is measured in years,

write 6 months as a fraction of 1 year:

$\frac{6}{12} = \frac{1}{2} = 0.5$

Substitute: $P = 1500$, $r = 0.07$, and $t = 0.5$

Use a calculator.

So, $I = 1500 \times 0.07 \times 0.5$

$\quad\quad = 52.5$

Hannah pays $52.50 interest.

b) Hannah pays back principal and interest:

$1500 + $52.50 = $1552.50

In *Example 1b* and *2b*, the principal + interest is the **amount**.

Practice

Use a calculator when you need to.

1. Write each percent as a decimal.

    a) 5%   **b)** 7%   **c)** 3%   **d)** 1.25%

    e) 3.5%   **f)** $3\frac{1}{4}$%   **g)** $5\frac{3}{4}$%   **h)** $2\frac{1}{2}$%

2. Calculate the simple interest paid on each deposit.

	Deposit ($)	Annual Interest Rate (%)	Time (years)
a)	300	3	1
b)	550	4	2
c)	800	2	3

3. Calculate the simple interest charged for each loan.

	Loan ($)	Annual Interest Rate (%)	Time (years)
a)	4 000	7	3
b)	10 000	9	2
c)	2 960	5	5

Calculator Skills

Copy these boxes.

Replace each box with the numbers 5, 6, 8, and 9 to make the product closest to 5000.

☐ ☐ ☐

×   ☐

4. Calculate the simple interest and the amount.
   a) $2500 invested at an annual interest rate of 6% for 6 months
   b) $6000 invested at an annual interest rate of 6% for 3 years
   c) $700 invested at an annual interest rate of 8% for 3 months

5. Mark borrows $2000 at a rate of 7% per year for 4 years.
   Mark pays back the money in equal monthly payments over the years.
   a) What simple interest does Mark pay?
   b) What is each monthly payment?

6. Shemaine has a $500 savings bond. She has had the bond for 1.5 years. The bond paid 2.5% interest per year. Shemaine cashes her bond. What amount will she receive?

7. Mumtaz borrows $3350 from a friend. She will pay back the money after 8 months, including interest at a rate of 6.25% per year. What amount will Mumtaz pay after 8 months?

8. **Assessment Focus** Marie won $1 000 000 on a lottery.
   She invested the money in her friend's business.
   Marie's friend paid Marie simple interest at a rate of 4% per year.
   Use this information to pose a problem.
   Solve your problem.
   Show your work.

9. Harpreet borrowed $2575 to start a business.
   He will pay back the money in 18 months.
   He pays simple interest at a rate of $12\frac{1}{4}$% per year.
   What amount will Harpreet pay in 18 months?

Take It Further

10. What is the interest rate when you borrow $600 for 1 year and pay $45.00 interest?

Reflect

What is simple interest?
How is simple interest earned?
How is it paid?
Use examples in your explanation.

Posing Math Problems

People see the objects and events of everyday life differently.

They bring different perspectives, and ask different questions.

A person with an historical perspective might ask how the object was invented, by whom, where, and so on.

A person with a scientific perspective might ask how the object works, with what materials it is made, or how it could be improved. Making sense of math often begins by seeing math in everyday experiences.

A mathematical perspective might ask questions about number, shape, patterns, and relationships.

Here are some questions or problems that you might pose about school desks, using the math strands as a guide:

Number

- How many desks are in the school?
- What is the ratio of desks to students in the school? Why?
- How many desks need replacing each year?
- What is the cost of replacing a student desk?

Geometry

- A desk manufacturer cuts many parts for the desk from flat sheets of material (metal or wood).
  Which figures are needed to create a desk?
- What is the best way to arrange and cut these figures from a flat sheet to save material?
- How can desks be stacked to take up the least space?
- How can desks be stacked differently if parts are disassembled and rearranged?
- How might you create a package to protect the desks during transportation?

Measurement

- What is the least shipping cost for delivering all the desks needed for a new school?
- How far do the desks travel from where they are made to the school?
- What is the operating cost for a transport truck?
- What is the volume of a desk?
- How much space is in a transport truck?
- How many trips would be necessary for a transport truck to deliver all the desks?

Patterning

- Based on the population growth of our school, how much money does the school need to set aside for purchasing desks in the next 10 years?
- How many desks will we need to purchase in the next 10 years?

Data Management

- A desk manufacturer needs to know how many desks to make in the next year. What would you suggest?
- What might be the typical costs for each year?
- How would you present this information in tables and graphs?

Many of these problems use information and skills from different math strands.

Choose an object or event. Create a series of mathematical problems related to it. Trade your problems with a classmate. Solve your classmate's problems.

Unit Review

Review any lesson with

online tutorial

What Do I Need to Know?

☑ An *equivalent ratio* can be formed by multiplying or dividing the terms
of a ratio by the same number.
For example:
$10:16 = (10 \times 2):(16 \times 2) = 20:32$
$10:16 = (10 \div 2):(16 \div 2) = 5:8$
5:8, 10:16, and 20:32 are equivalent ratios.

☑ A *proportion* is a statement that two ratios are equal.

☑ A *scale drawing* is a reduction or an enlargement of an object.
The scale may be written as a ratio.

☑ *Similar figures* have:
  • corresponding angles equal
  • corresponding sides in the same ratio
The ratio of the areas of similar figures is equal to the square of the ratio of
corresponding sides. For example, if the ratio of sides is 1:2, the ratio of
areas is 1:4.

☑ A *rate* is a comparison of two quantities with different units.
For example, 500 km in 4 h is a rate.
Divide 500 km by 4 h to get the unit rate of $\frac{500 \text{ km}}{4 \text{ h}}$, or 125 km/h.

☑ To calculate a *percent decrease*, divide the decrease by the original amount,
then write the quotient as a percent.

☑ To calculate a *percent increase*, divide the increase by the original amount,
then write the quotient as a percent.

☑ *Simple interest* is charged on money borrowed,
or earned on money invested.
The formula for simple interest is: $I = Prt$, where P is the principal,
r the annual interest rate as a decimal, and t the time in years.

What Should I Be Able to Do?

For extra practice, go to page 489.

2.1

1. Silva and Renate were given money in the ratio of 5 to 3. Silva's share was $60. How much did Renate receive?

2. At the 2004 Athens Olympics, the ratio of Canada's gold medals to Greece's gold medals was 1:2. Together, Canada and Greece won 9 gold medals.
   a) How many gold medals did Canada win?
   b) How many gold medals did Greece win?

3. The St. Croix junior hockey team won 3 out of every 4 games it played. The team played 56 games. How many games did it lose?

4. A punch recipe calls for orange juice and pop in the ratio of 2:5. The recipe requires 1 L of pop to serve 7 people.
   a) How much orange juice is needed for 7 people?
   b) About how much orange juice and pop do you need to serve 15 people? 20 people? Justify your answers.

2.2

5. A picture of a bug has a scale of 15:1. The length of the bug in the picture is 2.5 cm. What is the actual length of the bug?

6. A scale on a map is 1 cm represents 2.5 km. Vivian drove 45 km. What is this distance on the map?

2.3 **7.** Pierre baby-sits for $5.50/h. He earned $1155 during the summer. How many hours did Pierre baby-sit?

8. A speed walker walks 2.5 km in 20 min.
   a) What is her average speed in kilometres per minute?
   b) What is her average speed in kilometres per hour?

9. Kieran ran 8 laps of the track in 18 min. Jevon ran 6 laps of the track in 10 min. Who had the greater average speed? How do you know?

2.4 **10.** Forty percent of an average person's body is muscle.
   a) Ali has a mass of 73 kg. What is the mass of his muscle?
   b) What mass of your body is muscle? How do you know?
   c) Draw a diagram to illustrate your answers.

11. In a shipment of 2000 tomato sauce jars, 0.85% were broken.
   a) Estimate how many jars were broken.

b) Calculate how many jars were broken.

c) Draw a diagram to illustrate your answer in part b.

2.5 **12.** Joline collects hockey cards. She needs 5 cards to complete a set. This is 20% of the set. How many cards are in the set? Justify your answer.

13. The water in the dam was 15 m deep. After a storm, the depth of the water was 15% greater. What was the new depth of water?

14. Jehane had a mark of 39 on a test. This mark was 60%. What was the total possible mark?

15. About 310 000 t of Atlantic cod were caught in 1991. In 2001, the mass of cod caught was 87% less. How much cod was caught in 2001?

16. A ball bounced 64% of the distance through which it had fallen. The bounce was 72 cm high. From what height was the ball dropped?

2.4 **17.** A new queen size bed sheet 2.5 measures 210 cm by 240 cm. The length and width each shrinks by 2% after the first washing.
a) What are the dimensions of the sheet after washing?
b) What is the percent decrease in the area of the sheet?

2.6 **18.** A portable CD player is on sale. It is advertised as, "Save $20. You pay $49.99."
a) What is the regular price?
b) What is the percent discount?

19. A gym suit sells for $89.99 at Aerobics for All. For August, there is a 25% discount as a *Back to School* special. Calculate the total price, including taxes.

20. A family sold a home for $350 000 and paid 5% commission.
a) How much was the commission?
b) How much did the family receive for the house?
c) What if the family had sold its house through a realtor who charged only $2\frac{3}{4}$% commission. How much would the family have saved?

2.7 **21.** Find the simple interest paid on an investment of $7500 at 2% per year for 3 months.

22. Jamie borrowed $1500 from her mother. She promised to repay the money in $1\frac{1}{2}$ years, with simple interest at $7\frac{3}{4}$% per year.
a) What simple interest does Jamie pay?
b) What amount must Jamie pay in $1\frac{1}{2}$ years?
c) What if Jamie repaid the money in monthly instalments. What would each monthly payment be?

Practice Test

1. Fuel for an outboard motor has a gas to oil ratio of 50:1.
   How much oil is needed for 25.5 L of gas?

2. Spyri travels from Edmonton to Grande Prairie. On the map, this distance is 9.1 cm. The scale on the map is 1:5 000 000.
   a) What is the actual distance between these two cities?
   b) Spyri's car uses 1 L of gas to travel 7.5 km.
      At this rate, how much gas will she need for this trip?
   c) Spyri estimates that her average speed will be 70 km/h.
      How long will it take her to complete the trip?
   d) What if Spyri's average speed was 10 km/h greater.
      How much time would she save?

3. Ben received 1 mark for every correct answer on a multiple-choice test.
   He answered 75 questions correctly out of a total of 135 questions.
   On the test, 55% is a pass. Did Ben pass the test?
   Give reasons for your answer.

4. The Grade 8 class at Wheatly Middle School sold 77 boxes of almond chocolates.
   This is 22% of all the boxes of chocolates sold.
   a) Estimate the total number of boxes sold.
   b) Calculate how many boxes were sold.

5. Joginder borrowed $3500 at an annual rate of simple interest of 12%.
   Calculate the amount Joginder repays after 2 years.

6. The price of a house increased by 20% from 2002 to 2003.
   The price of the house decreased by 20% from 2003 to 2004.
   Will the price of the house at the beginning of 2002 be equal to its price at the end of 2004?
   If your answer is yes, show how you know.
   If your answer is no, when is the house less expensive?

Alison is 18 years old.

She plans to go to college for 4 years to study computer animation.
Alison has two choices:

- Live at home and buy a car to travel 30 km each way to college every day.
- Live in residence and pay $15 000 room and board each year.

Here are the list prices of new and used cars.

Vehicles	New ($)	Used ($)	Gas Consumption in City (L/100 km)
Hyundai Accent	12 895	10 495	8.0
Kia Spectra	14 995	9 995	10.0
Chevrolet Aveo	13 480	11 195	8.8

If Alison buys a car, she will have to pay GST and PST.
She will also have to pay for regular maintenance, insurance,
and a vehicle plate fee.
A new car should not need as many repairs as a used car.
Interview a family member or friend.
Find the approximate costs of car repairs, maintenance, and fuel.
Should Alison buy a car or live
in residence?
Justify your answer.

Check List

Your work should show:

✓ the methods you used to calculate and compare costs of buying a new or used car, and living in residence

✓ correct calculations of unit rates, and all costs, including taxes

✓ your conclusions and justification for your decision

✓ clear explanations, with correct use of mathematical language

Reflect on the Unit

How are ratio, rate, and percent related?
How can you use your knowledge of percents outside the classroom?
Include examples in your explanation.

Geometry and Measurement

These students are setting up a tent.
How do the students know how to set up the tent?
How is the shape of the tent created?
How could students find the amount of material needed to make the tent?
Why might students want to know the volume of the tent?

What You'll Learn

- Recognize and sketch objects.
- Use nets to build objects.
- Develop and use a formula for the surface area of a triangular prism.
- Develop and use a formula for the volume of a triangular prism.
- Solve problems involving prisms.

Why It's Important

- We find out about our environment by looking at objects from different views. When we combine these views, we have a better understanding of these objects.
- We need measurement and calculation skills to design and build objects, such as homes and parks.

Key Words

- isometric diagram
- pictorial diagram
- triangular prism
- surface area
- volume

Skills You'll Need

Drawing Isometric and Pictorial Diagrams

An **isometric diagram** shows three dimensions of an object. It is drawn on isometric (triangular) dot paper. Vertical edges of an object are drawn as vertical line segments. Parallel edges of an object are drawn as parallel line segments.

Example 1

Make an isometric diagram of this object.

Solution

On isometric paper, join a pair of dots for each vertical edge.

Join a pair of dots diagonally for each horizontal edge that goes up to the right.

Join a pair of dots diagonally for each horizontal edge that goes up to the left.
Shade the faces so the object appears three-dimensional.

In a **pictorial diagram**, the depth of an object is drawn to a smaller scale than the length and width. This gives the object a three-dimensional appearance.

Example 2

Make a pictorial diagram of this cylinder.

Solution

The top and bottom of a cylinder are circles.
In a pictorial diagram, a circular face is an oval.
One-half of the bottom circular face is
drawn as a broken curve. This indicates that
this half cannot be seen.
Draw vertical line segments to join the
top and bottom ovals.

1. Make an isometric diagram of each object.
   a) a rectangular prism with dimensions
      3 units by 4 units by 5 units
   b) a square pyramid

   **Remember to look up any terms
   that you are unsure of in the *Glossary*.**

2. Make a pictorial diagram of each object.
   a) a rectangular prism with dimensions 3 units by 4 units by 5 units
   b) a regular tetrahedron

Calculating the Surface Area and Volume of a Rectangular Prism

The surface area of a rectangular prism is the sum of the areas of all its faces.
Since opposite faces are congruent,
this formula can be used to find the surface area:
Surface area = 2 × area of base + 2 × area of side face + 2 × area of front face
Using symbols, we write: $SA = 2lw + 2hw + 2lh$
Since the congruent faces occur in pairs, this formula can be written as:
$SA = 2(lw + hw + lh)$
In this formula,
l represents length,
w represents width,
and h represents height.

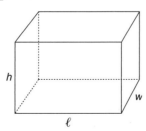

The volume of a rectangular prism is the space occupied by the prism.
One formula for the volume is: Volume = area of base × height
Using symbols, we write: $V = lwh$

Example 3

A rectangular prism has dimensions 4 m by 6 m by 3 m.

a) Calculate the surface area.   **b)** Calculate the volume.

Solution

Draw and label a pictorial diagram.

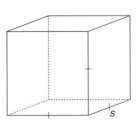

a) Use the formula for the surface area
   of a rectangular prism:
   $SA = 2(lw + hw + lh)$
   Substitute: $l = 6$, $w = 4$, and $h = 3$
   $SA = 2(6 \times 4 + 3 \times 4 + 6 \times 3)$ In the brackets, multiply then add.
   $\quad\; = 2(24 + 12 + 18)$
   $\quad\; = 2(54)$
   $\quad\; = 108$

Area and surface area are measured in square units (m²).

   The surface area is 108 m².

b) Use the formula for the volume of a rectangular prism:
   $V = lwh$
   Substitute: $l = 6$, $w = 4$, and $h = 3$
   $V = 6 \times 4 \times 3$
   $\quad = 72$

Volume is measured in cubic units (m³).

   The volume is 72 m³.

A cube is a regular polyhedron with 6 square faces.
Since all the faces of a cube are congruent,
we can simplify the formulas for surface area
and volume. Each edge length is s.
The area of each square face is: $s \times s = s^2$
So, the surface area of a cube is: $SA = 6s^2$
The volume of a cube is: $V = s \times s \times s = s^3$

✓ Check

3. Find the surface area and volume of each rectangular prism.
   Include a labelled pictorial diagram for each rectangular prism.
   a) 12 cm by 6 cm by 8 cm   **b)** 7 mm by 7 mm by 4 mm
   c) 2.50 m by 3.25 m by 3.25 m   **d)** 5 cm by 5 cm by 5 cm

Calculating the Area of a Triangle

The area of a triangle is calculated with either of these formulas:

Area = base × height ÷ 2, or

Area = one-half × base × height

Using symbols,

we write: $A = \frac{bh}{2}$ or $A = \frac{1}{2}bh$

These formulas are equivalent. Dividing bh by 2 is the same as multiplying bh by $\frac{1}{2}$.

where b is the length of the base and h is the corresponding height.

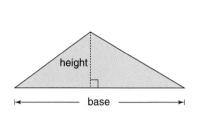
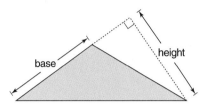

Example 4

The side lengths of △PQR are 12 cm, 5 cm, and 13 cm.

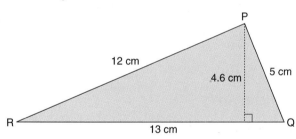

a) The height from P to QR is about 4.6 cm.

Use this to calculate the area of △PQR.

b) Triangle PQR is a right triangle with ∠P = 90°.

Use this to calculate the area of △PQR a different way.

Solution

a) Use the formula: $A = \frac{bh}{2}$

Substitute: $b = 13$ and $h = 4.6$

$A = \frac{13 \times 4.6}{2}$

$= \frac{59.8}{2}$

$= 29.9$

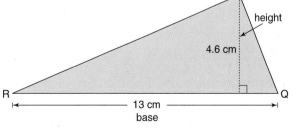

The area is 30 cm² to the nearest square centimetre.

b) Since △PQR is a right triangle,
the two sides that form the right angle
are the base and height.
∠QPR = 90°, so QP = 5 cm is the height;
and PR = 12 cm is the base.
Use the formula: $A = \frac{bh}{2}$
Substitute: $b = 12$ and $h = 5$

$A = \frac{12 \times 5}{2}$

$= \frac{60}{2}$

$= 30$

The area is 30 cm².

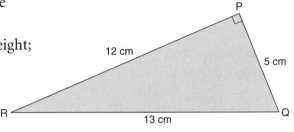

**Any triangle has 3 sets of base and height.
All sets produce the same area.**

✓ Check

4. Calculate the area of each triangle.

a)

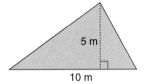

5 m
10 m

b)

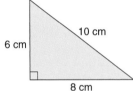

6 cm
10 cm
8 cm

c)

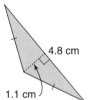

4.8 cm
1.1 cm

Converting among Units of Measure

1 m = 100 cm

100 cm
1 m

The area of a square with side length 1 m is:
A = 1 m × 1 m
  = 1 m²

The area of a square with side length 100 cm is:
A = 100 cm × 100 cm
  = 10 000 cm²
So, 1 m² = 10 000 cm², or 10⁴ cm²

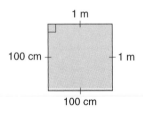

1 m
100 cm
1 m
100 cm

The volume of a cube with edge length 1 m is:

$V = 1 \text{ m} \times 1 \text{ m} \times 1 \text{ m}$
$\quad = 1 \text{ m}^3$

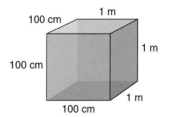

The volume of a cube with edge length 100 cm is:

$V = 100 \text{ cm} \times 100 \text{ cm} \times 100 \text{ cm}$
$\quad = 1\,000\,000 \text{ cm}^3$
So, $1 \text{ m}^3 = 1\,000\,000 \text{ cm}^3$, or 10^6 cm^3

The volume of a cube with edge length 10 cm is:

$V = 10 \text{ cm} \times 10 \text{ cm} \times 10 \text{ cm}$
$\quad = 1000 \text{ cm}^3$
Since $1 \text{ cm}^3 = 1 \text{ mL}$,
then $1000 \text{ cm}^3 = 1000 \text{ mL}$
$\qquad\qquad\quad = 1 \text{ L}$

Example 5

Convert.

a) 0.72 m² to square centimetres

b) 1.05 m³ to cubic centimetres

Solution

a) 0.72 m² to square centimetres
$1 \text{ m}^2 = 10\,000 \text{ cm}^2$
So, $0.72 \text{ m}^2 = 0.72 \times 10\,000 \text{ cm}^2$
$\qquad\qquad\quad = 7200 \text{ cm}^2$

To multiply by 10 000, move the decimal point 4 places to the right.

b) 1.05 m³ to cubic centimetres
$1 \text{ m}^3 = 1\,000\,000 \text{ cm}^3$, or 10^6 cm^3
So, $1.05 \text{ m}^3 = 1.05 \times 10^6 \text{ cm}^3$ ◄——— This answer is in scientific notation.
$\qquad\qquad\quad = 1\,050\,000 \text{ cm}^3$ ◄——— This answer is in standard form.

✓ Check

5. Convert. Write your answers in standard form and in scientific notation, where appropriate.

a) 726.5 cm to metres

b) 4300 cm² to square metres

c) 980 000 cm³ to cubic metres

d) 4 280 000 cm³ to litres

e) 8.75 m to centimetres

f) 1.36 m² to square centimetres

g) 14.98 m³ to cubic centimetres

h) 9.87 L to cubic centimetres

When we draw a view of an object, we show internal line segments only where the depth or thickness of the object changes.
Here is an object made with 7 linking cubes.

Here are the views:

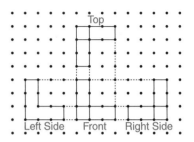

The broken lines show how the views are aligned.

Explore

Work in a group of 3.
Each student needs 8 linking cubes and isometric dot paper.
Each student chooses *one* of these views.

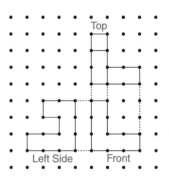

You use 8 cubes to build an object that matches the view you chose.
Sketch your object on isometric paper.
Use your cubes to build, then sketch, a different object with the view you chose.

Reflect & Share

Compare your objects with those of other members of your group.
Does any object match all 3 views?
If not, build and sketch one that does.
What helped you decide the shape of the object?
Are any other views needed to identify the object? Explain.

Each view of an object provides information about the shape of the object. When an object is built with linking cubes, the top, front, and side views are often enough to identify and build the object. These views are drawn with the top view above the front view, and the side views beside the front view, as they were at the top of page 102. In this way, matching edges are adjacent.

Example

Which object has these views?

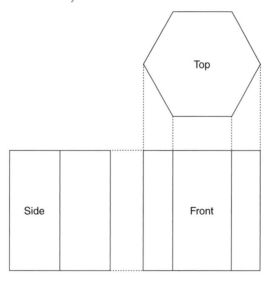

Solution

The top view is a regular hexagon.
The side view is 2 congruent rectangles.
The front view is 3 rectangles, 2 of which are congruent.
This object is a hexagonal prism.

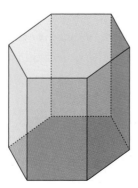

You will need linking cubes, isometric dot paper, and grid paper.

1. Match each view A to D with each object H to L.

Name each view: top, bottom, front, back, left side, or right side

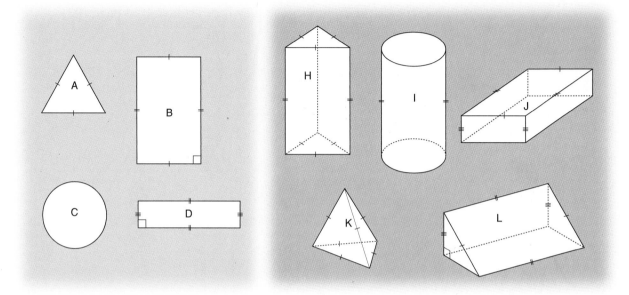

2. Sketch a different view of two of the objects in question 1.

3. Use these 4 clues and linking cubes to build an object.
Draw the object on isometric paper.
Clue 1: There are 6 cubes in all. One cube is yellow.
Clue 2: The green cube shares one face with
each of the other 5 cubes.
Clue 3: The 2 red cubes do not touch each other.
Clue 4: The 2 blue cubes do not touch each other.

4. **a)** Build the object for the set of views below.
b) Sketch the object on isometric dot paper.

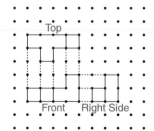

5. a) Build an object with each number of linking cubes.
For 3 or more cubes, do *not* make a rectangular prism.
i) 2 **ii)** 3 **iii)** 4 **iv)** 5 **v)** 6

b) For each object you build, count the number of its faces, edges, and vertices. Record your results in a table.

c) Look for a pattern in the table in part b.
For any object, how are the numbers of faces, edges, and vertices related?

d) Build an object with 7 linking cubes.
Check that the relationship in part c is true.

The relationship in part c is called *Euler's formula*. It is named for a Swiss mathematician, Leonhard Euler, who lived in the 18th century.

6. Assessment Focus

a) Use these views to build an object.

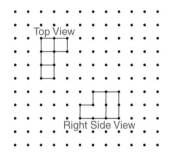

b) Sketch the object on isometric dot paper.

c) Draw the other views of the object.

Number Strategies

Write each number in scientific notation.
- 3 590 000
- 40 400 000
- 398 759

7. a) Use these views to build an object. A shaded region has no cubes.

b) Sketch the object. Explain your work.

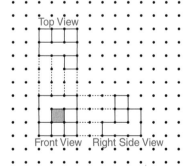

Reflect

How do views help to show a three-dimensional object?
Use an example to explain.

Sketching and Folding Nets

Focus | Sketch nets and use them to build objects.

A net is a pattern that can be folded to make an object.

Here is a net and the rectangular prism it forms.

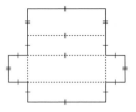 

A polyhedron can have several different nets.

Explore

Work with a partner.
You will need scissors, tape, and 1-cm grid paper.
For each set of views below:
➤ Identify the object. Draw a net of the object.
➤ Cut out your net. Check that it folds to form the object.
➤ Describe the object.

Set A

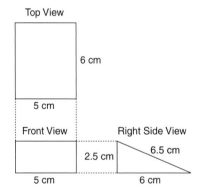

Top View

6 cm

5 cm

Front View   Right Side View

2.5 cm   6.5 cm

5 cm   6 cm

Remember that an internal line segment on a view shows that the depth changes.

Set B

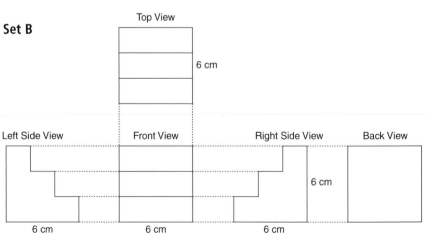

Top View

6 cm

Left Side View   Front View   Right Side View   Back View

6 cm

6 cm   6 cm   6 cm

Reflect & Share

Compare your nets with those of another pair of classmates.
How did you know how many faces to draw?
How do you know which faces share a common edge?
How do the faces on the net compare to the views for each object?
Could you have drawn a different net for the same object? Explain.

Connect

Some faces are not visible from a particular view.
Look at the views of an object, at the right.

The top view shows that four congruent
triangular faces meet at a point.
The front and side view shows an
isosceles triangular face.
The bottom view shows the base is a square.
This object is a square pyramid.
We can use a ruler, protractor, and compass to
draw the net.

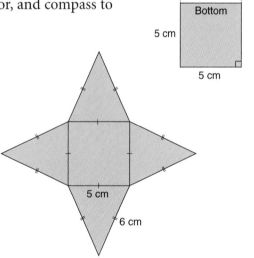

In the centre of the paper,
draw a 5-cm square
for the base.
On each side of the base,
draw an isosceles
triangle with two
equal sides of 6 cm.

Other arrangements of the five faces may
produce a net.
Each outer edge must match another
outer edge, and no faces must overlap.
The net at the right is constructed so that
the 5-cm square base is attached
to only one triangle.

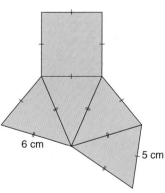

Each net folds to form a square pyramid.

Many views of an object are needed to
create its net, especially if the object
is made with linking cubes.

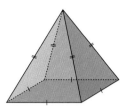

Example

These four views show
an object made
with linking cubes.
Use the views to draw
a net for the object.

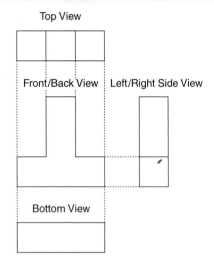

Solution

Use 1-cm dot paper.
Start with the simplest view, which is the bottom view.
Draw the bottom face.
Draw the front face and back face above
and below the bottom face.
The top view and side views show a change in depth.
A new face is drawn for each depth change.
A side square is drawn for each depth change.

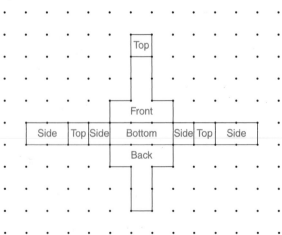

Each side view is drawn to
touch the square that shows
the change in depth.

The net in the *Example* folds to form
this object with 10 faces, including
the bottom face.

The symmetry of the object made it
easier to draw a net. Other
arrangements of the 10 faces may
be folded to form the object.

Practice

1. Each set of views below represents an object.
   a) Identify each object.
   b) Draw a net for the object.
   c) Cut out the net. Build the object.
   d) Describe the object.

   i)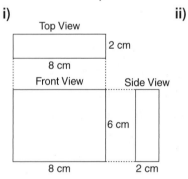

   ii)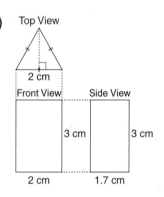

2. Each set of views represents an object.
   Draw 2 different nets for each object.
   Build the object.

   a)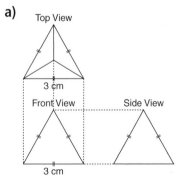

   b)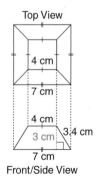

3. Choose one set of views from question 2.
   Describe the steps you used to draw the net.

4. A chocolate box has the shape of a prism with a base that is a rhombus. Each side length of the rhombus is 3.6 cm. The angles between adjacent sides of the base are 60° and 120°. The prism is 10.8 cm long.

a) Draw a net for this box.

b) How is your net different from the cardboard net from which the box is made? Explain.

5. These views represent an object. Each shaded region is an opening in a face. Draw a net for the object. Build the object.

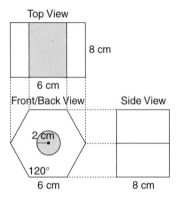

Top View

8 cm

6 cm

Front/Back View   Side View

2 cm

120°

6 cm   8 cm

6. Assessment Focus

These views represent an object.

a) Identify the object.

b) Draw two different nets.

c) Use the two different nets to build the object.

d) Is one net easier to draw or fold? Explain.

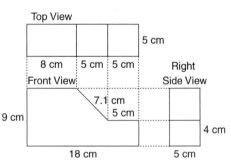

Top View

5 cm

8 cm   5 cm   5 cm   Right
Front View   Side View

7.1 cm
5 cm

9 cm   4 cm

18 cm   5 cm

Reflect

Describe how a set of views of an object relates to the figures on a net of the object. Is there one correct net for a set of views? Explain with an example.

Mid-Unit Review

LESSON

3.1 **1.** **a)** Use linking cubes.
Use the views below to build the object.
Remember that internal lines show where the depth of the object changes.

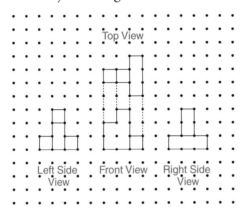

b) Sketch the object on isometric dot paper.

2. **a)** Use the views below to describe the object. Remember that a shaded area shows an opening in the face.

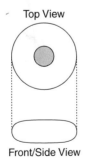

Top View

Front/Side View

b) Sketch a pictorial diagram of the object.

3.2 **3.** Each set of views that follows represents an object.

a) Identify each object.
b) Draw a net for each object.
c) Cut out the net. Build the object.
d) Describe the object.

i)

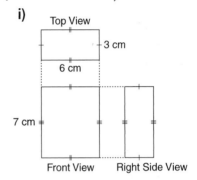

ii)

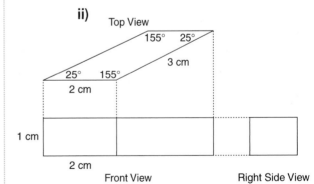

iii)

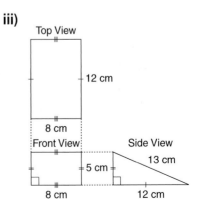

Surface Area of a Triangular Prism

Focus | Develop a formula for finding the surface area of a triangular prism.

A triangular prism is formed when a triangle is translated in the air so that each side of the triangle is always parallel to its original position.

The two triangular faces are the bases of the prism.

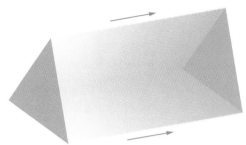

Explore

Work with a partner.
You will need 1-cm grid paper.
➤ For each triangular prism below:
  Draw a net.
  Find the surface area of the prism.

The surface area of an object is the sum of the areas of its faces.

4 cm   3 cm
5 cm
6 cm

Prism A

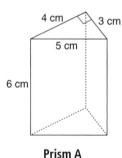

2.5 cm   3.2 cm
6 cm
4 cm

Prism B

➤ Write a formula you can use to find the surface area of any triangular prism.

Reflect & Share

Compare your nets and formula with those of another pair of classmates.
Did you write the same formula?
If not, do both formulas work? Explain.
Did you write a word formula or use variables? Explain.

Here is a triangular prism and its net.
Both the prism and net are drawn to scale.

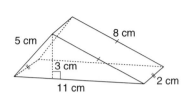

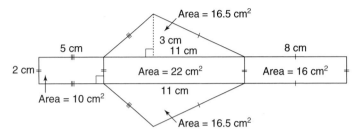

The two triangular faces of the prism are congruent.
Each triangular face has base 11 cm and height 3 cm.
So, the area of one triangular face is: $\frac{1}{2} \times 11 \times 3$
The surface area of a triangular prism can be expressed
using a word formula:

The area of a rectangle is base × height.

SA = sum of the areas of three rectangular faces +
$\qquad$ 2 × area of one triangular face

Use this formula to find the surface area of the prism above.

Use the order of operations. Multiply before adding.

$SA = (5 \times 2) + (11 \times 2) + (8 \times 2) + 2 \times \frac{1}{2} \times 11 \times 3$
$\qquad = 10 + 22 + 16 + 33$
$\qquad = 81$

The surface area of the prism is 81 cm².

We can use variables to write a formula for the surface area of a
triangular prism.

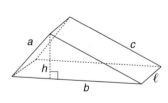

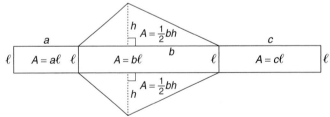

To avoid confusion between the height of a triangle and the height of
the prism, we now use *length* instead of *height* to describe the edge
that is perpendicular to the base.
For the triangular prism above:
The length of the prism is *l*.
Each triangular face has side lengths *a*, *b*, and *c*.

a, b, c, h, and l are variables.

The height of a triangular face is *h* and its base is *b*.

The surface area of the prism is:

SA = sum of the areas of the 3 rectangular faces + 2 × area of one triangular face

$SA = al + bl + cl + 2 \times \frac{1}{2}bh$

$SA = al + bl + cl + bh$

Example

Find the surface area of the prism below.

Each dimension has been rounded to the nearest whole number.

Write the surface area in square centimetres and in square metres.

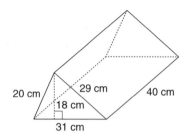

Solution

Identify the variable that represents each dimension.
Sketch, then label the prism with these variables.
The length of the prism is:

$l = 40$

The 3 sides of a triangular face are:

$a = 20, b = 31, c = 29$

The height of a triangular face is:

$h = 18$

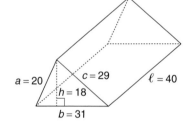

Substitute for each variable in the formula for surface area.

$SA = al + bl + cl + bh$

$= (20 \times 40) + (31 \times 40) + (29 \times 40) + (31 \times 18)$

$= 800 + 1240 + 1160 + 558$

$= 3758$

The surface area of the prism is 3758 cm².

To convert square centimetres to square metres, divide by 10 000.

$3758 \text{ cm}^2 = \frac{3758}{10\ 000} \text{ m}^2$

$= 0.3758 \text{ m}^2$

The surface area of the prism is 0.3758 m².

1. Calculate the area of each net.

a)

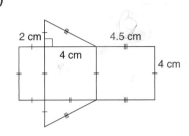

b)

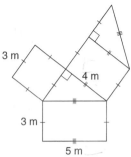

2. Calculate the surface area of each prism.
Draw a net first if it helps.
Write the surface area in square metres.

a)

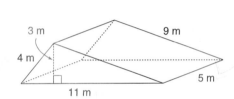

b)

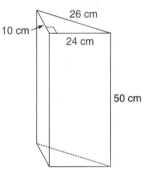

3. Calculate the surface area of each prism.
The shaded region indicates that the face is missing.
Write the surface area in square centimetres.

When a face of a prism is missing, the prism is a shell, not a solid.

a)

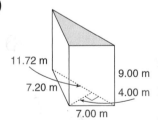

b)

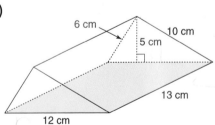

4. a) What area of wood, in square metres, is needed to make the ramp at the left?
The ramp does not have a base.

b) Suppose the ramp is built against a stage.
The vertical face that is a rectangle is against the stage.
How much wood is needed now?

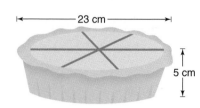

23 cm

5 cm

5. A plastic container company designs a container with a lid to hold one piece of pie.
   a) Design the container as a triangular prism. Explain your choice of dimensions.
   b) Calculate the area of plastic in your design.

6. The total area of the 3 rectangular faces of an equilateral triangular prism is 72 cm².
   a) No dimension can be 1 cm. What are the possible whole-number dimensions of the edges?
   b) Sketch the prism with the greatest length. Include dimensions. Explain your choice.

7. ⬤ **Assessment Focus** How much metal, in square metres, is needed to build this water trough?

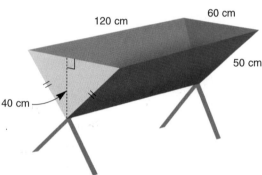

120 cm

60 cm

50 cm

40 cm

8. A right triangular prism has a base with perimeter 12 cm and area 6 cm².
   a) Find the whole-number dimensions of the base.
   b) The length of the prism is 6 cm. Calculate the surface area of the prism.
   c) Sketch the prism. Include its dimensions.

Take It Further

9. Use the variables below to sketch and label a triangular prism. The lengths have been rounded to the nearest whole number. Calculate the surface area of the prism.
   $a = 7$ cm, $b = 17$ cm, $c = 11$ cm, $h = 3$ cm, $l = 12$ cm

Reflect

Write to explain how to find the surface area of a triangular prism. Include an example and a diagram.

Volume of a Triangular Prism

Focus Develop and use a formula for finding the volume of a triangular prism.

Recall that the area of a triangle is one-half the area of a rectangle that has the same base and height. That is,

the area of $\triangle$DEC = $\frac{1}{2}$ the area of rectangle ABCD

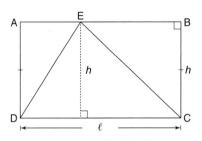

Explore

Work in a group of 4.

You will need 4 identical cereal boxes, one for each group member, and markers.

➤ Find the volume of your cereal box, which is a rectangular prism.

➤ Use a ruler to draw a triangle on one end of the cereal box. The base of each triangle should be along one edge of the box. The third vertex of the triangle should be on the opposite edge. Make sure you draw different triangles.
What is the volume of a triangular prism with this triangle as its base, and with length equal to the length of the cereal box?

➤ Compare your answer with those of other members of your group. What do you notice?

➤ Work together to write a formula for the volume of a triangular prism.

Reflect & Share

How is the volume of a triangular prism related to the volume of a rectangular prism?

Compare your formula for the volume of a triangular prism with that of another group.

Did you use variables in your formula?

If not, work together to write a formula that uses variables.

Connect

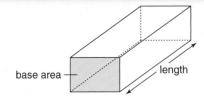

The volume of a rectangular prism is:
V = base area $\times$ length

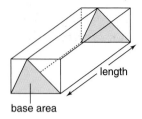

base area

Suppose we draw a triangle on the base of the prism so that the base of the triangle is one edge, and the third vertex of the triangle is on the opposite edge.

The volume of a triangular prism with this base, and with length equal to the length of the rectangular prism, is one-half the volume of the rectangular prism.

Since the base area of the triangular prism is one-half the base area of the rectangular prism, the volume of a triangular prism is also: $V = \text{base area} \times \text{length}$
The base is a triangle, so the base area is the area of a triangle.

We can use variables to write a formula for the volume of a triangular prism.

For the triangular prism below:
The length of the prism is l.
Each triangular face has base b and height h.
The volume of the prism is:

When we use this formula, we must identify what each variable represents.

$V = \text{base area} \times \text{length}$
$V = \frac{1}{2} bh \times l$
$V = \frac{1}{2} bhl$

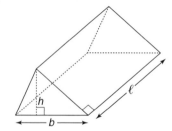

Example

How much water can the water trough hold?
Give the answer in litres.

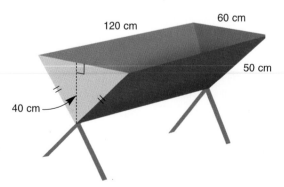

Solution

Capacity is the amount a container will hold, commonly measured in litres (L) or millilitres (mL). Volume is the amount of space an object occupies, commonly measured in cubic units.

The amount of water the trough can hold is the capacity of the triangular prism.

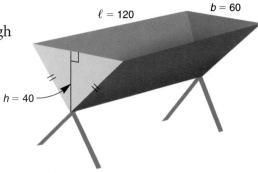

Sketch the prism.
Identify the variable that represents each dimension.
The base of the triangle is: $b = 60$
The height of the triangle is: $h = 40$
The length of the prism is: $l = 120$
Substitute for each variable into the formula for volume.
$V = \frac{1}{2}bhl$
$V = \frac{1}{2} \times 60 \times 40 \times 120$
$V = 144\ 000$
The volume of the trough is 144 000 cm³.
$1000\ \text{cm}^3 = 1\ \text{L}$
So, 144 000 cm³ = 144 L
The trough can hold 144 L of water.

Practice

1. The base area and length for each triangular prism are given. Find the volume of each prism.

a)
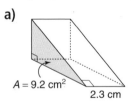
$A = 9.2\ \text{cm}^2$
2.3 cm

b)
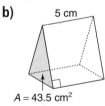
5 cm
$A = 43.5\ \text{cm}^2$

c)
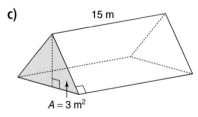
15 m
$A = 3\ \text{m}^2$

2. Find the volume of each triangular prism.

a)
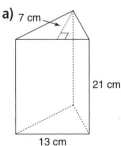
7 cm
21 cm
13 cm

b)
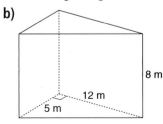
8 m
12 m
5 m

c)
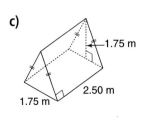
1.75 m
2.50 m
1.75 m

3. Calculate the volume of each prism.

a)

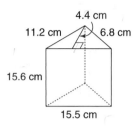
11.2 cm
4.4 cm
6.8 cm
15.6 cm
15.5 cm

b)

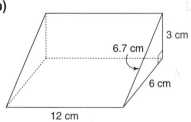
3 cm
6.7 cm
6 cm
12 cm

Use an isometric diagram or a pictorial diagram for the sketch of the prism.

4. Find possible values for *b*, *h*, and *l* for each volume of a triangular prism. Sketch one possible prism for each volume.
a) 5 cm^3 b) 9 m^3 c) 8 m^3 d) 18 cm^3

5. What is the volume of glass in this prism?

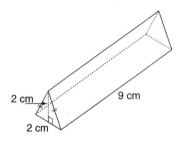
2 cm
9 cm
2 cm

6. Any face can be used as the base of a rectangular prism.
Can any face be used as the base of a triangular prism? Explain.

7. The volume of a triangular prism is 30 cm^3.
Each triangular face has an area of 4 cm^2.
How long is the prism?

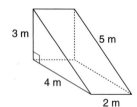
3 m
5 m
4 m
2 m

8. a) Calculate the surface area and volume of this triangular prism.
b) What do you think happens to the surface area and volume when the length of the prism is doubled? Justify your answer. Calculate the surface area and volume to check your ideas.
c) What do you think happens to the surface area and volume when the base and height of the triangular faces are doubled? Justify your answer.
Calculate the surface area and volume to check your ideas.
d) What do you think happens to surface area and volume when all the dimensions are doubled?
Justify your answer.
Calculate the surface area and volume to check your ideas.

9. **Assessment Focus** Jackie uses this
   form to build a concrete pad.
   a) How much concrete will Jackie
      need to mix to fill the form?
   b) Suppose Jackie increases the equal
      sides of the form from 3 m to 6 m.
      How much more concrete will Jackie nee̶
      Include a diagram.

10. A chocolate company produces different sizes
    of chocolate bars that are packaged in
    equilateral triangular prisms.
    Here is the 100-g chocolate bar.

a) Calculate the surface area and volume of the box.
b) The company produces a 400-g chocolate bar.
   It has the same shape as the 100-g bar.
   i) What are the possible dimensions for the 400-g box?
   ii) How are the dimensions of the two boxes related?

Take It Further

11. The volume and surface area of a prism, with a base that is not
    a triangle or a rectangle, can be found by dividing the prism
    into smaller prisms.
    Find the volume and surface area of each prism.
    a)

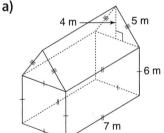

b)

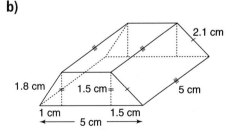

Reflect

Describe the relationships among the dimensions, faces,
and volume of a triangular prism.
Include an example in your description.

eatures of Word Problems

In *Unit 2*, you wrote problem statements as questions.
Math problems include information to help you understand
and solve the problem.
Here are some features of math word problems.

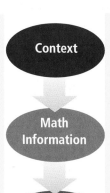

Context – may describe who, what, when, where, and why, like the setting of a story.

Math Information – may include words, numbers, figures, drawings, tables, graphs, and/or models.

Problem Statement – tells you to do something with the information. It may be a question or an instruction. Key words are sometimes used to suggest how the response is to be communicated.

Suppose you have an old refrigerator and are thinking about buying a new one.

The new refrigerator costs $900. It costs $126 a year to run your old fridge, and $66 a year to run the new, more energy-efficient fridge.

Is it more economical to buy the new fridge? Justify your answer.

To solve the problem, we need to compare the costs of buying the new fridge and running it with the costs of running the old fridge.
We can use a table and a graph.
In the second column of the table, add $126 each year.
In the third column of the table, add $66 each year.

Cumulative Costs		
Year	Old Fridge Cost ($)	New Fridge Cost ($)
1	126	900 + 66 = 966
2	252	1032
3	378	1098
4	504	1164
5	630	1230
6	756	1296
7	882	1362

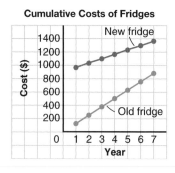

Cumulative Costs of Fridges

Here is part of the table and a graph.
Continue the table until the costs are equal.
Draw a graph for the data.
Use the table and graph to solve the problem.

 **Check**

With a partner, identify the context, math information, and problem statement for each of questions 1 to 4. Then solve the problem.

CARPET SALE

Regular $9.99
per square metre

Now on Sale

For **20%** off

1. Hori is buying carpet for his living room.
   It is rectangular with dimensions 4 m by 5 m.
   How much will Hori save if he buys the carpet at the sale price shown at the left?

2. Suppose you are in charge of setting up the cafeteria for a graduation dinner. One hundred twenty-two people will attend.
   The tables seat either 8 or 10 people.
   You do not want empty seats.
   How many of each size table will you need to use to make sure everyone has a seat? List all the combinations.

3. To win a contest, you have to find a mystery number.
   The mystery number is described this way:
   Sixteen more than $\frac{2}{3}$ of the mystery number is equal to two times the mystery number.
   What is the mystery number?

4. Alicia and Chantelle are playing a game.
   There are six tiles in a box: three red and three blue.
   A player picks two tiles without looking.
   Alicia gets a point if the tiles do not match; Chantelle gets a point if they do match. The tiles are returned to the box each time.
   What is the probability that both tiles have the same colour?
   Is this a fair game? Explain.

5. Write a word problem using the numbers 48, 149, and 600.
   Remember to include a context, math information, and a problem statement.
   Trade problems with a classmate.
   Identify the features of your classmate's problem.
   Solve the problem.

Unit Review

What Do I Need to Know?

☑ **Euler's Formula**

For any polyhedron, the numbers of faces, edges, and vertices are related by this formula:

vertices + faces − edges = 2

☑ **Surface Area of a Triangular Prism**

Surface area = sum of the areas of 3 rectangular faces +
2 × area of one triangular face

$SA = al + bl + cl + bh$

The side lengths of a triangular face are a, b, and c.

The height of a triangular face is h.

The length of the prism is l.

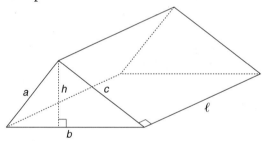

☑ **Volume of a Triangular Prism**

Volume = area of triangular base × length of prism

$V = \frac{1}{2}bhl$

The base and height of a triangular face are b and h, respectively.

The length of the prism is l.

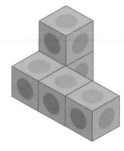

LESSON

3.1 **1.** You will need linking cubes and isometric dot paper.

a) Build the object that matches the views below.

b) Sketch the object on isometric dot paper.

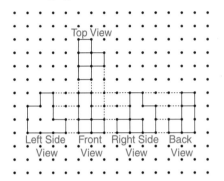

2. Sketch as many objects as possible that have two different rectangles as two of its views.

3.2 **3.** This set of views represents an object.

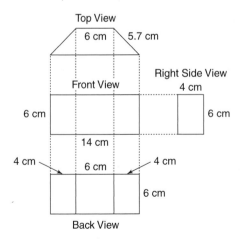

a) Identify the object.

b) Draw a net for the object on 1-cm grid paper.

c) Cut out the net. Build the object.

d) Describe the object.

3.3 **3.4** **4.** Here is a net of a triangular prism.

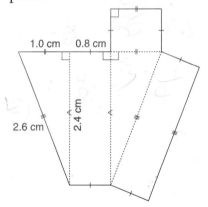

a) Calculate the surface area of the prism in square centimetres.

b) Calculate the volume of the prism in cubic centimetres.

5. a) Calculate the surface area of this prism. Sketch a net first, if it helps.

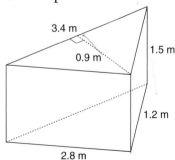

b) Calculate the volume of the prism.

3.4 **6.** The horticultural society is building a triangular flower bed at the intersection of two streets. The edges of the bed are raised 0.25 m. How much soil is needed to fill this flower bed? Justify your answer.

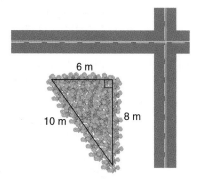

7. Find the possible values of *b*, *h*, and *l* for a triangular prism with volume 21 m³.
How many different ways can you do this? Sketch a diagram of one possible prism.

8. Alijah volunteers with the horticultural society.
He wants to increase the size but not the depth of the flower bed in question 6.
   a) How can Alijah change the dimensions so that:
   • the flower bed remains triangular, and
   • the area of the ground covered by the bed doubles?
   b) Sketch the new flower bed. Label its dimensions.
   c) How does the change in size affect the volume of soil needed? Explain.

9. The bucket on the front of a lawn tractor is a triangular prism.

   a) Find the volume of soil, in cubic metres, the bucket can hold. What assumptions do you make?
   b) Suppose the dimensions of the triangular faces are doubled.
   How much more soil do you expect the new bucket to hold? Explain.
   c) Calculate the new volume. Sketch the new bucket and include the new dimensions.

10. A tent has the shape of a triangular prism.
Its volume is 25 m³.
   a) Find possible dimensions for this prism.
   b) Choose one set of dimensions from part a. Sketch the tent. Label the dimensions.
   c) A larger tent has volume 100 m³. It is also a triangular prism.
   i) What could the dimensions of this larger tent be? Justify your answer.
   ii) How are the dimensions of the two tents related?

Practice Test

1. You will need linking cubes and isometric dot paper.
   a) Build the object that has these views.
   b) Sketch the object on isometric dot paper.
   c) Draw a net for the object. Cut out and fold the net to make the object.

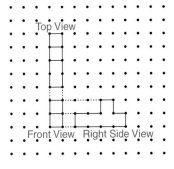

Top View

Front View   Right Side View

2. Calculate the surface area and volume of this prism.

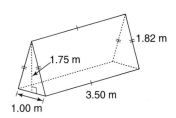

1.82 m
1.75 m
3.50 m
1.00 m

3. Look at the triangular prism in question 2.
   Suppose the base and height of the triangular faces are tripled.
   a) How does this affect the volume of the prism? Explain.
   b) Sketch the larger prism.
   c) Calculate the volume of the larger prism.

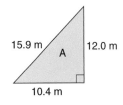

15.9 m   A   12.0 m
10.4 m

11.2 m
B   7.8 m
16 m

4. The triangular faces at the left are the bases of four triangular prisms. All the prisms have the same length.
   a) Which prism has the greatest volume? Explain.
   b) Which prism has the least surface area? Explain.

5. The volume of a triangular prism is 210 cm³.
   a) The length of the prism is 7 cm.
      What are the possible base and height of the triangular faces?
   b) On 1-cm grid paper, draw two possible triangular faces for this prism. Measure to find the lengths of any sides you do not know.
   c) Calculate the surface area of each prism whose face you drew in part b.

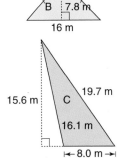

15.6 m   C   19.7 m
16.1 m
8.0 m

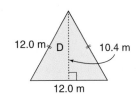

12.0 m   D   10.4 m
12.0 m

Suppose your class is responsible for building a circus tent.
The organizers have given you the views below.

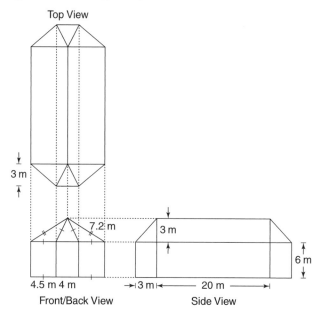

Top View

3 m

7.2 m

3 m

6 m

4.5 m 4 m

3 m 20 m

Front/Back View

Side View

Work in a group.

Part 1

Prepare a presentation for the organizers.
Your presentation must include:

- a 3-D sketch
- a net for the tent
- a list of steps for building the net
- a model of the tent

Part 2

Some members of a local service club will perform in the circus. They need a prop built for their act. It is a triangular prism in which one performer can hide, and which can also be used as a ramp for a bicycle jump. Prepare an estimated cost to build this prop. Your estimate must include:

- a diagram of the prop with appropriate dimensions
- an explanation of the dimensions you chose
- calculations for the amount of materials used
- cost of materials if the building material is $16.50/m^2$

Check List

Your work should show:

✓ a correctly constructed model

✓ all diagrams, sketches, and calculations in detail

✓ clear explanations of your choices, procedures, and results

✓ a description of situations at home where volume and capacity are used

Part 3

The volume or capacity of the tent depends on its dimensions. Research situations where volume and capacity are used in your home. Write a report on your findings.

Reflect on the Unit

Write a paragraph on what you have learned about representing three-dimensional objects and triangular prisms. Try to include something from each lesson in the unit.

Golden Rectangles

Work with a partner.

Similar polygons have the same shape, but different sizes.
These two rectangles are similar.

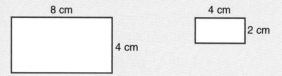

Similar rectangles have the same base:height ratio.
That is, 8 cm:4 cm = 4 cm:2 cm

In this *Investigation*, you will draw rectangles and compare
the ratios of their side lengths.

As you complete the *Investigation*, include all your work in a
report that you will hand in.

Materials:

• 1-cm grid
paper

Part 1

➤ Use 1-cm grid paper.
Turn your paper so the longer side is horizontal.
Draw a 1-cm square in the bottom left hand corner.
Label this square ABCD.
Draw another 1-cm square directly above the first.
You should now have a rectangle that measures 2 cm by 1 cm.
Label this rectangle ABEF.
Make a larger rectangle by drawing a square along a longer
side of rectangle ABEF. Label this 3-cm by 2-cm rectangle AGHF.

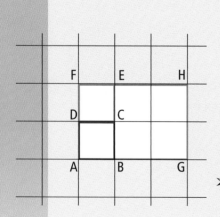

➤ Continue to make larger rectangles by drawing squares
along a longer side of the previous rectangle.
Continue to label the rectangles with letters.
Is there a pattern in the way you draw the rectangles? Explain.

➤ Stop drawing rectangles when you have no more room
on the paper.
Record the base and height of each rectangle in a table.
For this *Investigation*, the base is always the longer side.

➢ Write the ratio of the base to the height in fraction form. Calculate the value of the ratio by dividing the base by the height. Use a calculator if you need to.
Round each number to 3 decimal places when necessary. Include these numbers in the table.

Rectangle	Base	Height	Ratio $\frac{b}{h}$	Ratio Value
ABCD	1	1	$\frac{1}{1}$	1.0
ABEF	2	1	$\frac{2}{1}$	2.0

➢ What patterns do you see in the table?

➢ Extend the patterns. What are the dimensions of the next 3 rectangles? How do you know?

➢ Calculate the ratio of $\frac{b}{h}$ for each of the next 3 rectangles. What do you notice about the value of the ratio as the rectangles get larger?

➢ Predict the value of the ratio for the 50th rectangle. What do you think the ratio will be for the 100th rectangle? Explain.

➢ Are the rectangles in the table similar? Explain.

Part 2

➢ Look for rectangles around the room.
Some examples might be found in notebook paper, textbooks, and windows. Include these rectangles in your table.

➢ Measure the base and height of each rectangle you include.
What is the ratio of $\frac{b}{h}$ for each?
What is the value of this ratio?

Take It Further

Find some pictures of structures or buildings.
Look for rectangles in these structures.
Find the values of $\frac{b}{h}$ for each rectangle. Does any rectangle have the same ratio as the last rectangle in your table?

Fractions and Decimals

Look at the figures at the right. How is each figure made from the previous figure? The equilateral triangle in *Step 1* has side length 1 unit. What is the perimeter of each figure?

What patterns do you see? Suppose the patterns continue. What will the perimeter of the figure in *Step 4* be?
For any figure after *Step 1*, how is its perimeter related to the preceding figure?

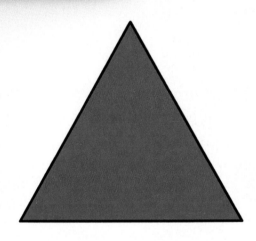

Step 1

What You'll Learn

- Compare and order fractions.
- Add and subtract fractions.
- Multiply a fraction by a whole number and by a fraction.
- Divide a whole number by a fraction, and a fraction by a fraction.
- Convert between fractions and decimals.

Why It's Important

You use fractions and decimals when you shop, measure, and work with a percent; and in sports, recipes, and business.

Step 2

Step 3

Key Words

- reciprocals
- terminating decimal
- repeating decimal

Skills You'll Need

Multiply a Fraction by a Whole Number

We multiply a fraction by a whole number to find that fraction of the whole number.

Example

a) Find $\frac{1}{3}$ of 27.

b) Multiply. $\frac{3}{4} \times 6$

Solution

a) $\frac{1}{3}$ of 27 is

$\frac{1}{3} \times 27 = 27 \times \frac{1}{3}$

Think: 27 times $\frac{1}{3}$ is 27 thirds.

$\frac{1}{3} \times 27 = \frac{27}{3}$

$= 9$

b) $\frac{3}{4} \times 6 = \frac{18}{4}$

$= \frac{9}{2}$

In *Example* part a, note that $27 \times \frac{1}{3}$ is the same as $27 \div 3$.
To find $\frac{1}{3}$ of a number, we multiply by $\frac{1}{3}$ or divide by 3.
$\frac{1}{3}$ and 3 are **reciprocals**.

A unit fraction has numerator 1.

To multiply a number by a unit fraction, we can divide by the reciprocal instead.

✓ Check

1. Find.

a) $\frac{1}{5}$ of 25

b) $\frac{1}{4}$ of 64

c) $\frac{1}{8}$ of 40

2. Multiply.

a) $\frac{3}{2} \times 20$

b) $\frac{5}{3} \times 5$

c) $\frac{7}{9} \times 4$

3. There are 660 students in Parkside School from Kindergarten to Grade 8.

a) Three-quarters of the students are boys. How many boys attend the school?

b) One-third of the students are in K to Grade 4. How many students are in Grades 5 to 8?

Focus | Use equivalent fractions to compare fractions.

Some photographers use a manual camera with a shutter speed dial. The numbers on the dial show how long the shutter stays open when a person takes a picture.

This setting opens the shutter for 2 s.

This setting opens the shutter for $\frac{1}{2}$ of 1 s.

This setting opens the shutter for $\frac{1}{4}$ of 1 s.

This pattern continues.
How long is the shutter open when the setting is 8? 15? 30?
Does a setting of 8 allow more or less light than a setting of 15?
Suppose a setting of 125 does not allow enough light.
Which setting might allow enough light?

Explore

Work in a group.

This square has side length 1 unit.

Your teacher will give you a copy of this square. What fraction of the whole square is each piece? Order the fractions from least to greatest.

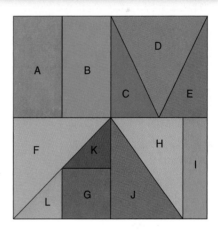

Reflect & Share

Share your results with those of another group of classmates.
How did you write each piece as a fraction of the whole?
What strategies did you use to order the fractions?
Could you use these strategies to order any set of fractions?
Explain.

One way to compare fractions is to use equivalent fractions.
Write each fraction with the same denominator,
then compare the numerators.

For example, to order $\frac{5}{8}$, $\frac{4}{5}$, and $\frac{3}{4}$:

Write equivalent fractions for each fraction until all the fractions
have the same denominator:

**These are proper
fractions. In a proper
fraction, the numerator
is less than the
denominator.**

$$\frac{5}{8} = \frac{10}{16} = \frac{15}{24} = \frac{20}{32} = \frac{25}{40}$$

(×2, ×3, ×4, ×5)

$$\frac{4}{5} = \frac{8}{10} = \frac{12}{15} = \frac{16}{20} = \frac{20}{25} = \frac{24}{30} = \frac{28}{35} = \frac{32}{40}$$

$$\frac{3}{4} = \frac{6}{8} = \frac{9}{12} = \frac{12}{16} = \frac{15}{20} = \frac{18}{24} = \frac{21}{28} = \frac{24}{32} = \frac{27}{36} = \frac{30}{40}$$

Now, each fraction has denominator 40.

Compare the numerators: $25 < 30 < 32$

So, $\frac{25}{40} < \frac{30}{40} < \frac{32}{40}$

So, $\frac{5}{8} < \frac{3}{4} < \frac{4}{5}$

In order from least to greatest: $\frac{5}{8}$, $\frac{3}{4}$, $\frac{4}{5}$

A simpler way to find the common denominator is to find
the lowest common multiple of the denominators.
We can use this method to order improper fractions.

Example

Write these fractions in order from least to greatest:
$\frac{5}{3}$, $\frac{3}{2}$, $\frac{8}{5}$

Solution

$\frac{5}{3}$, $\frac{3}{2}$, $\frac{8}{5}$

Find the lowest common denominator.

**In an improper
fraction, the
numerator is greater
than the denominator.**

Since the denominators have no common factors,
list the multiples of 3, 2, and 5.

3: 3, 6, 9, 12, 15, 18, 21, 24, 27, **30**, ...

2: 2, 4, 6, 8, 10, 12, 14, 16, 18, 20, 22, 24, 26, 28, **30**, ...

5: 5, 10, 15, 20, 25, **30**, ...

The lowest common denominator is 30.

$$\overset{\times 10}{\frac{5}{3}} = \frac{50}{30} \qquad \overset{\times 15}{\frac{3}{2}} = \frac{45}{30} \qquad \overset{\times 6}{\frac{8}{5}} = \frac{48}{30}$$

So, in order from least to greatest: $\frac{3}{2}, \frac{8}{5}, \frac{5}{3}$

Use this method to find the lowest common denominator when the denominators have no common factors.

In the *Example*, notice that the lowest common multiple of 3, 2, and 5 is their product: $3 \times 2 \times 5 = 30$
When two or more numbers have no common factors, their lowest common multiple is their product.

Practice

1. In each pair, which fraction is greater? How do you know?

a) $\frac{1}{2}, \frac{2}{5}$ b) $\frac{2}{3}, \frac{5}{6}$ c) $\frac{1}{2}, \frac{2}{3}$ d) $\frac{3}{4}, \frac{2}{5}$

e) $\frac{1}{4}, \frac{1}{3}$ f) $\frac{2}{3}, \frac{3}{4}$ g) $\frac{3}{4}, \frac{5}{8}$ h) $\frac{2}{5}, \frac{3}{10}$

2. Order the fractions in each set from least to greatest.

a) $\frac{3}{8}, \frac{4}{5}, \frac{1}{2}$ b) $\frac{7}{10}, \frac{6}{8}, \frac{3}{5}$ c) $\frac{5}{2}, \frac{6}{3}, \frac{7}{4}$ d) $\frac{10}{3}, \frac{7}{5}, \frac{13}{6}$

3. Use the fractions $\frac{19}{10}, \frac{11}{3}, \frac{9}{4}$.

a) Order the fractions from least to greatest.

b) Write each fraction as a mixed number.

c) Order the mixed numbers from least to greatest.

d) Which method was easier: ordering the improper fractions or ordering the mixed numbers? Explain.
    When would you use the method of ordering mixed numbers?

4. Maria stated that $\frac{5}{6}$ is between $\frac{4}{5}$ and $\frac{6}{7}$.
Do you agree? Give reasons for your answer.

5. Find the fraction that is halfway between each pair of numbers.
Use a number line if it helps.

a) 0 and 1   b) 1 and 2   c) 0 and $\frac{1}{2}$

d) $\frac{1}{2}$ and 1   e) 1 and $\frac{3}{2}$ f) $\frac{3}{2}$ and 2

The Farey sequence is named for the British geologist and mathematician, John Farey, who lived from 1766 to 1826.

6. A Farey sequence is a list of certain fractions that follow a pattern.

1st Farey sequence: $\frac{0}{1}, \frac{1}{1}$

2nd Farey sequence: $\frac{0}{1}, \frac{1}{1}, \frac{1}{2}$

3rd Farey sequence: $\frac{0}{1}, \frac{1}{1}, \frac{1}{2}, \frac{1}{3}, \frac{2}{3}$

4th Farey sequence: $\frac{0}{1}, \frac{1}{1}, \frac{1}{2}, \frac{1}{3}, \frac{2}{3}, \frac{1}{4}, \frac{3}{4}$

a) Look at the lists above. Extend the pattern. Write the 5th Farey sequence.

b) Order the fractions in the 5th Farey sequence from least to greatest.

7. **Assessment Focus** The fraction $\frac{11}{2}$ is halfway between 5 and 6. The fraction $\frac{23}{4}$ is halfway between $\frac{11}{2}$ and 6.

a) How many more fractions can you find between 5 and 6? List all the fractions you find.

b) Have you found *all* the fractions between 5 and 6? How do you know?

Show your work.

8. a) Use the digits 3, 4, 5, and 6 to write as many proper and improper fractions as you can. Each numerator and denominator is a single digit.

b) Order the fractions in part a from least to greatest.

c) Which fractions in part a are:

   i) less than $\frac{1}{2}$?

   ii) between $\frac{1}{2}$ and 1?

   iii) greater than 1?

Mental Math

Use whole numbers.

List all the possible dimensions of a rectangular prism with volume 72 cm³.

Take It Further

9. In each pair, which fraction is greater? How do you know?

a) $\frac{22}{32}$ or $\frac{43}{65}$

b) $\frac{91\,919}{99\,999}$ or $\frac{919}{999}$

Reflect

Name two ways you can compare and order fractions. Which way do you prefer? Explain.

Adding Fractions

Explore

Work on your own.

Copy these diagrams.

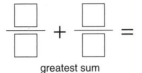
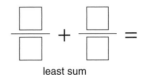

greatest sum   least sum

Use the numbers 1, 2, 4, and 8 to make
the greatest sum and the least sum.
In each case, use each number once.

Reflect & Share

Share your results with a classmate.
Do both of you have the same answers?
If not, which is the greatest sum? The least sum?
What strategies did you use to add?

Connect

Recall how to add fractions with the same denominator.
For example, to add $\frac{3}{12}$ and $\frac{4}{12}$,
add the numerators: $\frac{3}{12} + \frac{4}{12} = \frac{7}{12}$

We can illustrate this sum with a diagram.

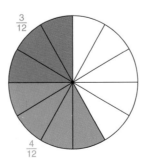

To add fractions that do not have the same denominator,
we find and use a common denominator.

Example 1

Add. $\frac{5}{12} + \frac{5}{6}$

Solution

$\frac{5}{12} + \frac{5}{6}$

Use equivalent fractions to write the fractions with a common denominator.
Since 6 is a factor of 12, the lowest common multiple of 12 and 6 is 12.

Use 12 as the common denominator.

$$\overset{\times 2}{\underset{\times 2}{\frac{5}{6} = \frac{10}{12}}}$$

$\frac{5}{12} + \frac{5}{6} = \frac{5}{12} + \frac{10}{12}$ Add the numerators.

A fraction is in simplest form when the numerator and denominator have no common factors.

$= \frac{15}{12}$

$= \frac{15 \div 3}{12 \div 3}$ To reduce to simplest form, divide the numerator and denominator by their greatest common factor, 3.

$= \frac{5}{4}$

Since 5 > 4, this is an improper fraction.

To write the fraction as a mixed number:

$\frac{5}{4} = \frac{4}{4} + \frac{1}{4}$

$= 1 + \frac{1}{4}$

$= 1\frac{1}{4}$ ◄——————— This is a mixed number.

We can also use common denominators to add more than two fractions.

Example 2

Add. $\frac{2}{3} + \frac{4}{5} + \frac{3}{4}$

Solution

$\frac{2}{3} + \frac{4}{5} + \frac{3}{4}$

Use equivalent fractions to write the fractions with a common denominator.

The denominators 3, 5, and 4 have no common factors.
So, their lowest common multiple is their product: $3 \times 5 \times 4 = 60$
Write each fraction with denominator 60.

$$\overset{\times 20}{\frac{2}{3}} = \frac{40}{60} \qquad \overset{\times 12}{\frac{4}{5}} = \frac{48}{60} \qquad \overset{\times 15}{\frac{3}{4}} = \frac{45}{60}$$

$$\frac{2}{3} + \frac{4}{5} + \frac{3}{4} = \frac{40}{60} + \frac{48}{60} + \frac{45}{60}$$

$$= \frac{133}{60} \quad \text{Since } 133 > 60, \text{ this is an improper fraction.}$$

$$\text{It can be written as a mixed number.}$$

$$= \frac{120}{60} + \frac{13}{60}$$

$$= 2 + \frac{13}{60}$$

$$= 2\frac{13}{60}$$

Practice

Write all sums in simplest form.

1. Add.

a) $\frac{4}{9} + \frac{1}{3}$ b) $\frac{1}{2} + \frac{1}{3}$ c) $\frac{2}{3} + \frac{1}{6}$ d) $\frac{3}{4} + \frac{1}{6}$

e) $\frac{2}{5} + \frac{1}{3}$ f) $\frac{2}{5} + \frac{1}{10}$ g) $\frac{1}{12} + \frac{1}{4}$ h) $\frac{3}{8} + \frac{1}{4}$

2. Add.

a) $\frac{3}{8} + \frac{3}{2}$ b) $\frac{7}{4} + \frac{4}{5}$ c) $\frac{7}{6} + \frac{5}{7}$ d) $\frac{13}{10} + \frac{4}{3}$

e) $\frac{5}{8} + \frac{2}{3}$ f) $\frac{4}{5} + \frac{4}{7}$ g) $\frac{9}{4} + \frac{4}{9}$ h) $\frac{8}{5} + \frac{11}{6}$

3. Damara and Baldwin had to shovel snow from their driveway.
Damara shovelled about $\frac{3}{10}$ of the driveway.
Baldwin shovelled about $\frac{2}{3}$ of the driveway.
About what fraction of the driveway was cleared of snow?

4. a) Write each fraction as the sum of two fractions
with the same denominator.

   i) $\frac{1}{2}$ ii) $\frac{3}{4}$ iii) $\frac{9}{10}$

b) Write each fraction in part a as the sum of two fractions
with different denominators.

5. **Assessment Focus** Write each fraction as the sum of two or more fractions in as many different ways as you can.

   a) $\frac{4}{5}$ b) $\frac{7}{10}$ c) $\frac{2}{9}$

   Show your work.

To add mixed numbers, add the fractions, then add the whole numbers.

6. Add.

   a) $3\frac{1}{3} + 4\frac{1}{4}$ b) $2\frac{1}{2} + 1\frac{9}{10}$ c) $1\frac{3}{4} + 2\frac{3}{5}$

   d) $\frac{7}{8} + 1\frac{2}{3}$ e) $2\frac{3}{5} + \frac{2}{3}$ f) $5\frac{2}{5} + 1\frac{7}{8}$

7. Two students, Galen and Mai, worked on a project.
   Galen worked for $3\frac{2}{3}$ h.
   Mai worked for $2\frac{4}{5}$ h.
   What was the total time spent on the project?

8. Add.

   a) $\frac{1}{2} + \frac{2}{3} + \frac{3}{4}$ b) $\frac{1}{3} + \frac{3}{4} + \frac{2}{5}$ c) $\frac{5}{6} + \frac{4}{5} + \frac{4}{3}$

   d) $\frac{5}{4} + \frac{3}{5} + \frac{1}{6}$ e) $\frac{7}{10} + \frac{7}{5} + \frac{7}{2}$ f) $\frac{5}{12} + \frac{6}{5} + \frac{3}{4}$

Math Link

History
Unit fractions are often called Egyptian fractions because the ancient Egyptians investigated fractions in this form.

9. Each fraction below is written as the sum of two unit fractions. Which sums are correct? How do you know?

   a) $\frac{7}{10} = \frac{1}{5} + \frac{1}{2}$ b) $\frac{5}{12} = \frac{1}{3} + \frac{1}{4}$ c) $\frac{5}{6} = \frac{1}{3} + \frac{1}{3}$

   d) $\frac{7}{12} = \frac{1}{2} + \frac{1}{6}$ e) $\frac{11}{8} = \frac{1}{2} + \frac{1}{9}$ f) $\frac{2}{15} = \frac{1}{10} + \frac{1}{30}$

   g) $\frac{7}{15} = \frac{1}{5} + \frac{1}{3}$ h) $\frac{2}{5} = \frac{1}{3} + \frac{1}{15}$ i) $\frac{4}{15} = \frac{1}{5} + \frac{1}{15}$

10. Write each fraction as the sum of two different unit fractions.

    a) $\frac{3}{4}$ b) $\frac{5}{12}$ c) $\frac{7}{10}$

Take It Further

11. Find this sum. Explain your method.

    $\frac{1}{2} + \frac{1}{3} + \frac{1}{4} + \frac{1}{5} + \frac{1}{6} + \frac{2}{3} + \frac{2}{4} + \frac{2}{5} + \frac{2}{6} + \frac{3}{4} + \frac{3}{5} + \frac{3}{6} + \frac{4}{5} + \frac{4}{6} + \frac{5}{6}$

Reflect

Choose two improper fractions. Add them.
Write each improper fraction as a mixed number.
Add the mixed numbers.
Which method is more efficient for finding the sum of two improper fractions? Why?

Explore

Work with a partner.
You will need 1-cm grid paper and coloured pencils.

Use these rules to create a rectangular design on grid paper:

- The design must have line symmetry or rotational symmetry.
- One-half of the grid squares must be red. One-third of the grid squares must be blue. The remaining grid squares must be green.
- The rectangle must have the fewest squares possible.

What fraction of the squares are green? How do you know?
How many squares did you use? Explain.
Describe your design.

Reflect & Share

Compare your design with that of another pair of classmates.
If the designs are different, do both of them obey the rules?
Explain.
Compare your designs with those of other classmates.
How many different designs are possible?

Connect

To subtract fractions, we use a strategy similar to that
for adding fractions.
When the denominators are different,
we find a common denominator first.

Example 1

Subtract. $\frac{4}{5} - \frac{3}{10}$

Solution

$\frac{4}{5} - \frac{3}{10}$

Since 5 is a factor of 10, the lowest common denominator is 10.

$$\overset{\times 2}{\frac{4}{5}} = \frac{8}{10}$$
$$\underset{\times 2}{}$$

$$\frac{4}{5} - \frac{3}{10} = \frac{8}{10} - \frac{3}{10}$$

$$= \frac{5}{10} \qquad \text{This is not in simplest form.}$$

$$= \frac{5 \div 5}{10 \div 5} \qquad \text{5 is a factor of the numerator and denominator.}$$

$$= \frac{1}{2}$$

To subtract mixed numbers, we subtract the fractions, then subtract the whole numbers. We must check the fractions to see which is greater. When the second fraction is greater than the first fraction, we cannot subtract directly.

Example 2

Subtract. $3\frac{1}{5} - 1\frac{3}{4}$

Solution

Estimate:
Since $\frac{1}{5} < \frac{1}{2}$
and $\frac{3}{4} > \frac{1}{2}$;
$\frac{1}{5} < \frac{3}{4}$

Method 1

Subtract the whole numbers and the fractions separately.

$3\frac{1}{5} - 1\frac{3}{4}$

Subtract the fractions: $\frac{1}{5} - \frac{3}{4}$

But $\frac{1}{5} < \frac{3}{4}$, so we cannot subtract $\frac{3}{4}$ from $\frac{1}{5}$.

Write $3\frac{1}{5}$ as $2 + 1\frac{1}{5}$, or $2 + \frac{6}{5}$.

The problem can be written $2\frac{6}{5} - 1\frac{3}{4}$.

Then, subtract the fractions.

$\frac{6}{5} - \frac{3}{4}$

The denominators 5 and 4 have no common factors.

So, their lowest common denominator is $4 \times 5 = 20$.

$\overset{\times 4}{\frown}$
$\dfrac{6}{5} = \dfrac{24}{20}$
$\underset{\times 4}{\smile}$

$\overset{\times 5}{\frown}$
$\dfrac{3}{4} = \dfrac{15}{20}$
$\underset{\times 5}{\smile}$

$$\dfrac{6}{5} - \dfrac{3}{4} = \dfrac{24}{20} - \dfrac{15}{20}$$
$$= \dfrac{9}{20}$$

Now, subtract the whole numbers that remain: $2 - 1 = 1$

So, $3\dfrac{1}{5} - 1\dfrac{3}{4} = 1\dfrac{9}{20}$

Method 2

Change both fractions to improper fractions, then subtract.

$3\dfrac{1}{5} = 3 + \dfrac{1}{5}$
$\quad = \dfrac{15}{5} + \dfrac{1}{5}$
$\quad = \dfrac{16}{5}$

$1\dfrac{3}{4} = 1 + \dfrac{3}{4}$
$\quad = \dfrac{4}{4} + \dfrac{3}{4}$
$\quad = \dfrac{7}{4}$

The denominators have no common factors.

So, their lowest common denominator is $4 \times 5 = 20$.

$\overset{\times 4}{\frown}$
$\dfrac{16}{5} = \dfrac{64}{20}$
$\underset{\times 4}{\smile}$

$\overset{\times 5}{\frown}$
$\dfrac{7}{4} = \dfrac{35}{20}$
$\underset{\times 5}{\smile}$

$$\dfrac{16}{5} - \dfrac{7}{4} = \dfrac{64}{20} - \dfrac{35}{20}$$
$$= \dfrac{29}{20}$$

To write the fraction as a mixed number:

$\dfrac{29}{20} = \dfrac{20}{20} + \dfrac{9}{20}$
$\quad = 1 + \dfrac{9}{20}$
$\quad = 1\dfrac{9}{20}$

So, $3\dfrac{1}{5} - 1\dfrac{3}{4} = 1\dfrac{9}{20}$

In *Example 2*, the answer was written as a mixed number because the question was written with mixed numbers.
In general, the answer should be written in the same form as the question.

Practice

1. Subtract.

   a) $\frac{7}{2} - \frac{5}{4}$

   b) $\frac{7}{8} - \frac{3}{4}$

   c) $\frac{13}{6} - \frac{8}{12}$

   d) $\frac{5}{3} - \frac{2}{6}$

   e) $\frac{7}{5} - \frac{4}{10}$

   f) $\frac{5}{3} - \frac{2}{9}$

   g) $\frac{7}{2} - \frac{2}{4}$

   h) $\frac{3}{2} - \frac{9}{7}$

Number Strategies

Round each amount to the nearest dime, then add.
Round each amount to the nearest dollar, then add.

- $198.85
- $201.79

2. Subtract.

   a) $\frac{11}{12} - \frac{5}{6}$

   b) $\frac{7}{10} - \frac{1}{2}$

   c) $\frac{3}{4} - \frac{3}{5}$

   d) $\frac{7}{8} - \frac{1}{3}$

   e) $\frac{2}{3} - \frac{7}{12}$

   f) $\frac{7}{5} - \frac{2}{3}$

   g) $\frac{9}{5} - \frac{1}{2}$

   h) $\frac{4}{5} - \frac{1}{3}$

3. A sports store placed an order for shoes.
   Three-eighths of the order was basketball shoes;
   one-quarter was running shoes; and the rest were golf shoes.
   What fraction of the order was golf shoes?

4. Subtract.

   a) $\frac{10}{3} - \frac{3}{4}$

   b) $\frac{8}{5} - \frac{2}{3}$

   c) $\frac{7}{4} - \frac{3}{5}$

   d) $\frac{17}{10} - \frac{5}{6}$

   e) $\frac{7}{2} - \frac{3}{5}$

   f) $\frac{13}{6} - \frac{2}{5}$

   g) $\frac{7}{3} - \frac{3}{2}$

   h) $\frac{7}{3} - \frac{5}{8}$

Make sure the first fraction is greater than the second.

5. **Assessment Focus** Copy this diagram.
   Use the numbers 2, 3, 4, and 5 as numerators or denominators.

   $$\frac{\square}{\square} - \frac{\square}{\square} =$$

   a) Write as many different subtraction statements as possible.
      Use each number once each time.
   b) Which statement has the greatest difference?
   c) Which statement has the least difference?
   Show your work.

6. Subtract.

   a) $3\frac{3}{4} - 1\frac{1}{5}$

   b) $3\frac{2}{5} - 1\frac{5}{8}$

   c) $3\frac{7}{10} - 2\frac{1}{3}$

   d) $3\frac{1}{3} - 2\frac{7}{10}$

   e) $4\frac{2}{9} - 1\frac{1}{6}$

   f) $4\frac{1}{6} - 1\frac{2}{9}$

7. a) Subtract.

   i) $3 - \frac{4}{5}$

   ii) $4 - \frac{3}{7}$

   iii) $5 - \frac{5}{6}$

   b) Which methods did you use in part a? Explain your choice.

8. Write each fraction as the difference of two proper fractions with different denominators.

a) $\frac{1}{2}$　　　b) $\frac{3}{4}$　　　c) $\frac{1}{10}$　　　d) $\frac{1}{6}$　　　e) $\frac{1}{4}$

9. a) One-half of the books in Kevin's backpack are novels.
He also has 3 science books, 2 history books, and 1 geography book.
How many books are in Kevin's backpack?

b) In Raji's locker, one-third of the books are novels and one-third are science books. She also has 2 geography books, 3 history books, and 1 social studies book.
How many books are in Raji's locker?

10. Here are some subtraction statements with unit fractions having denominators that are consecutive whole numbers:

$$\frac{1}{1} - \frac{1}{2}; \ \frac{1}{2} - \frac{1}{3}; \ \frac{1}{3} - \frac{1}{4}; \ \frac{1}{4} - \frac{1}{5}$$

a) Find the difference of the fractions in each pair above.

b) Write more pairs of fractions like these.
Find each difference.

c) What patterns do you see?

Take It Further

11. There are some pennies on a table.
One-quarter of the pennies show heads.
Two pennies are turned over.
Now, one-third of the pennies show heads.
How many pennies are on the table?
How do you know?

12. Two matching pitchers contain grapefruit juice.
Pitcher A is $\frac{1}{3}$ full. Pitcher B is $\frac{2}{5}$ full.
Each pitcher is then filled with water.
The contents of both pitchers are poured into one large bowl.
What fraction of the liquid is grapefruit juice?

Reflect

Which fractions or mixed numbers are easy to subtract?
Which are more difficult?
Give an example in each case.

Focus | Use an area model to multiply fractions.

You have used area models to multiply 2 whole numbers, and
to multiply a whole number and a fraction.
We extend the area model to multiply 2 fractions.

Explore

Work with a partner.
One-quarter of a cherry pie was left after dinner.
Trevor ate one-half of the leftover pie for lunch the next day.
What fraction of the pie did he have for lunch?
What if Trevor had eaten only one-quarter of the leftover pie.
What fraction of the pie would he have eaten?

Reflect & Share

How did you solve the problems?
Compare your solutions and strategies with those
of another pair of classmates.
Was one strategy more efficient than another? Explain.

Connect

Use an area model to find the product of fractions.
For example:
Sandi cut $\frac{2}{3}$ of the grass on a rectangular lawn.
Akiva cut $\frac{1}{2}$ of the remaining grass.
What fraction of the lawn did Akiva cut?

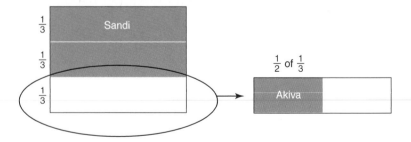

Sandi cut $\frac{2}{3}$. So, $\frac{1}{3}$ remains to be cut.
Akiva cut $\frac{1}{2}$ of $\frac{1}{3}$.

From the rectangle, $\frac{1}{2}$ of $\frac{1}{3} = \frac{1}{6}$

$\frac{1}{6}$ is the area of part of a rectangle,
with one side length $\frac{1}{2}$ of the original length
and the other side length $\frac{1}{3}$ of the original length.

Since the area of a rectangle is length $\times$ width, we can write
$\frac{1}{2}$ of $\frac{1}{3}$ as the multiplication statement: $\frac{1}{2} \times \frac{1}{3} = \frac{1}{6}$

So, Akiva cut $\frac{1}{6}$ of the lawn.

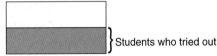

Example

One-half of the Grade 8 students tried out for the school's volleyball team. Three-quarters of the students were successful.

a) What fraction of the Grade 8 students are on the team?

b) Draw an area model and write a multiplication statement to show your answer.

Solution

a) Three-quarters of one-half of the Grade 8 students are on the team.
Draw a rectangle.
Show $\frac{1}{2}$ of the rectangle.

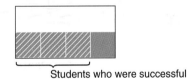

Divide $\frac{1}{2}$ of the rectangle into quarters.
Shade $\frac{3}{4}$.

Use broken lines to divide the whole rectangle into equal parts.
There are 8 equal parts.
Three parts are shaded.
So, $\frac{3}{8}$ of the Grade 8 students are on the team.

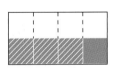

b) $\frac{3}{4}$ of $\frac{1}{2}$ is $\frac{3}{8}$.
So, $\frac{3}{4} \times \frac{1}{2} = \frac{3}{8}$

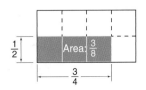

1. Draw each rectangle on grid paper.
   Use the rectangle to find each product.

   a) $\frac{1}{2} \times \frac{3}{4}$

   b) $\frac{3}{4} \times \frac{2}{3}$

   c) $\frac{2}{5} \times \frac{1}{2}$

   d) $\frac{5}{6} \times \frac{1}{2}$

   e) $\frac{3}{5} \times \frac{7}{8}$

   f) $\frac{4}{5} \times \frac{3}{4}$

2. Draw a rectangle on grid paper to find each product.

   a) $\frac{3}{4} \times \frac{5}{8}$

   b) $\frac{4}{9} \times \frac{2}{5}$

   c) $\frac{3}{4} \times \frac{2}{3}$

   d) $\frac{6}{7} \times \frac{2}{3}$

   e) $\frac{2}{3} \times \frac{1}{3}$

   f) $\frac{4}{5} \times \frac{4}{5}$

3. Write 3 multiplication statements using proper fractions. Make sure they are different from any products you have found so far. Draw a rectangle to illustrate each product.

4. **Assessment Focus**

   a) Draw an area model to find each product.

   i) $\frac{3}{4} \times \frac{2}{5}$

   ii) $\frac{2}{4} \times \frac{3}{5}$

   iii) $\frac{1}{4} \times \frac{3}{8}$

   iv) $\frac{3}{4} \times \frac{1}{8}$

   v) $\frac{3}{5} \times \frac{4}{6}$

   vi) $\frac{3}{6} \times \frac{4}{5}$

   b) Compare the area models. What patterns do you see?
      Write some other products similar to those in part a.
      Show your work.

5. Why is $\frac{5}{8}$ of $\frac{3}{12}$ equal to $\frac{3}{8}$ of $\frac{5}{12}$?
   Draw area models to explain your answer.

Calculator Skills

Find a perfect square that is the sum of a two-digit perfect square and a one-digit perfect square. How many can you find?

Reflect

When you use an area model to multiply two fractions, how do you decide how to draw the rectangle? Include an example in your explanation.

4.5 Multiplying Fractions

Explore

Work with a partner.
You will need grid paper.
Copy this diagram on grid paper.

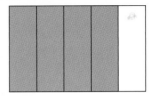

Find $\frac{2}{3} \times \frac{4}{5}$.
What patterns do you notice in the numbers?
How can you use patterns to multiply $\frac{2}{3} \times \frac{4}{5}$?
Use your method to calculate $\frac{7}{8} \times \frac{3}{10}$.
Use an area model to check.

Reflect & Share

Compare your strategies with those of another pair of classmates.
Do you think your strategy will work with all fractions? Explain.

Connect

Here is an area model to show:
$\frac{4}{7} \times \frac{2}{5} = \frac{8}{35}$

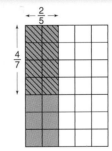

The product of the numerators is:
$4 \times 2 = 8$
The product of the denominators is:
$7 \times 5 = 35$
That is, $\frac{4}{7} \times \frac{2}{5} = \frac{4 \times 2}{7 \times 5}$
$= \frac{8}{35}$

So, to multiply two fractions, multiply the numerators and
multiply the denominators.

We can use this method to multiply proper fractions and
improper fractions.
Example 1 illustrates that the product expression may be simplified
before multiplying.

Example 1

Multiply.

a) $\frac{4}{9} \times \frac{3}{8}$

b) $\frac{7}{5} \times \frac{8}{3}$

Solution

a) $\frac{4}{9} \times \frac{3}{8} = \frac{4 \times 3}{9 \times 8}$

Notice that the numerator and denominator have common factors 3 and 4.

Divide the numerator and denominator by these factors.

$$\frac{4}{9} \times \frac{3}{8} = \frac{\overset{1}{\cancel{4}} \times \overset{1}{\cancel{3}}}{\underset{3}{\cancel{9}} \times \underset{2}{\cancel{8}}}$$

$$4 \div 4 = 1 \qquad 3 \div 3 = 1$$
$$9 \div 3 = 3 \qquad 8 \div 4 = 2$$

$$= \frac{1 \times 1}{3 \times 2}$$

$$= \frac{1}{6}$$

b) $\frac{7}{5} \times \frac{8}{3}$

There are no common factors in the numerators and denominators.

So, $\frac{7}{5} \times \frac{8}{3} = \frac{56}{15}$

Here is an area model to illustrate the product of the improper fractions in *Example 1b*.

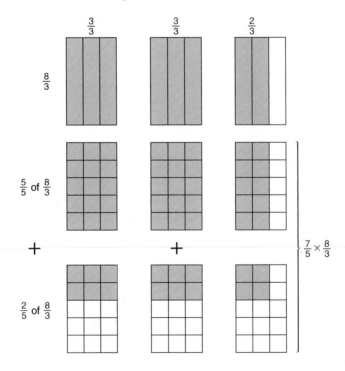

Each small square is $\frac{1}{15}$.

The number of shaded squares is: $15 + 15 + 10 + 6 + 6 + 4 = 56$

So, $\frac{56}{15}$ are shaded.

$\frac{7}{5} \times \frac{8}{3} = \frac{56}{15}$

To multiply two mixed numbers, change them to improper fractions first.

Example 2

Multiply. $2\frac{1}{4} \times 3\frac{2}{5}$

Solution

$2\frac{1}{4} \times 3\frac{2}{5}$

$2\frac{1}{4} = 2 + \frac{1}{4}$ $\qquad\qquad\qquad$ $3\frac{2}{5} = 3 + \frac{2}{5}$

$\quad = \frac{8}{4} + \frac{1}{4}$ $\qquad\qquad\qquad\quad = \frac{15}{5} + \frac{2}{5}$

$\quad = \frac{9}{4}$ $\qquad\qquad\qquad\qquad\quad = \frac{17}{5}$

$2\frac{1}{4} \times 3\frac{2}{5} = \frac{9}{4} \times \frac{17}{5}$

$\qquad\qquad\ = \frac{153}{20}$

As a mixed number: $\frac{153}{20} = 7\frac{13}{20}$

So, $2\frac{1}{4} \times 3\frac{2}{5} = \frac{153}{20}$, or $7\frac{13}{20}$

We can apply the rules for multiplying two fractions to multiply three or more fractions.
You will do this in *Practice* question 7.

Practice

1. Multiply. Check each product by drawing an area model.
    a) $\frac{3}{8} \times \frac{5}{6}$ $\qquad\qquad$ b) $\frac{4}{5} \times \frac{1}{2}$ $\qquad\qquad$ c) $\frac{3}{10} \times \frac{3}{4}$

2. Multiply.
    a) $\frac{3}{5} \times \frac{2}{3}$ $\qquad\qquad$ b) $\frac{1}{2} \times \frac{5}{10}$ $\qquad\qquad$ c) $\frac{1}{6} \times \frac{1}{4}$
    d) $\frac{13}{8} \times \frac{3}{2}$ $\qquad\qquad$ e) $\frac{5}{4} \times \frac{11}{10}$ $\qquad\qquad$ f) $\frac{7}{3} \times \frac{7}{8}$

3. Paula has $\frac{7}{8}$ of a tank of gas.

She estimates she will use $\frac{2}{3}$ of the gas to get home.

What fraction of a tank of gas does she use?

4. a) Find each product.

    i) $\frac{3}{4} \times \frac{4}{3}$ **ii)** $\frac{1}{5} \times \frac{5}{1}$

    iii) $\frac{7}{2} \times \frac{2}{7}$ **iv)** $\frac{5}{6} \times \frac{6}{5}$

    v) $\frac{8}{3} \times \frac{3}{8}$ **vi)** $\frac{12}{11} \times \frac{11}{12}$

b) What do you notice about the products in part a?

Write 3 more pairs of fractions that have
the same product.

What is special about these fractions?

5. Multiply.

   a) $1\frac{3}{4} \times 2\frac{1}{2}$ **b)** $3\frac{2}{3} \times 2\frac{1}{5}$ **c)** $4\frac{3}{8} \times 1\frac{1}{4}$

   d) $3\frac{3}{4} \times 3\frac{3}{4}$ **e)** $4\frac{3}{10} \times \frac{4}{5}$ **f)** $\frac{7}{8} \times 2\frac{3}{5}$

6. Play this game with a partner.

Your teacher will give you a copy of this spinner.

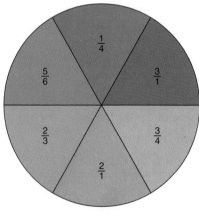

You will need an open paperclip as a pointer and
a sharp pencil to keep the pointer in place.

For each turn, players spin twice.

Player A adds the fractions.

Player B multiplies the same two fractions.

The player with the greater result gets one point.

The first person to get 12 points wins.

Is this game fair?

Give reasons for your answer.

7. Use your knowledge of exponents and multiplying fractions to evaluate each power.

a) $\left(\frac{2}{9}\right)^2$ b) $\left(\frac{6}{5}\right)^3$ c) $\left(\frac{3}{10}\right)^4$ d) $\left(\frac{5}{2}\right)^4$

8. **Assessment Focus** In question 4, each product is 1.

a) Write a pair of fractions that has each product.

     i) 2   **ii)** 3   **iii)** 4   **iv)** 5

b) Write a pair of fractions that has the product 1. Change only one numerator or denominator each time to write a pair of fractions that has each product.

     i) 2   **ii)** 3   **iii)** 4   **iv)** 5

c) How could you write a pair of fractions that has the product 10?

Show your work.

9. The product of two fractions is $\frac{2}{3}$.
One fraction is $\frac{3}{5}$.
What is the other fraction?
How do you know?

Take It Further

10. Amar baked a cake. John ate $\frac{1}{6}$ of the cake.
Susan ate $\frac{1}{5}$ of what was left.
Chan ate $\frac{1}{4}$ of what was left after that.
Cindy ate $\frac{1}{3}$ of what was left after that.
Luigi ate $\frac{1}{2}$ of what was left after that.
How much of the original cake was left?

Number Strategies

A lunch counter offers 2 soups, 4 sandwiches, and 3 drinks.

How many different possible combinations are there for a person who wants a soup, a sandwich, and a drink?

Reflect

Look at your answers to all the questions.
Some products were in simplest form after you multiplied.
Some products were not in simplest form.
How can you tell if a product of two fractions will be in simplest form after you multiply? Use examples in your explanation.

Mid-Unit Review

LESSON

4.1 **1.** Order these fractions from least to greatest.
$\frac{2}{3}, \frac{1}{2}, \frac{5}{8}, \frac{1}{4}, \frac{3}{4}$

2. Paola has read $\frac{3}{4}$ of her novel. Rafferty has read $\frac{5}{7}$ of the same novel.
Who has read more?
How do you know?

3. Which fractions below are:
a) between 0 and $\frac{1}{2}$?
b) between $\frac{1}{2}$ and 1?
$\frac{2}{5}, \frac{1}{4}, \frac{2}{3}, \frac{3}{8}, \frac{7}{12}, \frac{8}{10}, \frac{1}{3}, \frac{5}{6}$
How do you know?

4.2 **4.** Find each sum. What patterns do you see in the fractions and their sums?
a) $\frac{1}{2} + \frac{2}{1}$ b) $\frac{2}{3} + \frac{3}{2}$
c) $\frac{3}{4} + \frac{4}{3}$ d) $\frac{4}{5} + \frac{5}{4}$

5. Takoda and Wesley are collecting shells on the beach in identical pails. Takoda estimates she has filled $\frac{7}{12}$ of her pail. Wesley estimates he has filled $\frac{4}{10}$ of his pail. Suppose the children combine their shells. Will one pail be full? Explain.

6. Add.
a) $3\frac{1}{4} + 1\frac{3}{8}$ b) $2\frac{3}{4} + 2\frac{3}{4}$
c) $4\frac{3}{10} + 1\frac{1}{8}$ d) $2\frac{2}{3} + 1\frac{5}{8}$
e) $3\frac{2}{3} + 1\frac{5}{9}$ f) $1\frac{3}{5} + 2\frac{1}{6}$

7. Is this a magic square? How do you know?

$\frac{3}{8}$	$\frac{1}{6}$	$\frac{11}{24}$
$\frac{5}{12}$	$\frac{1}{3}$	$\frac{1}{4}$
$\frac{5}{24}$	$\frac{1}{2}$	$\frac{7}{24}$

4.3 **8.** Subtract.
a) $\frac{3}{5} - \frac{1}{2}$ b) $\frac{4}{3} - \frac{2}{7}$
c) $\frac{8}{5} - \frac{3}{4}$ d) $\frac{8}{5} - \frac{3}{2}$

9. Subtract.
a) $3\frac{7}{10} - 2\frac{1}{5}$ b) $4\frac{2}{5} - 1\frac{3}{8}$
c) $2\frac{3}{4} - 1\frac{9}{10}$ d) $4\frac{3}{8} - 3\frac{7}{10}$

10. Farrah has run $\frac{7}{10}$ of a race. Malcom has run $\frac{6}{9}$ of the race.
a) Who has run farther?
b) How much farther?

4.4 **11.** Draw a rectangle on grid paper to find each product.
a) $\frac{7}{8} \times \frac{1}{2}$ b) $\frac{1}{2} \times \frac{3}{4}$
c) $\frac{3}{4} \times \frac{2}{3}$ d) $\frac{2}{3} \times \frac{4}{5}$

12. Multiply.
a) $\frac{4}{10} \times \frac{2}{3}$ b) $\frac{7}{5} \times \frac{3}{8}$
c) $2\frac{2}{3} \times 3\frac{1}{10}$ d) $2\frac{2}{9} \times 2\frac{2}{9}$

4.5 **13.** Aiko says that $\frac{2}{3}$ of her stamp collection are Asian stamps. One-fifth of her Asian stamps are from India. What fraction of Aiko's stamp collection is from India?

Using Models to Divide Fractions and Whole Numbers

Focus Use a number line to divide fractions and whole numbers.

When you first studied division, you learned two ways: sharing and grouping.
For example, 20 ÷ 5 can be thought of as:

- Sharing 20 items equally among 5 sets
- Grouping 20 items into sets of 5

Recall that multiplication and division are inverse operations.
We know: 20 ÷ 5 = 4
So, we also know: 4 × 5 = 20

Explore

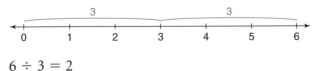

Work with a partner.
Suppose you have 5 cups of concentrate.

➤ A recipe for a bowl of punch calls for $\frac{1}{4}$ cup of concentrate.
How many bowls of punch can you make?

➤ A different recipe calls for $\frac{3}{4}$ cup of concentrate for
one bowl of punch.
How many bowls of punch could you make if you used
this recipe?
Draw a diagram to illustrate your answers.

Reflect & Share

Compare your answers with those of another pair of classmates.
Did you solve the problems the same way?
If not, explain your method to your classmates.

Connect

Before we divide a whole number by a fraction,
think about how we divide whole numbers.

➤ To find how many 3s are in 6, group 6 into 3s.
We can show this on a number line.
There are 2 groups of 3 in 6.

6 ÷ 3 = 2

➤ To find how many thirds are in 6, divide 6 into thirds.

There are 18 thirds in 6.
Write this as a division statement.
$6 \div \frac{1}{3} = 18$ Notice: $6 \times 3 = 18$

➤ Use the same number line to find how many two-thirds are in 6.

Arrange 18 thirds into groups of two-thirds.
There are 9 groups of two-thirds.
We write: $6 \div \frac{2}{3} = 9$

➤ Use the number line again to find how many five-thirds are in 6;
that is, $6 \div \frac{5}{3}$.
Arrange 18 thirds into groups of five-thirds.

There are 3 groups of five-thirds.
There are 3 thirds left over.
Think: What fraction of $\frac{5}{3}$ is $\frac{3}{3}$?

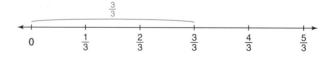

From the number line, $\frac{3}{3}$ is $\frac{3}{5}$ of $\frac{5}{3}$.
So, $6 \div \frac{5}{3} = 3\frac{3}{5}$

We can also use a number line to divide a fraction by a whole
number. This is illustrated in the *Example* that follows.

Example

Divide.

a) $\frac{1}{5} \div 4$ **b)** $\frac{3}{5} \div 4$

Solution

a) To find $\frac{1}{5} \div 4$, mark $\frac{1}{5}$ on a number line.

Divide the interval 0 to $\frac{1}{5}$ into 4 equal parts.

$\frac{1}{5}$ is $\frac{4}{20}$
$\frac{1}{10}$ is $\frac{2}{20}$

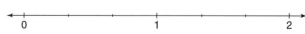

Each part is $\frac{1}{20}$.

So, $\frac{1}{5} \div 4 = \frac{1}{20}$

b) To find $\frac{3}{5} \div 4$, mark $\frac{3}{5}$ on a number line.

Divide the interval 0 to $\frac{3}{5}$ into 4 equal parts.

To do this, divide the fifths into twentieths.

Arrange the 12 twentieths into 4 equal groups.

$\frac{1}{5}$ is $\frac{4}{20}$
$\frac{3}{5}$ is $\frac{12}{20}$

There are $\frac{3}{20}$ in each group.

So, $\frac{3}{5} \div 4 = \frac{3}{20}$

Practice

1. Use a number line to find each quotient.

    a) i) $2 \div \frac{1}{3}$ **ii)** $2 \div \frac{2}{3}$

    b) i) $3 \div \frac{1}{4}$ **ii)** $3 \div \frac{2}{4}$ **iii)** $3 \div \frac{3}{4}$

    c) i) $\frac{4}{8} \div 2$ **ii)** $\frac{4}{8} \div 4$ **iii)** $\frac{4}{8} \div 8$

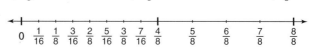

Number Strategies

Order the numbers in each set from least to greatest.

- $1.2, \frac{11}{2}, 2.12, 125\%$
- $2^7, 22^1, 20^4, 40^2, 7^2$

2. Find each quotient. Use number lines to illustrate the answers.

   a) $2 \div \frac{1}{2}$ **b)** $3 \div \frac{1}{3}$ **c)** $3 \div \frac{2}{3}$

   d) $4 \div \frac{1}{4}$ **e)** $4 \div \frac{2}{4}$ **f)** $4 \div \frac{3}{4}$

3. Find each quotient. Use number lines to illustrate the answers.

   a) $\frac{1}{2} \div 2$ **b)** $\frac{1}{3} \div 3$ **c)** $\frac{2}{3} \div 3$

   d) $\frac{1}{4} \div 4$ **e)** $\frac{2}{4} \div 4$ **f)** $\frac{3}{4} \div 4$

4. Use a number line to find each quotient.

   a) $\frac{4}{5} \div 3$ **b)** $2 \div \frac{3}{8}$ **c)** $\frac{1}{2} \div 5$

   d) $6 \div \frac{3}{4}$ **e)** $4 \div \frac{2}{3}$ **f)** $\frac{5}{8} \div 2$

5. Ioana wants to spend $\frac{3}{4}$ of an hour studying each subject. She has 3 h to study. How many subjects can she study?

6. Why is $\frac{2}{3} \div 4$ not the same as $4 \div \frac{2}{3}$? Use number lines in your explanation.

7. **Assessment Focus** Copy these boxes.

$$\square \div \frac{\square}{\square}$$

   a) Write the digits 2, 4, and 6 in the boxes to find as many division statements as possible.

   b) Which statement in part a has the greatest quotient? The least quotient? How do you know?

   Show your work.

Take It Further

8. The numbers $\frac{9}{2}$ and 3 share this property: their difference is equal to their quotient. That is, $\frac{9}{2} - 3 = \frac{3}{2}$ and $\frac{9}{2} \div 3 = \frac{3}{2}$ Find other pairs of numbers with this property. Describe any patterns you see.

Reflect

When you divide a whole number by a proper fraction, is the quotient greater than or less than the whole number? Include an example in your explanation.

Focus Develop algorithms to divide fractions.

You have used grouping to divide 4 by $\frac{2}{3}$: $4 \div \frac{2}{3}$

You have used sharing to divide $\frac{2}{3}$ by 4: $\frac{2}{3} \div 4$

You will now investigate dividing a fraction by a fraction: $\frac{2}{3} \div \frac{1}{4}$

Explore

Work with a partner.

Use this number line to find how many quarters are in $\frac{2}{3}$;
that is, find $\frac{2}{3} \div \frac{1}{4}$.

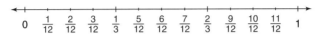

Look at the quotient.
Try to find a method to calculate the quotient
without using a number line.
Use a different division problem to check your method.

Reflect & Share

Compare your method with that of another pair of classmates.
Does your method work with their problem? Explain.
Does their method work with your problem? Explain.

Connect

Here are two ways to divide fractions.

➤ Use common denominators.
To divide: $\frac{3}{5} \div \frac{1}{4}$
Write each fraction with a common denominator.
Since 5 and 4 have no common factors,
their common denominator is $5 \times 4 = 20$.

$$\frac{3}{5} \overset{\times 4}{\underset{\times 4}{=}} \frac{12}{20} \qquad\qquad \frac{1}{4} \overset{\times 5}{\underset{\times 5}{=}} \frac{5}{20}$$

When the denominators are the same, divide the numerators.

$$\frac{3}{5} \div \frac{1}{4} = \frac{12}{20} \div \frac{5}{20}$$

This means: How many five-twentieths are in $\frac{12}{20}$?

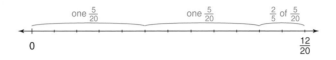

From the number line, this is: $12 \div 5 = 2\frac{2}{5}$

So, $\frac{3}{5} \div \frac{1}{4} = 2\frac{2}{5}$

➤ Use multiplication.

Recall that another way to divide by 4 is to multiply by $\frac{1}{4}$.

$12 \div 4 = 3$ and $12 \times \frac{1}{4} = 3$

Since 4 can be written as $\frac{4}{1}$,

dividing by 4 is the same as dividing by $\frac{4}{1}$.

So, we can write $12 \div \frac{4}{1} = 3$ and $12 \times \frac{1}{4} = 3$.

Similarly, another way to divide by $\frac{1}{4}$ is to multiply by 4.

$3 \div \frac{1}{4} = 12$ and $3 \times \frac{4}{1} = 12$

We can use the same pattern to divide two fractions.

The fraction $\frac{1}{4}$ is the reciprocal of the fraction $\frac{4}{1}$.

That is, $\frac{3}{5} \div \frac{1}{4} = \frac{3}{5} \times \frac{4}{1}$

$$= \frac{12}{5}$$

Up until now, we have divided a fraction by a lesser fraction. We can use the same methods when we divide a fraction by a greater fraction or when we divide mixed numbers.

Example

Divide.

a) $\frac{3}{4} \div \frac{5}{6}$

b) $1\frac{7}{8} \div 1\frac{1}{4}$

Solution

a) $\frac{3}{4} \div \frac{5}{6}$

Use multiplication.

Dividing by $\frac{5}{6}$ is the same as multiplying by $\frac{6}{5}$.

$\frac{3}{4} \div \frac{5}{6}$ can be written as

$$\frac{3}{4} \times \frac{6}{5} = \frac{3 \times \overset{3}{6}}{\underset{2}{4} \times 5}$$

$$= \frac{3 \times 3}{2 \times 5}$$

$$= \frac{9}{10}$$

b) $1\frac{7}{8} \div 1\frac{1}{4}$

Change the mixed numbers to improper fractions.

$$1\frac{7}{8} = \frac{8}{8} + \frac{7}{8} \qquad\qquad 1\frac{1}{4} = \frac{4}{4} + \frac{1}{4}$$
$$= \frac{15}{8} \qquad\qquad\qquad\quad = \frac{5}{4}$$

So, $1\frac{7}{8} \div 1\frac{1}{4} = \frac{15}{8} \div \frac{5}{4}$

Use common denominators.

Since 4 is a factor of 8, the lowest common denominator is 8.

Multiply the numerator and denominator by 2: $\frac{5}{4} = \frac{10}{8}$

$\frac{15}{8} \div \frac{5}{4} = \frac{15}{8} \div \frac{10}{8}$ Since the denominators are the same,
divide the numerators.

$\qquad\quad = \frac{15}{10}$ Reduce to simplest form.

$\qquad\quad = \frac{15 \div 5}{10 \div 5}$

$\qquad\quad = \frac{3}{2}$, or $1\frac{1}{2}$

Practice

1. Use a copy of each number line to illustrate each quotient.

  a) $\frac{5}{6} \div \frac{1}{3}$

  b) $\frac{3}{4} \div \frac{1}{3}$

2. Use multiplication to find each quotient.

  a) $\frac{8}{5} \div \frac{3}{4}$ **b)** $\frac{9}{10} \div \frac{5}{3}$ **c)** $\frac{7}{2} \div \frac{4}{3}$ **d)** $\frac{1}{2} \div \frac{7}{6}$

3. Use common denominators to find each quotient.

  a) $\frac{7}{12} \div \frac{1}{4}$ **b)** $\frac{3}{5} \div \frac{11}{10}$ **c)** $\frac{5}{2} \div \frac{1}{3}$ **d)** $\frac{5}{6} \div \frac{9}{8}$

4. Divide.

  a) $1\frac{9}{10} \div 2\frac{2}{3}$ **b)** $2\frac{3}{4} \div 2\frac{1}{3}$ **c)** $3\frac{1}{2} \div 1\frac{4}{5}$ **d)** $1\frac{3}{8} \div 1\frac{3}{8}$

5. Divide.

    a) $\frac{5}{3} \div \frac{3}{5}$ **b)** $\frac{4}{9} \div \frac{4}{9}$ **c)** $\frac{1}{6} \div \frac{5}{2}$ **d)** $1\frac{3}{4} \div 2\frac{9}{10}$

6. a) Find each quotient.

    i) $\frac{3}{4} \div \frac{5}{8}$ **ii)** $\frac{5}{8} \div \frac{3}{4}$ **iii)** $\frac{7}{12} \div \frac{2}{5}$

    iv) $\frac{2}{5} \div \frac{7}{12}$ **v)** $\frac{5}{3} \div \frac{4}{5}$ **vi)** $\frac{4}{5} \div \frac{5}{3}$

    b) In part a, what patterns do you see in the division statements and their quotients?

    Write two more pairs of division statements that follow the same pattern.

7. **Assessment Focus**

    a) Copy the boxes below.

    Write the digits 2, 3, 4, and 5 in the boxes to make as many different division statements as you can.

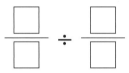

    b) Which division statement in part a has the greatest quotient? The least quotient? How do you know?

    Show your work.

8. Which statement has the greatest value?

    Give reasons for your answer.

    a) $3\frac{1}{5} \times \frac{1}{2}$ **b)** $3\frac{1}{5} \times \frac{2}{3}$ **c)** $3\frac{1}{5} \div \frac{2}{3}$

    d) $3\frac{1}{5} \div \frac{2}{1}$ **e)** $3\frac{1}{5} + \frac{2}{3}$ **f)** $3\frac{1}{5} + \frac{3}{2}$

9. Write as many division questions as you can that have $\frac{5}{6}$ as their quotient.

Reflect

When you divide two fractions, how can you tell, before you divide, if the quotient will be:

- greater than 1?
- less than 1?
- equal to 1?

Use examples in your explanation.

4.8

Converting between Decimals and Fractions

Focus Use a calculator and patterns to convert between decimals and fractions.

Recall that we can write a fraction as a division.
For example, $\frac{5}{2}$ can be written as $5 \div 2$.

Explore

Use the *Glossary* if you have forgotten what a unit fraction is.

Work with a partner.
You will need a calculator.

➤ Write all the unit fractions with denominators from 1 to 10.
Write each fraction as a decimal.
Use a calculator to check your answers.

➤ Choose 3 different proper fractions.
Write each fraction as a decimal.
Trade decimals with your partner.
Order the decimals from least to greatest.

Reflect & Share

Compare your fractions and decimals with those
of another pair of classmates.
Sort the decimals into two sets. Which attributes did you use?

Connect

Recall these two types of decimals.
- These are **terminating decimals**: 0.5, 0.76, 0.435
  Each decimal has a definite number of decimal places.
- These are **repeating decimals**:
  0.333...; 0.454 545...; 0.811 111...
  Some digits in each decimal repeat forever.

This is a repeating decimal.

➤ To write a fraction as a decimal,
divide the numerator by the denominator.
For example, $\frac{4}{11} = 4 \div 11 = 0.363\ 636\ 36...$.
We write $\frac{4}{11} = 0.\overline{36}$, with a bar over the digits that repeat.
When we use a calculator to divide, the calculator may round
the last digit and display 0.363 636 364.

➤ To write a terminating decimal as a fraction, look at these patterns.

$$0.3 = \frac{3}{10}$$

$$0.03 = \frac{3}{100} \qquad\qquad 0.33 = \frac{33}{100}$$

$$0.003 = \frac{3}{1000} \qquad\qquad 0.333 = \frac{333}{1000}$$

The number of digits after the decimal point tells the power of 10 in the denominator:

**0.333 is
333 thousandths.**

$$0.333 = \frac{333}{1000}$$

10^3 in the denominator

3 digits after the decimal point

**0.4567 is
4567 ten-thousandths.**

$$0.4567 = \frac{4567}{10\,000}$$

10^4 in the denominator

4 digits after the decimal point

Example 1

Write each decimal as a fraction in simplest form.

a) 0.365   **b)** 0.0054

Solution

a) 0.365

There are 3 digits after the decimal point.

In fraction form, the denominator is 10^3, or 1000.

**0.365 is
365 thousandths.**

$$0.365 = \frac{365}{1000}$$

Write in simplest form.

5 is a common factor of 365 and 1000.

So, divide numerator and denominator by 5.

$$\frac{365}{1000} = \frac{365 \div 5}{1000 \div 5} = \frac{73}{200}$$

$$0.365 = \frac{73}{200}$$

b) 0.0054

There are 4 digits after the decimal point.

In fraction form, the denominator is 10^4, or 10 000.

**0.0054 is
54 ten-thousandths.**

$$0.0054 = \frac{54}{10\,000}$$ 2 is a common factor of 54 and 10 000.

$$= \frac{27}{5000}$$ Divide numerator and denominator by 2.

We use place value to order decimals.

Example 2

Order these decimals from least to greatest.
$0.\overline{45}, 0.4\overline{5}, 0.4, 0.45$

Solution

All the decimals have 0 in the ones place and 4 in the tenths place.
Compare digits in the hundredths place and beyond.

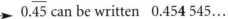

$0.\overline{45}$ can be written   0.454 545...
$0.4\overline{5}$ can be written   0.455 555...
0.4 can be written   0.4**00 000**
0.45 can be written   0.45**0 000**

0.4 has 0 in the hundredths place. It is the least.
0.45 has 0 in the thousandths place.
$0.\overline{45}$ has 4 in the thousandths place.
$0.4\overline{5}$ has 5 in the thousandths place. It is the greatest.

In order from least to greatest: $0.4, 0.45, 0.\overline{45}, 0.4\overline{5}$

Practice

1. a) Write each fraction as a decimal.
    i) $\frac{2}{3}$ ii) $\frac{3}{4}$ iii) $\frac{4}{5}$ iv) $\frac{5}{6}$ v) $\frac{6}{7}$
   b) How can you tell which fractions in part a repeat
      and which terminate?

2. Write each decimal as a fraction, in simplest form.
   a) 0.73   b) 0.765   c) 0.8765   d) 0.0006

3. For each fraction, write an equivalent fraction with a
   denominator that is a power of 10.
   Then, write the fraction as a decimal.
   a) $\frac{1}{2}$ b) $\frac{2}{5}$ c) $\frac{3}{4}$ d) $\frac{13}{25}$ e) $\frac{19}{50}$

4. Write each fraction as a decimal.
   a) $\frac{2}{7}$ b) $\frac{3}{11}$ c) $\frac{2}{9}$ d) $\frac{5}{17}$ e) $\frac{5}{13}$

5. In question 4d, the calculator display is not long enough to show the repeating digits. How could you find the repeating digits?

6. Write $\frac{1}{5}$ as a decimal.
Use this decimal to write each number below as a decimal.

a) $\frac{4}{5}$ b) $\frac{7}{5}$ c) $1\frac{4}{5}$ d) $2\frac{1}{5}$

7. a) How many fractions can you write that are equivalent to the decimal 0.76?

b) Have you written all possible fractions? Explain.

 8. **Assessment Focus** Here is the Fibonacci sequence:
1, 1, 2, 3, 5, 8, 13, 21, 34, 55, 89, …
We can write consecutive terms as fractions:
$\frac{1}{1}, \frac{2}{1}, \frac{3}{2}, \frac{5}{3}, \frac{8}{5}, \frac{13}{8}$, and so on

a) Write each fraction above as a decimal.
What patterns do you see?

b) Continue to write consecutive terms as decimals.
Write about what you find out.

9. In each set, write the first three fractions as decimals.
Look for a pattern.
Use the pattern to write the remaining fractions as decimals.

a) $\frac{1}{7}, \frac{2}{7}, \frac{3}{7}, \frac{4}{7}, \frac{5}{7}, \frac{6}{7}$

b) $\frac{1}{9}, \frac{2}{9}, \frac{3}{9}, \frac{4}{9}, \frac{5}{9}, \frac{6}{9}, \frac{7}{9}, \frac{8}{9}$

c) $\frac{1}{11}, \frac{2}{11}, \frac{3}{11}, \frac{4}{11}, \frac{5}{11}, \frac{6}{11}, \frac{7}{11}, \frac{8}{11}, \frac{9}{11}, \frac{10}{11}$

10. Order the decimals in each set from least to greatest.

a) $1.01, 0.1, 0.01, 0.\overline{1}$

b) $1.\overline{3}, 0.\overline{3}, 2.3, 0.3, 0.35$

c) $0.46, 0.64, 1.4\overline{6}, 1.06, 0.\overline{6}$

Reflect

When you look at a decimal, how can you tell if it repeats or terminates?
Use examples in your explanation.

Focus   Relate division by 0.1, 0.01, and 0.001 to multiplication by powers of 10.

Recall that when you multiply a decimal by 10, the digits move 1 place to the left on a place-value chart, or the decimal point shifts 1 place to the right.

What happens when you multiply a decimal by 100? By 1000?

Explore

Work on your own.
Use a calculator.

➤ Choose a 4-digit decimal.
  Divide it by 0.1, 0.01, and 0.001.
  What patterns do you notice?

➤ Choose a different 4-digit decimal.
  Use patterns to divide it by 0.1, 0.01, and 0.001.
  Check your answers with a calculator.

Reflect & Share

Compare your strategies for dividing with those of a classmate.
How could you use multiplication to divide by 0.1, 0.01, and 0.001?

Connect

➤ Dividing by $\frac{1}{10}$ is the same as multiplying by 10.

$0.1 = \frac{1}{10}$

So, $1.35 \div 0.1 = 1.35 \div \frac{1}{10}$
$\qquad\qquad\qquad = 1.35 \times 10$
$\qquad\qquad\qquad = 13.5$

➤ Dividing by $\frac{1}{100}$ is the same as multiplying by 100.

$0.01 = \frac{1}{100}$

So, $1.35 \div 0.01 = 1.35 \div \frac{1}{100}$
$\qquad\qquad\qquad = 1.35 \times 100$
$\qquad\qquad\qquad = 135$

➤ Dividing by $\frac{1}{1000}$ is the same as multiplying by 1000.

$0.001 = \frac{1}{1000}$

So, $1.35 \div 0.001 = 1.35 \div \frac{1}{1000}$

$= 1.35 \times 1000$

$= 1350$

➤ Recall how to divide by powers of 10, such as 10 and 100.

$1.35 \div 10 \ = 0.135$

$1.35 \div 100 = 0.0135$

We can think of dividing by 10 as multiplying by $\frac{1}{10} = 0.1$.

So, $1.35 \div 10 = 1.35 \times 0.1$

$= 0.135$

And, dividing by 100 is the same as multiplying by $\frac{1}{100} = 0.01$.

So, $135 \div 100 = 1.35 \times 0.01$

$= 0.0135$

We can use these patterns to mentally divide by multiples of 0.1, 0.01, and 0.001.

Example

Divide.

a) $0.275 \div 0.2$

b) $1.863 \div 0.03$

Solution

a) $0.275 \div 0.2 = 0.275 \div \frac{2}{10}$

$= 0.275 \times \frac{10}{2}$

$= \frac{2.75}{2}$

$= 1.375$

b) $1.863 \div 0.03 = 1.863 \div \frac{3}{100}$

$= 1.863 \times \frac{100}{3}$

$= \frac{186.3}{3}$

$= 62.1$

Practice

Use mental math.

1. Predict the quotient when you divide each number by 100, 10, 1, 0.1, 0.01, and 0.001.

a) 547   b) 879   c) 34.5   d) 6.52

e) 6542.12   f) 0.234   g) 8.9   h) 10.01

2. Write each quotient.

a) $\dfrac{147}{1000}$ b) $\dfrac{147}{0.01}$ c) $\dfrac{9.64}{0.1}$ d) $\dfrac{12.30}{0.001}$

e) $\dfrac{0.345}{0.01}$ f) $\dfrac{12.3}{10}$ g) $\dfrac{23.45}{0.01}$ h) $\dfrac{0.123}{0.001}$

3. Find the missing divisor in each division statement.

a) $\dfrac{4.3}{?} = 4.3$ b) $\dfrac{54}{?} = 5.4$ c) $\dfrac{65.4}{?} = 6540$

d) $\dfrac{43.45}{?} = 434.5$ e) $\dfrac{785.03}{?} = 7850.3$ f) $\dfrac{0.0345}{?} = 3.45$

g) $\dfrac{0.003\,45}{?} = 0.345$ h) $\dfrac{345.6}{?} = 3456$ i) $\dfrac{0.593}{?} = 59.3$

4. Find the missing dividend in each division statement.

a) $\dfrac{?}{10} = 234$ b) $\dfrac{?}{0.1} = 34.5$ c) $\dfrac{?}{0.01} = 12.23$

d) $\dfrac{?}{0.001} = 12\,000$ e) $\dfrac{?}{0.01} = 1320$ f) $\dfrac{?}{0.001} = 50$

g) $\dfrac{?}{0.1} = 0.725$ h) $\dfrac{?}{0.1} = 72.5$ i) $\dfrac{?}{100} = 0.1456$

Mental Math

Copy this statement.

$5 + 15 \div 3 + 2 \times 2 - 13$

Insert one pair of brackets to make the answer 8.

5. **Assessment Focus** A student says that when you divide two numbers, the quotient is always less than the dividend. Is this true? Use examples to explain your answer.

6. Find each quotient.

a) $356.2 \div 0.2$ b) $127.5 \div 0.03$ c) $0.448 \div 0.4$

d) $0.0525 \div 0.005$ e) $63.6 \div 0.06$ f) $211.4 \div 0.007$

7. A rectangle has an area of 15.5 cm². Find the length and perimeter of the rectangle for each width.

a) 10 cm   b) 1 cm   c) 0.1 cm   d) 0.01 cm   e) 0.001 cm

Take It Further

Similar figures have corresponding sides in the same ratio.

8. a) Draw each rectangle on grid paper.
   i) 4 cm by 4 cm   ii) 6 cm by 4.4 cm   iii) 8.6 cm by 4.8 cm
   b) Calculate the area of each rectangle.
   c) What if you divide the area of the first rectangle by 0.1; the second by 0.01; the third by 0.001. Would each new rectangle be larger or smaller than the original rectangle? Explain.
   d) Find the area of each new rectangle in part c. Is each new rectangle similar to the original rectangle? Give reasons for your answer.

Reflect

Explain how you divide by 0.1, 0.01, and 0.001 mentally. Use examples in your explanation.

Providing Math Information

When you designed your own math problem, you wrote a problem statement in the form of a question. You need to include the math information required to solve the problem. To know what information is needed, it is helpful to work backward from the problem statement.

Problem Statement

↓

Math Information

Start with the problem statement.
For example, here is a problem from *Unit 2*:
How many desks are in the school?

To answer this, you might need to know this information:
- How many classrooms are in the school?
- How many desks are in each classroom?
- How many other desks are in the school?

You can investigate to find the information or you can make up your own information.
The problem might be written this way:
There are 12 classrooms.
There are 30 desks in each classroom.
There are 25 more desks in the library resource room.
How many desks are in the school?

Some problems give more information than is needed.
It is helpful to highlight the math information needed to solve the problem and to ~~cross out~~ information that is not needed.

Ensure that all information needed to solve the problem is provided. Keep in mind the strategies at the left that can be used to solve a problem.

Strategies

- Make a table.
- Use a model.
- Draw a diagram.
- Solve a simpler problem.
- Work backward.
- Guess and check.
- Make an organized list.
- Use a pattern.
- Draw a graph.
- Use logical reasoning.

✓ Check

Reading and Writing in Math

1. a) In this problem, what extra information is needed to solve the problem?
Shazi bought some 30¢ candy bars and some 60¢ candy bars. She bought 10 candy bars in total. How many candy bars did she buy at each price?

b) Make up the information you need to solve the problem in part a.
Then solve the problem.

2. a) In this problem, what information is not needed to solve the problem?
A bicycle dealer put together a shipment of bicycles and tricycles.
Tricycles cost $25 more than bicycles.
The dealer used 50 seats and 130 wheels.
How many bicycles and how many tricycles did she put together?

b) Solve the problem in part a.

3. a) In this problem, what extra information is needed to solve the problem?
The local hockey league has two divisions.
Each division has 6 teams.
How many games are played during the season?

b) Try to find out about a local hockey league in your area.
Write a problem about the league.
Solve the problem.

4. Write your own word problem.
Identify what math information is needed to solve it.
Write the information needed to answer the problem, either by investigating to find the information or by making up your own information.
Solve the problem.
Trade problems with a classmate.
Solve your classmate's problem.

To write a word problem:
- Start with a problem statement. Work backward.
- Think about what is needed to solve the problem.
- Think about what math information is required.
- Write the word problem.

Unit Review

Review any lesson with

What Do I Need to Know?

☑ **To add or subtract two fractions:**

Use equivalent fractions to make the denominators the same,

then add or subtract the numerators. For example,

$$\frac{5}{4} + \frac{3}{5}$$

$$= \frac{25}{20} + \frac{12}{20}$$

$$= \frac{37}{20}, \text{ or } 1\frac{17}{20}$$

$$\frac{7}{3} - \frac{3}{8}$$

$$= \frac{56}{24} - \frac{9}{24}$$

$$= \frac{47}{24}, \text{ or } 1\frac{23}{24}$$

☑ **To multiply two fractions:**

Multiply the numerators and multiply the denominators.

$$\frac{2}{3} \times \frac{1}{5} = \frac{2 \times 1}{3 \times 5} = \frac{2}{15}$$

$$\frac{2}{3} \times \frac{1}{5} = \frac{2}{15}$$

☑ **To divide a whole number by a fraction:**

Write the whole number as a fraction, then multiply.

For $4 \div \frac{2}{3}$, write: $\frac{4}{1} \div \frac{2}{3}$ as $\frac{4}{1} \times \frac{3}{2} = \frac{12}{2} = 6$

☑ **To divide a fraction by a whole number:**

Write the whole number as a fraction, then use common denominators.

$$\frac{2}{3} \div 4 = \frac{2}{3} \div \frac{12}{3}$$

$$= \frac{2}{12}$$

$$= \frac{1}{6}$$

Since the denominators are the same,

divide the numerators.

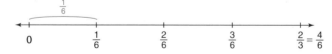

☑ **To divide two fractions:**

Method 1:

Use common denominators.

$$\frac{4}{5} \div \frac{3}{2} = \frac{8}{10} \div \frac{15}{10}$$

$$= \frac{8}{15}$$

Method 2:

Use multiplication.

$\frac{4}{5} \div \frac{3}{2}$ is the same as $\frac{4}{5} \times \frac{2}{3} = \frac{8}{15}$

LESSON

4.1 **1.** Name a fraction between each pair of fractions.

a) $\frac{1}{4}$ and $\frac{1}{2}$ b) $\frac{1}{2}$ and $\frac{3}{4}$

c) $\frac{1}{3}$ and $\frac{3}{4}$ d) $\frac{3}{5}$ and $\frac{7}{8}$

2. Fletcher completed $\frac{3}{5}$ of the test questions. Lalo completed $\frac{2}{3}$ of the test questions.

a) Who completed more questions?

b) How many questions might have been on the test? Explain.

3. Order the fractions in each set from least to greatest.

a) $\frac{2}{3}, \frac{4}{5}, \frac{5}{6}, \frac{3}{4}, \frac{1}{4}$ b) $\frac{7}{10}, \frac{1}{3}, \frac{3}{7}, \frac{3}{8}, \frac{2}{5}$

4. Order the fractions in each set from greatest to least.

a) $\frac{1}{2}, \frac{3}{4}, \frac{7}{6}, \frac{7}{8}$

b) $\frac{4}{3}, \frac{3}{4}, \frac{1}{6}, \frac{4}{10}, \frac{3}{12}$

c) $\frac{4}{5}, \frac{4}{6}, \frac{4}{10}, \frac{2}{3}, \frac{2}{4}$

4.2 **5.** Add.

a) $\frac{3}{8} + \frac{3}{4}$

b) $\frac{5}{6} + \frac{2}{7}$

c) $\frac{3}{2} + \frac{5}{3} + \frac{9}{10}$

4.3 **6.** Subtract.

a) $\frac{9}{10} - \frac{3}{4}$ b) $\frac{19}{12} - \frac{1}{2}$

c) $\frac{8}{9} - \frac{1}{8}$ d) $\frac{7}{5} - \frac{7}{6}$

4.2
4.3 **7.** Add or subtract as indicated.

a) $2\frac{2}{3} + 1\frac{1}{2}$ b) $3\frac{1}{3} - 1\frac{7}{10}$

c) $2\frac{1}{6} + 4\frac{7}{8}$ d) $3\frac{1}{2} - 2\frac{3}{4}$

8. A flask contains $2\frac{1}{2}$ cups of juice. Ping drinks $\frac{3}{8}$ cup of juice. Preston drinks $\frac{7}{10}$ cup of juice. How much juice is in the flask now?

4.4
4.5 **9.** Multiply. Use an area model to illustrate each product.

a) $\frac{2}{3} \times 15$ b) $\frac{7}{10} \times \frac{5}{8}$

c) $5 \times \frac{3}{2}$ d) $\frac{2}{3} \times \frac{3}{8}$

e) $\frac{4}{5} \times \frac{3}{10}$ f) $\frac{9}{8} \times \frac{1}{5}$

g) $\frac{10}{3} \times \frac{5}{2}$ h) $\frac{11}{6} \times \frac{7}{4}$

10. Twenty-five Grade 8 students are going on a school trip. They pre-order sandwiches. Three-quarters of the students order a turkey sandwich, while $\frac{1}{4}$ of the students order a roasted vegetable sandwich. Of the $\frac{3}{4}$ who want turkey, $\frac{2}{5}$ do not want mayonnaise. What fraction of the students do not want mayonnaise?

4.6 **11.** Divide. Sketch a number line to show each quotient.

a) $1 \div \frac{1}{3}$ b) $2 \div \frac{3}{4}$

c) $3 \div \frac{4}{5}$ d) $4 \div \frac{5}{6}$

12. A glass holds $\frac{2}{3}$ cup of milk. A jug holds 8 cups of milk. How many glasses can be filled from the milk in the jug?

13. Divide. Sketch a number line to show each quotient.

a) $\frac{3}{10} \div 2$ b) $\frac{8}{5} \div 3$

c) $\frac{13}{2} \div 4$ d) $\frac{5}{4} \div 3$

14. Jaiden estimates that he takes $1\frac{1}{4}$ h to knit a square for a blanket.
How many squares can Jaiden knit in 25 h?

15. When you divide a fraction by a whole number, is the quotient greater than or less than 1? Include examples in your explanation.

4.6
4.7
16. Divide.

a) $6 \div \frac{2}{3}$ b) $\frac{3}{4} \div \frac{1}{4}$

c) $\frac{1}{2} \div \frac{1}{4}$ d) $\frac{2}{3} \div \frac{3}{8}$

e) $\frac{4}{5} \div \frac{3}{10}$ f) $\frac{9}{4} \div \frac{3}{2}$

g) $\frac{12}{5} \div \frac{5}{12}$ h) $\frac{11}{7} \div \frac{11}{7}$

4.7
17. Divide.

a) $\frac{5}{4} \div \frac{1}{3}$ b) $\frac{3}{8} \div \frac{9}{5}$

c) $\frac{5}{2} \div \frac{5}{4}$ d) $\frac{7}{10} \div \frac{10}{3}$

18. Divide.

a) $1\frac{3}{4} \div 2\frac{1}{8}$ b) $3\frac{5}{6} \div 2\frac{1}{5}$

c) $3\frac{1}{2} \div 1\frac{3}{8}$ d) $2\frac{1}{5} \div 4\frac{2}{5}$

19. When you divide a fraction by its reciprocal, is the quotient less than 1, greater than 1, or equal to 1? Use examples in your explanation.

4.5
4.6
4.7
20. Find each product and quotient. What patterns do you see?

a) i) $\frac{3}{1} \times \frac{1}{2}$ ii) $\frac{3}{1} \div \frac{2}{1}$

b) i) $\frac{3}{4} \times \frac{2}{3}$ ii) $\frac{3}{4} \div \frac{3}{2}$

c) i) $\frac{4}{5} \times \frac{3}{4}$ ii) $\frac{4}{5} \div \frac{4}{3}$

d) i) $\frac{5}{6} \times \frac{2}{3}$ ii) $\frac{5}{6} \div \frac{3}{2}$

4.2
4.3
4.5
4.7
21. Evaluate.

a) $\frac{9}{8} - \frac{3}{4}$

b) $\frac{9}{8} + \frac{3}{4}$

c) $\frac{4}{3} \times \frac{5}{2}$

d) $\frac{17}{10} \div \frac{2}{5}$

4.8
22. Write each decimal as a fraction.

a) 0.25   b) 0.75

c) 0.32   d) 0.005

23. Write each fraction as a decimal.

a) $\frac{1}{8}$ b) $\frac{3}{5}$

c) $\frac{123}{250}$ d) $\frac{19}{20}$

24. Write each fraction as a decimal.

a) $\frac{2}{3}$ b) $\frac{3}{7}$

c) $\frac{3}{13}$ d) $\frac{4}{11}$

25. The tenths and hundredths digits of a decimal can be any digit from 0 to 9.

a) Write all the decimals that are greater than $\frac{1}{3}$ and less than $\frac{3}{4}$.

b) Order the decimals in part a from least to greatest.

4.9
26. Use mental math to divide.

a) $57.8 \div 0.01$

b) $0.882 \div 0.2$

c) $1.374 \div 0.003$

Practice Test

1. Evaluate.
   a) $\frac{7}{5} + \frac{3}{4}$
   b) $\frac{13}{10} - \frac{2}{3}$
   c) $\frac{3}{7} \times \frac{4}{9}$
   d) $\frac{5}{2} \div \frac{7}{6}$

2. Which statement has the greatest value? How do you know?
   a) $\frac{7}{3} \times \frac{3}{4}$
   b) $\frac{7}{3} - \frac{3}{4}$
   c) $\frac{7}{3} \div \frac{3}{4}$
   d) $\frac{7}{3} + \frac{3}{4}$

3. Multiply a fraction by its reciprocal. What is the product?
   Use an example and an area model to explain.

4. a) Write $\frac{1}{7}$ as a decimal.
   b) What is the 2001st digit in the repeating decimal for $\frac{1}{7}$?
      Explain how you know.

5. Which number is added to the numerator and denominator of $\frac{2}{7}$
   to get a fraction that is equivalent to $\frac{1}{2}$? Show your work.

6. Three-fifths of the Grade 8 class are in the band.
   a) On Tuesday, only $\frac{1}{3}$ of these students went to band practice.
      What fraction of the class went to band practice on Tuesday?
   b) How many students might be in the class? How do you know?

7. Write each decimal as a fraction and each fraction as a decimal.
   a) $\frac{7}{8}$
   b) 0.64
   c) $\frac{5}{11}$
   d) 0.004

8. a) Choose a proper fraction.
      Add 1 to the numerator and to the denominator.
      Write the new fraction.
      Which fraction is greater?
   b) Repeat part a for 3 more different fractions.
      Is your answer about the greater fraction always the same?
      Explain.

Part 1

Recall that hatch marks show equal line segments.

The side length of this square is 1 unit. Write the area of each of the 4 figures as a fraction of the area of the square.

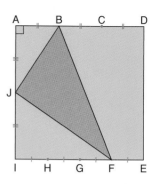

Show how you used multiplication of fractions to find the areas. Order the fractions from least to greatest.

Part 2

What fraction of each square is shaded green?

Square A

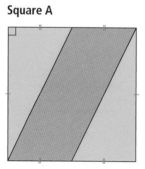

Square B

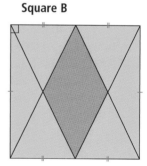

Square C

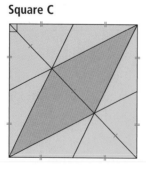

Square D

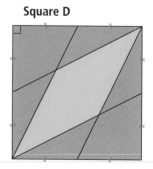

How did you use the addition or subtraction of fractions to find each fraction?

Part 3

Draw a large square with side length 1 unit.

Draw line segments to divide the square into different figures.

Find the area of each figure as a fraction of the area of the square.

Copy the square without the fractions.

Trade squares with a classmate.

For your classmate's square, write the area of each figure as a fraction of the area of the square.

Reflect on the Unit

What do you know about fractions and decimals that you did not know before this unit? Use examples in your explanation.

1. The bird with the most feathers is the whistling swan.
It has 25 216 feathers.
The ruby-throated hummingbird has the fewest feathers.
It has 940 feathers.
   a) How many more feathers does the whistling swan have than the ruby-throated hummingbird?
   b) About how many hummingbirds together would have the same number of feathers as one whistling swan? Explain.

2. Write each number as a product of prime factors.
Use exponents where possible.
   a) 38   **b)** 15
   c) 252   **d)** 105

3. According to *Guinness World Records 2005*, the greatest number of dominoes set up single-handed and toppled is 303 621 out of 303 628 by Ma Li Hua, in 2003.
   a) Write the number toppled in scientific notation.
   b) Write the number set up in scientific notation.
   c) What is the difference between the two numbers?
   Why can we not write this difference in scientific notation?

4. Copy each statement.
Insert brackets to make each statement true.
   a) $40 \div 5 + 3 \times 2^2 - 1 = 17$
   b) $40 \div 5 + 3 \times 2^2 - 1 = 19$
   c) $40 \div 5 + 3 \times 2^2 - 1 = 43$
   d) $40 \div 5 + 3 \times 2^2 - 1 = 15$

5. Twelve less than a number is 13.
Let x represent the number.
Then an equation is $x - 12 = 13$.
Solve the equation.
What is the number?

6. A primary class is going to the zoo. The ratio of adults to children must be 2:7. Twenty-eight children go on the zoo trip. How many adults are needed for supervision?

7. The ostrich runs at 65 km/h.
At this rate, how far can the ostrich run in 90 s?

8. There are 429 students registered at Woodside Public School.
On Wednesday, about 0.7% of the students were absent.
   a) How many students were absent?
   b) What percent of students were at school?

9. A salesperson earns commission at a rate of 8%. Last week, she earned $450 commission. What were her total sales for the week?

10. Calculate the simple interest and the amount.

  a) $500 invested at an annual interest rate of 2% for 1 year

  b) $2750 invested at an annual interest rate of 3.5% for 4 years

  c) $4500 invested at an annual interest rate of 6.25% for 18 months

3 **11.** Use 9 linking cubes.
Build an object.
Draw as many views as a classmate would need to build the object.
Trade views with a classmate.
Build your classmate's object.
Compare the object your classmate built from your views with the object you built.
Explain any differences.

12. Here is the net of a triangular prism.

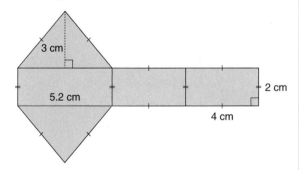

3 cm

5.2 cm

2 cm

4 cm

  a) Calculate the surface area of the prism.

  b) Calculate the volume of the prism.

13. The base of a triangular prism has base b and height h. The length of the prism is l.
What are the possible values of b, h, and l for a triangular prism:

  a) with volume 12 cm³?

  b) with volume 24 cm³?

4 **14.** Copy each pair of fractions.
Use $<$, $>$, or $=$ to compare the fractions.

  a) $\frac{2}{5} \square \frac{6}{15}$ **b)** $\frac{1}{9} \square \frac{2}{18}$

  c) $\frac{8}{10} \square \frac{3}{8}$ **d)** $\frac{2}{3} \square \frac{4}{5}$

  e) $\frac{3}{8} \square \frac{2}{5}$ **f)** $\frac{5}{6} \square \frac{6}{7}$

15. Add or subtract.

  a) $\frac{2}{5} + \frac{1}{4}$ **b)** $\frac{3}{8} + \frac{1}{2}$

  c) $\frac{7}{8} - \frac{1}{4}$ **d)** $\frac{1}{2} - \frac{1}{10}$

  e) $5\frac{7}{9} - 2\frac{1}{4}$ **f)** $3\frac{1}{3} + 1\frac{1}{8}$

16. Which statement has the least value?

  a) $\frac{2}{3} + \frac{1}{6}$ **b)** $\frac{2}{3} - \frac{1}{6}$

  c) $\frac{2}{3} \times \frac{1}{6}$ **d)** $\frac{2}{3} \div \frac{1}{6}$

  e) $\frac{1}{6} \div \frac{2}{3}$ **f)** $\frac{1}{6} \times \frac{2}{3}$

17. Write each fraction as a decimal.
Then order the decimals from least to greatest.

  a) $\frac{13}{50}$ **b)** $\frac{1}{4}$ **c)** $\frac{51}{200}$ **d)** $\frac{3}{11}$

18. Find each quotient.

  a) $\frac{3275}{0.1}$ **b)** $\frac{3275}{0.01}$

  c) $\frac{3275}{0.001}$ **d)** $\frac{3275}{0.5}$

  e) $\frac{3275}{0.05}$ **f)** $\frac{3275}{0.005}$

UNIT 5

Data Management

What would you like to know about people in your community? Here are some questions to ask.

- What careers are your classmates interested in?
- How do people spend their leisure time?
- How many people speak two languages? Three languages?

What other questions could you ask?

What You'll Learn

- Relate a census and sample.
- Identify bias in data-collection methods.
- Use databases and spreadsheets.
- Collect, display, and evaluate data on charts and graphs.
- Identify and describe trends in graphs.
- Understand and apply the measures of central tendency.

Why It's Important

Statistics Canada is the federal government department that collects data from every household in Canada. Private companies conduct surveys and publish the results.
You need to understand how to interpret the data you read.

Key Words

- population
- census
- sample
- *Census at School*
- inference
- histogram

Skills You'll Need

Using a Percent Circle to Draw a Circle Graph

The circle represents 100%.
A percent circle is divided into 10 congruent sectors.
Each large sector represents 10%.
Each sector is further divided into 10 parts.
Each part is 1% of the circle.

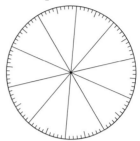

Example 1

Pitaq collected data about his classmates' favourite season.
He recorded the results in a table.

Classmates' Favourite Season

Season	Autumn	Winter	Spring	Summer
Number of Students	8	3	5	9

Use a percent circle to display the data on a circle graph.

Solution

Add the numbers in the table: $8 + 3 + 5 + 9 = 25$. There are 25 students.
Write the number of students who chose each season as a fraction of 25,
then as a percent.

Autumn: $\dfrac{8}{25} = \dfrac{32}{100} = 32\%$ Winter: $\dfrac{3}{25} = \dfrac{12}{100} = 12\%$

Spring: $\dfrac{5}{25} = \dfrac{20}{100} = 20\%$ Summer: $\dfrac{9}{25} = \dfrac{36}{100} = 36\%$

For autumn, 32% is three 10% sectors and two 1% parts.
For winter, 12% is one 10% sector and two 1% parts.
For spring, 20% is two 10% sectors.
For summer, 36% is three 10% sectors and six 1% parts.
Label each sector.
Write a title for the graph.

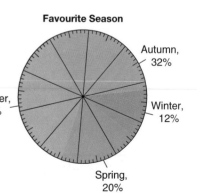

Favourite Season
Autumn, 32%
Winter, 12%
Spring, 20%
Summer, 36%

1. For each set of data, use a percent circle to draw a circle graph.

a)

Coins in Laura's Piggy Bank

Coin	Pennies	Nickels	Dimes	Quarters
Number	12	15	9	14

b)

Items in Lost and Found Box

Item	Hats	Socks	Sweaters	Gloves	Shoes
Number	8	16	6	8	2

Trends in Graphs

Data and graphs sometimes show a pattern or trend.

Example 2

Each line graph represents temperatures recorded in southern Ontario.

a) Describe any trends in each graph.

b) What were the temperatures on Sunday and Saturday?
How do you know?

Graph A
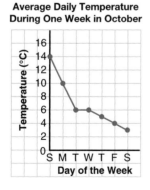

Graph B
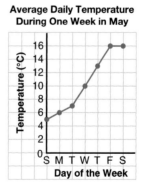

Solution

Graph A

a) The graph goes down to the right.

This means the temperature is decreasing.

The line segment from Tuesday to Wednesday is horizontal.

This means the temperature stayed the same during those two days.

b) The first **S** on the horizontal axis represents Sunday.
The corresponding value on the vertical axis is 14°C.
The temperature on Sunday was 14°C.
To find the temperature on Saturday:
Draw a vertical line, from the last **S** on the
horizontal axis, to meet the graph,
as shown.
Then draw a horizontal line to meet the
vertical axis.
The temperature was about 3°C on Saturday.

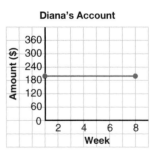

Graph B

a) The graph goes up to the right.
This means the temperature is increasing.
The line segment from Friday to Saturday is horizontal.
This means the temperature stayed the same during those two days.

b) The temperature on Sunday was 5°C.
The temperature on Saturday was 16°C.

✓ Check

2. Each of 3 people deposited $200 in a bank account at the beginning of January.
These graphs display the money in each account for the next 7 weeks.

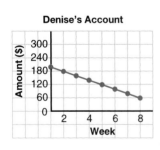
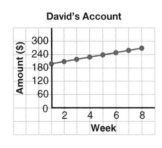
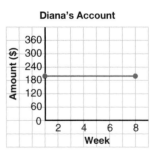

a) Who withdrew money from her or his account? How do you know?

b) Who deposited money to her or his account? How do you know?

c) Which account had no transactions? Explain.

d) How much money did each person have at the end?
How do you know?

5.1

Relating Census and Sample

Focus Collect data from a population and a sample.

Every five years, Statistics Canada (Stats Can) collects data from one in five households in Canada. It uses these data to help us better understand our country, including its people, natural resources, educational needs, society, and cultures.

Explore

Work in a group of 3.
Which kind of TV show is the most popular in your class?

- Have each student in your group name her or his favourite kind of TV show.
  Record the results in a tally chart.
  Make a conclusion based on these results.
- Combine your results with those of a second group of students.
  Make a conclusion.
- Combine results to get those of 3 groups of students.
  Make a conclusion.

Did your conclusions change after each additional group was included? Explain.

Would your conclusions likely change if you used the results for the whole class? Explain.

Reflect & Share

Record the results for your group on the board.
Compare them with the class results.
How did the results change as the number of responses increased?

Connect

One way to collect data is to survey all the people in a **population**.
This is called a **census** survey.
For example, Jared asked every student in each of the three Grade 8 classes in his school, "What is your favourite sport to play?"
Since Jared collected the data himself, the data are primary data.
Results from a census are accurate because the entire population provides input. For a large population, a census can be expensive and time-consuming.

Another way to collect data is to survey a portion or a **sample** of the population. For example, Alicia used the Internet to find the salaries of 10 professional basketball players. Since Alicia did not collect the data herself, the data are secondary data.

A sample survey is less costly and requires less time than a census. Results from a sample may not be as accurate as those from a census because not everyone in the population has input.

A sample is *biased* if it does not accurately represent the population. For example, if the population is 15 boys and 17 girls, a sample for this population should have approximately equal numbers of boys and girls. A sample of 8 boys is a biased sample.

A survey is *reliable* if the results can be duplicated in another survey. A biased sample may not produce reliable results.
For example, suppose 100 people at a hockey game were asked to name their favourite sport. The results would be biased and probably different from asking 100 randomly-selected people to name their favourite sport.

A survey is *valid* if the results represent the population.
Courtney surveys her friends in Grade 8 and finds that 75% have a DVD player at home. James surveys all the Grade 8 students in their school and finds that 38% have a DVD player at home.
Courtney's survey is not valid because her sample results do not represent the Grade 8 population.

Example 1

For each survey, state if the sample is biased or reliable.
a) To find out if the school cafeteria should change its menu, all the teachers were surveyed.
b) *Rock Music* magazine asks its readers to reply to a question about the favourite music group of teenagers.

Solution

a) The sample is biased because the cafeteria is also used by students and they should be included in the survey.
b) The sample is biased because the readers are most likely fans of rock music, and not all teenagers prefer rock music.

Example 2

The owners of the City Mall want to find out if shoppers would like a video arcade to be opened in the mall.

How might the survey results be affected in each situation? Explain.

a) A survey of all shoppers is taken on a weekday morning.

b) A survey of all shoppers is taken on a Saturday afternoon.

c) A survey of teenage shoppers is taken.

Solution

a) On a weekday morning, most customers might be retired people or parents with small children. The sample is not representative of the population. The data collected from the survey would most likely indicate that a video arcade should not be opened.

b) On a Saturday afternoon, the mall would have a mixed population in age and gender. The data collected from the survey would probably be representative of the population.

c) When only teenagers are surveyed, the sample is biased. The data collected from the survey would most likely indicate that a video arcade should be opened.

Practice

1. In each case, are the data collected from a census or a sample? Justify your answer.

   a) To find the favourite TV show of Grade 8 students in a school, fifteen of the 40 Grade 8 students in the school are surveyed.

   b) To find the favourite video game of Ontario 13-year-olds, all 13-year-old students in Ontario are surveyed.

   c) To find out if customers of a chain of coffee shops are happy with the service, some customers in every shop were surveyed.

2. For each survey, state if the sample is biased or reliable. Justify your answer.

   a) To find out if the arena should offer more public skating times, a survey is posted on a bulletin board at the arena and left for patrons to complete and return.

b) To find out the favourite breakfast of Grades 7 and 8 students, a survey of 300 randomly-selected Grades 7 and 8 students was conducted.

c) To find out about the exercise habits of Canadian teenagers, a fitness magazine asks its readers to send in information about the exercise habits of teenagers.

d) To find out if the soccer league should buy new uniforms for the players, 20 parents of the students in the soccer league were surveyed.

3. For each situation, explain why data are collected from a sample and not a census.

a) To find the mean cost of hockey equipment for teenagers in Canada

b) To find the number of Canadian families with a cell phone

c) To find the number of hours an AAA battery will last in a calculator

Number Strategies

Find each percent.
- 300% of 140
- 30% of 140
- 3% of 140
- 0.3% of 140

What patterns do you see in the answers?

4. Name two methods used to collect data.
Describe the advantages and disadvantages of each method.

5. Identify each population about which the data are to be collected.

a) The management team of a shopping mall in Brantford wants to know how to attract more people between the ages of 13 and 25 to the mall.

b) A juice company wants to find the actual volume of juice in a 1-L carton.

c) The board of education wants to find out which schools need renovations.

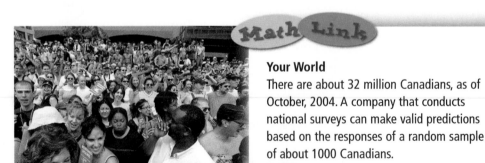

Math Link

Your World
There are about 32 million Canadians, as of October, 2004. A company that conducts national surveys can make valid predictions based on the responses of a random sample of about 1000 Canadians.

6. For each situation below:
   a) Might the sampling method provide biased data?
   b) If your answer to part a is yes, how can the sampling method be changed so the data collected represent the population?
      i) The student council wants to know if students will attend another school dance this month.
         The student council members survey all their friends to find out.
      ii) A sportswear store conducts a survey to find the most popular brands of athletic shoes.
         The first 300 people in the store who are wearing athletic shoes are surveyed.
      iii) To find out the number of hours per week that people in her city spend exercising, a newspaper journalist interviews all the clients at five different fitness centres in the city.
      iv) A company is hired to find out who Canadians think will win the next Grey Cup. To collect the data, the company puts an advertisement in the sports section of all major newspapers asking people to vote for their choice.

7. Should a census or sample be used to collect data about each topic? Explain your choice.
   a) the effectiveness of a new suntan lotion
   b) the popularity of a fruit-flavoured yogurt
   c) the number of students in Grades 6, 7, and 8 in your school with braces
   d) the number of your friends who like to play computer games

8. **Assessment Focus** Suppose you are the manager of a high school cafeteria.
   You want to create a new breakfast and lunch menu for the students.
   Describe at least 2 different methods that could be used to collect data about food items to offer on the menu.
   How can you ensure that the results are reliable? Explain.

Reflect

Explain the difference between a census and a sample.
When is a sample biased?

Using *Census at School* to Get Secondary Data

Focus	Search databases for information and to solve problems.

To collect data about students from 8 to 18 years old, Statistics Canada developed the ***Census at School*** website. You can use *Census at School* to find data about students under these headings.

- Which kids have the gadgets?
- Time spent travelling to school
- What's your favourite subject?
- What's for breakfast?
- Sports in your life
- Who do you look up to?

You can also find data about students in other countries that use *Census at School.* Your teacher can register your class, so you, too, can complete the survey and access the data.

To use *Census at School,* follow these steps:

1. Open the *Census at School* website.
You may be asked for your username and password.
Ask your teacher for these.

2. Under **Home Page**, click: Data and results

3. Under **Canadian results**, click on any topic that interests you from the list under *See summary data tables for 2003/04:*

4. Suppose you select: What's your favourite subject? A table similar to this appears.

What's your favourite subject?

	Elementary			Secondary		
	Girls	Boys	Total	Girls	Boys	Total
			number			
English	218	48	**266**	154	69	**223**
French	87	33	**120**	50	7	**57**
Art	601	241	**842**	182	84	**266**
Computers	138	254	**392**	37	100	**137**
Geography	18	14	**32**	34	23	**57**
History	59	63	**122**	50	45	**95**
Math	284	360	**644**	139	146	**285**
Music	172	76	**248**	106	45	**151**
Other	131	129	**260**	138	188	**326**
Physical education	759	1,206	**1,965**	186	357	**543**
Science	122	170	**292**	113	64	**177**
Social studies	66	47	**113**	53	17	**70**

Source: Census at School, Canada, 2003/04.

Source: Statistics Canada, *Census at School,* 2003/04,
http://www19.statcan.ca/04/04_002_e.htm
Date extracted: March 2005

Print these data for later use on page 205.

To find data from students in other countries, follow these steps:

5. Return to *Step 3*. Under **International results**,
click on <u>Random data selector</u>.

6. Under **Results and Data**, click: Phase 1 Results

7. Select any topic of interest listed on the page. Suppose you click:
Month of Birth – What is the most popular month to be born in?

8. Suppose you select All Data. A table similar to this appears. The table shows data for the United Kingdom.

Month of Birth

All Data	Counts All pupils	Counts Males	Counts Females	Percentage of all pupils	Percentage of male pupils	Percentage of female pupils
Total	58,322	27,793	30,529	100.00	100.00	100.00
January	4,584	2,150	2,434	7.86	7.74	7.97
February	4,465	2,165	2,300	7.66	7.79	7.53
March	4,978	2,359	2,619	8.54	8.49	8.58
April	4,784	2,247	2,537	8.20	8.08	8.31
May	5,065	2,422	2,643	8.68	8.71	8.66
June	4,990	2,406	2,584	8.56	8.66	8.46
July	5,020	2,393	2,627	8.61	8.61	8.60
August	4,898	2,282	2,616	8.40	8.21	8.57
September	5,199	2,509	2,690	8.91	9.03	8.81
October	5,000	2,461	2,539	8.57	8.85	8.32
November	4,669	2,222	2,447	8.01	7.99	8.02
December	4,670	2,177	2,493	8.01	7.83	8.17

Excludes non responses

Source: *Census at School*–United Kingdom
http://www.censusatschool.ntu.ac.uk/table8-0.asp
Date extracted: March 2005

Use *Census at School* to solve each problem.
Remember to use the <u>Random data selector</u> to obtain data from the United Kingdom, South Africa, New Zealand, and parts of Australia. Print your data.

1. a) In which month are most students in the United Kingdom born?
   b) Is this month the same for boys and girls? Explain.

2. What percent of students in Canada take more than 1 h to get to school?

3. What is the difference in percents of people in Canada with blue eyes and with brown eyes?

Inferring and Evaluating

Focus Make inferences and evaluate arguments.

Explore

Work in a group of 3.
Suppose your group is responsible for selecting one of these students to be on an *Aim for the Top* quiz team for your school board:

Marks of Top Students

Student	Math	Language Arts	History	Science	Geography
Suresh	93	89	90	97	87
Marco	96	91	86	94	90
Nella	98	80	91	95	94
Yoko	90	92	91	91	92

Which student would you choose to be on the team?
Justify your choice.
What would you say to convince other groups about your choice?

Reflect & Share

Share your reasons with another group that picked the same student.
Are the reasons the same? Explain.
Share your choice with a group that picked a different student.
Discuss the reasons for your choices.

Connect

Data and graphs may be used to make convincing arguments.

The school cafeteria collected data on the numbers of cups of hot chocolate purchased by girls and boys during the first 15 days of school in September.

Number of Cups of Hot Chocolate Purchased															Total
Girls	2	0	3	5	4	7	4	3	4	5	1	5	2	8	10	63
Boys	3	1	5	4	5	9	6	5	3	6	2	4	4	9	8	74

Amy used the data in this table to argue that boys drink more hot chocolate than girls.

Alice said there were not enough data to make that argument.

The students collected more data to test Amy's argument.

Total Hot Chocolate Sales Per Month						
	Sept.	Oct.	Nov.	Dec.	Jan.	Total
Girls	85	128	197	201	252	863
Boys	92	130	190	207	249	868
Total	177	258	387	408	501	1731

Amy said that her conclusion was correct because the total number of hot chocolates purchased by girls was 863.

This is less than 868, which is the total for boys.

Alice looked at the data and made this argument:

More students drink hot chocolate when the weather is cold.

Alice said her argument was correct because the monthly sales increased for the colder months.

Alice's argument is more convincing than Amy's because, in 5 months, the total sales for boys was only 5 more than the total for girls. The numbers are too close for Amy's argument to be convincing.

When we use data to predict a value in the future, or to estimate a value between given data, we make an **inference**.

When we use data to make a conclusion, we *infer*.

Example

This double-bar graph shows the hourly wages of Canadians in 1997 and 2003.

a) How can the graph be used to justify each argument?
  i) The percent of employees who earned under $8.00/h decreased from 1997 to 2003.
  ii) Approximately 50% of employees earned $16.00/h or more in 2003.

b) What can be inferred about the percent of employees who will earn hourly wages over $24.00 in 2009? Explain.

Solution

a) i) The bar for the interval "Under $8.00" is shorter for 2003 than it is for 1997. This means that there was a decrease in the percent of employees who earned less than $8.00/h, from 1997 to 2003.

  ii) The percent of employees who earned $16.00/h or more in 2003 is given by the lengths of the bars for the intervals from $16.00, and up.
  Estimate the lengths: 16% + 12% + 9% + 9% + 6% = 52%
  Since about 52% of employees earned $16.00/h or more in 2003, the argument is valid.

b) The graph shows an increase, from 1997 to 2003, in the percent of employees earning hourly wages in these intervals:
  $24.00−$27.99, $28.00−$34.99, and $35.00 and over
  If we assume this trend continues, we can infer that there will be an increase in the percent of employees who will earn hourly wages over $24.00 in 2009.

Practice

1. Elise, Mira, and Kim practised to compete in the 50-m butterfly stroke. Their practice times, in seconds, are given in the table. Which of the three swimmers would you choose to be on the competitive relay team to swim the butterfly stroke? Justify your choice.

Swimmer	1st Practice (s)	2nd Practice (s)	3rd Practice (s)	4th Practice (s)
Elise	45.4	45.3	45.8	46.2
Mira	47.9	43.2	44.7	45.0
Kim	45.2	48.3	43.1	44.3

2. Each of ten cats was given a bowl filled with
Purr-Fect cat food. Seven cats ate the food in the bowl.
Three cats did not.
Advertisers for Purr-Fect cat food claimed:
"70% of cats prefer Purr-Fect cat food."
a) Is this claim true? Explain.
b) What inferences would you make from the data? Explain.

3. a) What inferences can you make from this graph? Explain.

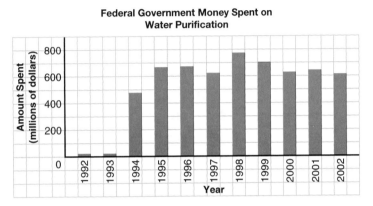

Federal Government Money Spent on Water Purification

Source: Statistics Canada, *Canada e-Book* (11-404-XIE) based on the 2001 Canada Yearbook (11-402-XPE).
http://142.206.72.67/01/01b/01b_graph/01b_graph_002d_1e.htm   Date extracted: April 11, 2005

b) How can the graph be used to justify each argument?
  i) The government has reduced its spending
     on water purification.
  ii) The government has increased its spending
     on water purification.

4. Liang asked 20 Grade 8 students:
"What is your favourite outdoor summer activity?"
He recorded these results:

Activity	Swimming	Biking	Canoeing	Skateboarding	Rollerblading
Number of Students	7	5	2	3	3

How could Liang use the results of his survey to make these
arguments? Justify your answers.
a) The majority of students surveyed chose swimming as their
   favourite outdoor activity.
b) Land activities are preferred over water activities.

5. Fifteen hundred students attend Bishop Ryan High School.
One hundred twenty students were asked,
"Do you think the school library hours should be extended?"
Forty students agreed that library hours should be extended,
10 students said they were against extended hours,
and 70 students had no opinion.
The Bishop Ryan High School newspaper reported:

"Survey indicates 80% of the student body who had an opinion, agree that school library hours should be extended."

a) Is this a valid inference? Justify your answer.
b) Write a convincing argument that could be justified by the survey results.
Explain your reasoning.

6. Police recorded this information about 100 car accidents:
- 40 were within 3 km of the driver's home
- 30 were on the highway
- 20 were in parking lots
- 10 were on country roads

A newspaper reporter wrote the following headline:

Majority of Car Accidents Happen Close to Home

a) How could the reporter use the data to convince readers that his statement is true?
b) What statement could you make to prove the reporter is wrong? Justify your statement.
c) Write two inferences that can be made from the data. Justify your inferences.

7. Assessment Focus

7. **Assessment Focus**

a) What does the table show?

Eats Fruits and Vegetables 5 to 10 Times Per Day

Age Group (years)	Females	Males
12–14	250 238	231 755
15–19	391 589	326 369
20–24	393 640	307 997
25–34	841 078	572 224

Source: Statistics Canada, *Canadian Community Health Survey*, 2003, CANSIM table 01050249.
http://www.statcan.ca/english/freepub/82-221-XIE/00604/tables/html/2187_03.htm   Date extracted: March 18, 2005

b) What inferences can you make about the eating habits of each group?

  i) females   **ii)** males

  Justify your answers.

c) What arguments can you make about the eating habits of each group?

  i) 12–14 year old males

  ii) 15–19 year old females compared to 20–24 year old females

    Justify your arguments.

d) **i)** What does this table show?

Eats Fruits and Vegetables more than 10 Times Per Day

Age Group (years)	Females	Males
12–14	31 289	40 472
15–19	59 418	63 755
20–24	61 520	53 264
25–34	98 376	69 277

Source: Statistics Canada, *Canadian Community Health Survey*, 2003, CANSIM table 01050249.
http://www.statcan.ca/english/freepub/82-221-XIE/00604/tables/html/2187_03.htm Date extracted: March 18, 2005

  ii) Do the inferences and arguments you made in parts b and c still hold? Explain.

Number Strategies

Insert brackets to make each statement true.

$4 \div 3 + 5 \times 2^2 + 4 = 16$

$4 \div 3 + 5 \times 2^2 + 4 = 48$

$4 \div 3 + 5 \times 2^2 + 4 = 112$

$4 \div 3 + 5 \times 2^2 + 4 = 24$

These data come from a Stats Can Community Health Survey in 2003.

Reflect

When you use a set of data to make an inference or a convincing argument, what sorts of things do you consider? Use an example in your explanation.

Focus Graph data, identify trends, and make inferences.

Explore

Work in a small group.

Each graph is incomplete.

Identify each type of graph.

Match each graph with a table below. Explain your choice.

On a copy of the graphs, insert the title and labels for each graph.

Make a list of some information you know from each graph.

Graph A

Graph B

Graph C

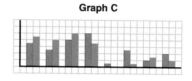

Graph D

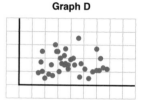

Table 1

Instruments Students Learn to Play

Instrument	Girls	Boys
Drums	7	9
Classical Guitar	5	8
Electric Guitar	8	10
Piano	10	7
Harp	1	0
Clarinet	5	1
Violin	2	3
Bagpipes	4	2

Table 2

Daily Maximum Temperatures for 2 Weeks in February

Feb.	1	2	3	4	5	6	7	8	9	10	11	12	13	14
Temp (°C)	−3	5	4	−1	0	6	3	5	−2	0	5	8	−4	2

Table 3

Fun Run Fundraiser – Participants' Age in Years and Distance Completed in Kilometres

Age	Distance	Age	Distance	Age	Distance	Age	Distance
28	30	23	8	36	27	42	20
63	20	60	11	35	22	62	11
46	35	47	6	36	16	15	9
37	15	48	16	65	13	35	15
38	12	51	12	60	27	25	15
29	25	40	19	57	10	18	10
33	16	31	10	68	12	17	17
54	5	20	5	32	20	18	25
43	15	30	25	38	15	26	7

Table 4

Homework Time Spent on Each Subject

Subject	Percent
French	15
Math	30
Spelling	25
History	10
Science	20

Reflect & Share

With your group, decide if the data could have been displayed another way. Draw each new display.

Connect

Here are some different types of graphs.

A circle graph displays data that represent parts of one whole.
Graph A in *Explore* is a circle graph.

A line graph displays data that change over time.
On a line graph, line segments join data points.
Graph B in *Explore* is a line graph.

A bar graph displays data that can be counted.
Two sets of data can be displayed on a double-bar graph.
Graph C in *Explore* is a double-bar graph.

A scatter plot displays two related sets of data that are measured or counted.
Graph D in *Explore* is a scatter plot.

Example

This table displays the life expectancy at birth for people born between 1920 and 1990.

Life Expectancy at Birth

Birth Year	Females (years)	Males (years)
1920	61	59
1930	62	60
1940	66	63
1950	71	66
1960	74	68
1970	76	69
1980	79	72
1990	81	75

Source: Statistics Canada, *Canadian Statistics summary tables.* Last modified 2005-02-17.
http://www40.statcan.ca/l01/cst01/health26.htm
Date extracted: April 11, 2005

a) A person is born in 1970.
How long can he or she expect to live?

b) Which types of graphs could be used to display the data? Explain.

c) Display the data using an appropriate graph. Justify your choice.

d) What trends do you see in the data? How does the graph show the trends?

e) How could the graph be used to predict the life expectancy of a female born in 2010? What assumptions are made?

f) Simon extended the data in the table. The life expectancy for a male increased by 75 − 59, or 16 years from 1920 to 1990. So, Simon predicted that a male born in 2060 would have a life expectancy of 75 + 16, or 91 years. Is this a valid argument?

Solution

a) From the table, a female born in 1970 can expect to live for 76 years, and a male for 69 years.

b) There are two sets of data. They could be displayed using a double-bar graph or double-line graph.

c) Since the data represent change over time, a line graph is an appropriate graph.
Plot each set of data on the same grid.
Use a scale of 1 square for every 5 years.

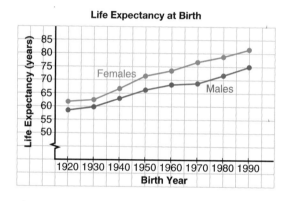

Remember to label each axis and give the graph a title.

d) Both lines go up to the right, so life expectancy is increasing for both males and females.
The line representing females is above the line representing males, so females have a greater life expectancy than males.

e) To predict the life expectancy of a female born in 2010, extend the last line segment for female life expectancy to 2010.

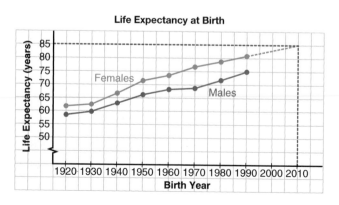

From the graph, the life expectancy of a female born in 2010 will be about 85 years. We assume that life expectancy will continue to increase at the same rate as it did from 1980 to 1990. This may not happen.

f) The argument is not valid.
We cannot assume that life expectancy will continue to increase at the same rate for the next 70 years.
Certain diseases may be cured, but the human body wears out.
The life expectancy will probably reach an upper limit.

Practice

1. The table shows the attendance at after-school football games.

Game	1	2	3	4	5	6	7	8	9	10	11
Attendance	235	197	203	185	163	149	126	118	102	85	71

a) Display the data.
b) What trend does the graph show? Explain.
c) What are some possible reasons for this trend?

Populations of Some Canadian Cities

City	1996	2001
St. John's	174 051	172 918
Sudbury	165 336	155 219
Saint John	125 705	122 678
Chicoutimi	160 454	154 938
Thunder Bay	126 643	121 986
Regina	193 652	192 800
Trois-Rivières	139 956	137 507

2. This table shows the populations of some Canadian cities in 1996 and in 2001.
a) Round the data to the nearest 1000. Display the data using an appropriate graph. Justify your choice.
b) What trends do you see in the data? How does the graph show the trends?
c) Predict the population of each city in 2006. Justify your numbers.

3. a) What do the data in the table represent?

Average Daily Temperature (°C)

Month	J	F	M	A	M	J	J	A	S	O	N	D
Vancouver	2	3	5	9	12	15	18	17	13	9	5	2
Hawaii	22	22	21	18	15	13	12	13	15	18	19	21

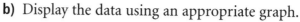

Evaluate.

- $\frac{4}{5} \div \frac{2}{3}$
- $\frac{4}{5} \times \frac{2}{3}$
- $\frac{4}{5} - \frac{2}{3}$
- $\frac{4}{5} + \frac{2}{3}$

b) Display the data using an appropriate graph.

c) Describe any trends in the graph.

d) When is the best time to visit Vancouver? Hawaii? Explain.

e) Who might be interested in these data? Why?

4. The table shows the ages in years and heights in centimetres of 15 students on a baseball team.

Age (years)	12	14	15	17	18	11	13	14
Height (cm)	134	161	158	185	199	157	161	183
Age (years)	18	16	15	16	14	15	17	
Height (cm)	207	172	189	169	175	166	185	

a) Display these data using an appropriate graph. Explain your choice.

b) Does there appear to be a relationship between age and height? Explain.

Average Annual Earnings of Canadians

Year	Females ($)	Males ($)
1993	22 300	34 700
1994	22 500	36 200
1995	23 000	35 400
1996	22 700	35 300
1997	22 900	36 200
1998	23 900	37 400
1999	24 200	37 800
2000	24 900	39 000
2001	25 100	39 100
2002	25 300	38 900

5. Assessment Focus

a) What do the data in the table represent?

b) Display the data using a suitable graph. Justify your choice.

c) Describe any trends. Justify your answers.

d) What do you think the average annual female earnings were in 1990? Explain.

e) Predict the average annual male earnings for 2005. Explain.

f) What inferences can you make from these data?

Show your work.

Reflect

Why do we need different types of graphs to display data? Give examples to support your answer.

Using a Spreadsheet to Create Graphs

Focus Display data on graphs using spreadsheets.

Spreadsheet software, such as *AppleWorks*, can be used to record, then graph, data.

Use these data for elementary students from *Census at School*.

What's your favourite subject?
Elementary Students

Subject	Girls	Boys
English	218	48
French	87	33
Art	601	241
Computers	138	254
Geography	18	14
History	59	63
Math	284	360
Music	172	76
Other	131	129
Phys-Ed	759	1206
Science	122	170
Social Studies	66	47

You found these data from *Census at School* on page 192.

To graph these data using *AppleWorks*, follow these steps.
Open *AppleWorks*. Choose Spreadsheet.
Enter the data into rows and columns in the spreadsheet.

To create a double-bar graph

1. Highlight the data. Include the column heads but do not include the table title.

2. Click the Make a Chart icon on the toolbar.
   A Chart Options dialogue box appears.
   Choose Bar, then click OK.

3. Right-click the graph. Choose Chart Options.
   Select the Labels tab.
   In the Title box, type **Favourite Subject, Elementary Students**.
   To insert labels, right-click the graph. Choose Chart Options.
   Click the Axes tab.

Select X axis. Type **Subject**. Ensure Grid Lines is not selected.
Select Y axis. Type **Number of Students**. Select Grid Lines.
Under Minimum: enter 0. Under Maximum: enter 1400.
Under Step Size: enter 200. Then click OK.
Your graph should look like the one below.

The double-bar graph shows a legend at the right. The legend shows what the two sets of bars represent.

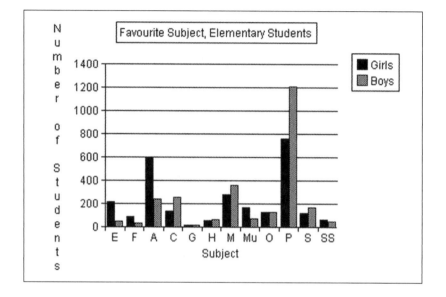

To create a circle graph

This table shows how Stacy budgets her money each month.

Stacy's Monthly Budget

Category	Amount ($)
Food	160
Clothing	47
Transportation	92
Entertainment	78
Savings	35
Rent	87
Other	28

1. Enter the data into rows and columns in a spreadsheet.
Highlight the data. Do not include the column heads or title.

2. Click the Make a Chart icon on the tool bar.
A Chart Options dialogue box appears.
Choose Pie, then click OK.

3. Right-click the graph. Choose Chart Options. Choose Series. Click Label Data. Click % in slice. Click OK.

4. To add a title, right-click the graph. Choose Chart Options. Select the Labels Tab. In the Title box, type **Stacy's Monthly Budget**. Then click OK. Your graph should look like the one below.

The circle graph shows a legend at the right. The legend shows what category each sector represents.

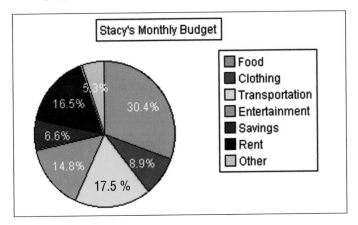

To create a double-line graph

Average Ticket Prices ($), Ontario

Year	Movie Theatres	Drive-Ins
1996–7	5.13	6.39
1997–8	5.38	6.79
1998–9	5.70	6.86
1999–0	6.14	6.92
2000–1	6.78	7.28
2001–2	—	—
2002–3	8.08	8.63

1. Enter the data into rows and columns in the spreadsheet.

2. Highlight the data. Include the column heads, but do not include the table title.

3. Click the Make a Chart icon on the tool bar.
A Chart Options dialogue box appears.
Choose Lines, then click OK.

4. Right-click the graph. Choose Chart Options.
Select the Labels tab.
In the Title box, type **Average Ticket Prices ($), Ontario**.
Then click OK. To insert labels, right-click the graph.
Choose Chart Options. Click the Axes tab.
Select X axis. Type **Year**. Ensure Grid Lines is not selected.
Select Y axis. Type **Amount ($)**. Select Grid Lines.
Then click OK.
Your graph should look like the one below.

The double-line graph shows a legend at the right. The legend shows what each line represents.

Note how *AppleWorks* draws the graph when data are missing. How would *you* draw this graph?

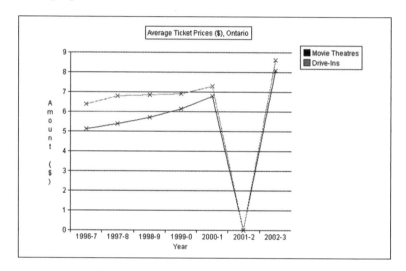

Check

1. Use the data on page 205 or the double-bar graph on page 206.
  a) Which subject is liked most by boys? How do you know?
  b) Which subject is liked least by girls? How do you know?
  c) Which subject is liked almost equally by boys and girls?
   How do you know?
  d) Write 3 more things you know from looking at the table
   or the graph.

2. Use the circle graph on page 207.
One category is about 3 times as great as another.
How is it easier to identify these categories from the graph
than the table?

3. Use the data on page 207 or the double-line graph on page 208.

    a) What trends does the graph show? Explain.

    b) Estimate the average ticket prices for 2001–2002. Justify your answer.

    c) Estimate the average ticket prices for 2004–2005. Explain your reasoning.

    d) When did movie tickets have the greatest increase? The least increase? How do you know?

    e) What else can you infer from the table or the graph? Explain.

4. These data are from *Census at School*.

What's for Breakfast?
Elementary Students (%)

Item	Girls	Boys
Milk	51.0	57.8
Fruit juice	44.1	40.8
Hot drink	12.1	12.4
Hot cereal	11.8	12.9
Cold cereal	47.0	55.6
Eggs	23.9	32.0
Toast	50.0	52.0
Muffin	18.6	17.2
Bagel	30.9	27.8
Cereal bar	7.7	8.6
Cheese	10.3	13.1
Yogurt	18.5	14.8
No breakfast	16.7	11.2
Other	23.3	25.1

Source: Statistics Canada, *Census at School*, 2003/04
http://www19.statcan.ca/04/04_006_e.htm
Date extracted: March 2005

    a) Use a spreadsheet to draw the most suitable graph for the data. Justify your choice of graph.

    b) Write 3 questions about your graph. Answer your questions.

    c) Compare your questions with those of a classmate. What else can you infer from the table or the graph?

Mid-Unit Review

LESSON

5.1 **1.** For each situation, explain why data are collected from a sample and not a census.

a) To find the mean cost of ski equipment

b) To find the number of Canadian families that have a DVD player

5.2 **2. a)** Write an inference based on this double-line graph. Justify your inference.

Time Students Took on a Math Test

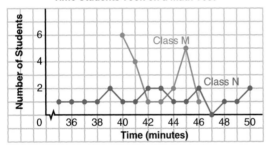

b) Write a convincing argument based on the graph. Justify your argument.

3. This table shows the sales, in thousands of dollars, for 6 months for 2 employees of Electronics Warehouse.

Month	Jamar ($1000s)	Laura ($1000s)
Jan.	117	124
Feb.	118	125
Mar.	119	128
Apr.	120	126
May	121	124
June	137	126

a) What do you infer from these data?

b) Write a convincing argument each employee could use to persuade the manager that she should be named the better salesperson. Justify each argument.

5.3 **4.** This table, taken from *Census at School*, shows the days of the week students were born.

Day of Birth

Day	Males	Females
Monday	3369	3602
Tuesday	3973	4335
Wednesday	4235	4571
Thursday	4278	4486
Friday	4131	4756
Saturday	4128	4671
Sunday	3670	4108

Source: Statistics Canada, *Census at School*, 2003/04
http://www.censusatschool.ntu.ac.uk/table9-0.asp
Date extracted: March 2005

a) Draw a double-bar graph to display the data.

b) What can you infer from these data?

c) What other type of graph could you draw to display these data? Justify your answer.

d) Ask each student in the class for the day of the week he or she was born. How do these data compare with the data in the table?

5.4 Applying Measures of Central Tendency

Focus Understand how mean, median, and mode are affected by outliers.

Explore

Recall that mean, median, and mode are measures of central tendency.

Work with a partner.

Students in a Grade 8 class measured their pulse rates.

Here are their results in beats per minute:

97, 69, 83, 66, 78, 8, 55, 82, 47, 52, 67, 76, 84,

64, 72, 80, 72, 70, 69, 80, 66, 60, 72, 88, 88

➤ Calculate the mean, median, and mode for these data. Which measure best represents the data? Explain.

➤ Remove any numbers that are significantly different from most of the data.

Calculate the mean, median, and mode again.

How are the three measures affected? Explain.

Reflect & Share

Compare your results with those of another pair of classmates.

How did you decide which numbers were significantly different?

Why do you think they are so different?

Connect

The *mean* of a set of data is the sum of the data values divided by the number of data values.

The mean is usually the best measure when no numbers in the data set are significantly different from the other numbers.

The *median* of a set of data is the middle number when the numbers are arranged in order. One-half the numbers are above the median and one-half are below the median. When there is an even number of data values, the median is the mean of the two middle values. The median is usually the best measure when there are numbers in the data set that are significantly different.

The *mode* of a set of data is the number that occurs most often. There may be no mode or there may be more than one mode.

The mode is usually the best measure when the data represent measures such as shoe sizes or other clothing sizes.

A number in a set of data that is significantly different from the other numbers is called an **outlier**.

Example

Here are the marks for an English test for students in a Grade 8 class. They are shown in a stem-and-leaf plot.

English Test Marks

Stem	Leaf
2	1 3 4 4 7 9 9 9
3	2 7 7 8 9
4	0
5	0 0 1 4 6 7 8 9
6	1
7	1
8	0
9	9

a) How many students were in the class? How do you know?

b) What are the outliers? Explain.

c) Calculate the mean, median, and mode test marks.

d) Calculate the mean, median, and mode without the outliers. What do you notice? Explain.

e) Which measure of central tendency best describes the average test mark for the Grade 8 class? Explain.

Solution

a) Count the entries in the stem-and-leaf plot to find the number of students in the class. There are 26 students.

b) There is only one number, 99, that is significantly different. The outlier is 99.

c) There are 26 marks. The mean is:

$(21 + 23 + 2 \times 24 + 27 + 3 \times 29 + 32 + 2 \times 37 + 38 + 39 + 40 + 2 \times 50 + 51 + 54 + 56 + 57 + 58 + 59 + 61 + 71 + 80 + 99) \div 26$

$= 1175 \div 26 \doteq 45.2$

The median is the mean of the 13th and 14th marks.
The 13th mark is 39. The 14th mark is 40.
So, the median is: $\frac{39 + 40}{2} = \frac{79}{2} = 39.5$
The mode is the mark that occurs most often. This is 29.

d) Without the outlier:

The mean is $\frac{1175 - 99}{25} = 43.04$

The median is the 13th mark.

The median is 39.

The mode is 29.

e) Since the outlier has a greater effect on the mean, it is not the best measure to describe the average. The mode is too low to best describe the average. So, the median best describes the average mark.

Practice

1. For each set of data:

   i) Calculate the mean, median, and mode.

   ii) Identify the outliers.

   iii) Calculate the mean, median, and mode without the outliers.

   How is each measure of central tendency affected when the outliers are not included?

a) marks on a set of exams:

   30, 66, 65, 72, 78, 93, 70, 68, 64, 90, 65, 68

b) weekly income: $625, $750, $800, $650, $725, $850, $625, $650, $625, $1250, $700, $625

c) waiting time, in minutes, at a fast-food restaurant:

   5, 5, 5, 6, 5, 7, 0, 5, 1, 7, 7, 5, 6, 5, 5, 5, 8, 5, 0, 5, 4, 5, 2, 7, 9

d) number of baskets scored by a basketball player in 10 games: 15, 7, 8, 6, 2, 7, 5, 7, 1, 8

2. In each case:

   i) Calculate the mean, median, and mode of the new data set formed.

   ii) Explain how the mean, median, and mode are affected by the change.

a) Add 5 to each number in question 1a.

b) Subtract $10 from each amount in question 1b.

c) Multiply each time in question 1c by 3.

d) Divide each number in question 1d by 2.

3. Is each conclusion correct? Explain.
  a) The mean cost of a medium pizza is $10. So, the prices of three medium pizzas could be $9, $10, and $11.
  b) The number of raisins in each of 30 cookies was counted. The mean number of raisins was 15. So, in 10 cookies, there would be 150 raisins.

4. The daily high temperatures for one week at Clearwater Harbour were: 27°C, 31°C, 23°C, 25°C, 28°C, 23°C, 28°C. The weather channel reported that the average temperature for Clearwater Harbour for that week was 23°C. Is this correct? Explain.

5. The masses, in tonnes, of household garbage collected in a municipality each weekday in April are:
285, 395, 270, 305, 320, 300, 290, 310, 315, 295, 310, 295, 305, 325, 315, 310, 305, 300, 325, 305, 305, 300
  a) Organize the data in a stem-and-leaf plot.
  b) Calculate the mean, median, and mode for the data.
  c) What are the outliers?
    Calculate the mean without the outliers.
    What do you notice? Explain.
  d) Which measure of central tendency best describes the data? Explain.

6. There are seven numbers in a set of data.
  The mean and the median are 7.
  The mode is 4 and occurs 3 times.
  Find the numbers. Explain your method.

7. Andrew has these marks:
  English 82%, French 75%, History 78%, Science 80%
  a) What mark will Andrew need in math if he wants his mean mark to be each percent?
    i) 80% **ii)** 81% **iii)** 82%
  b) Is it possible for Andrew to get a mean mark of 84% or higher?
    Justify your answer.

8. Celia received a mean mark of 80% in her first three exams. She then had 94% on her next exam.

Celia stated that her overall mean mark was 87% because the mean of 80 and 94 is 87. Is Celia's reasoning correct? Explain.

9. **Assessment Focus** A Grade 8 class wanted to find out if a TV advertisement was true. It claimed that *Full of Raisins* cereal guaranteed an average of 23 raisins per cup of cereal.

Each pair of students tested one box of cereal.

Each box contained 20 cups of cereal.

The number of raisins in each cup was counted.

a) Assume the advertisement is true.

How many raisins should there be in 1 box of cereal?

b) Here are the results for the numbers of raisins in 15 boxes of cereal: 473, 485, 441, 437, 489, 471, 400, 453, 465, 413, 499, 428, 419, 477, 467

   i) Display the data. Justify your choice of display.

   ii) How do the outliers affect the mean?

   iii) Was the advertisement true? Justify your answer.

Take It Further

10. There are 20 students in a Grade 8 class.

Eighteen students wrote a Social Studies test.

Here are their marks: 75, 56, 83, 61, 91, 42, 57, 56, 60, 87, 32, 42, 57, 67, 89, 43, 49, 81

Two students who were absent wrote the test the next day.

The mean mark for the 20 students was 63.

The range was 63.

a) What are the median mark and the mode mark for the 18 students?

b) What are possible marks for the two students who wrote the test late?

c) Suppose the teacher gives each student 3 extra marks. How do the mean, median, mode, and range change?

Mental Math

Divide each number by 0.1, 0.01, and 0.001.

- 385
- 3.84
- 0.286
- 25.7

What patterns do you see in the answers?

Reflect

What is an outlier? How are measures of central tendency affected by outliers? Use examples in your explanation.

Explore

Work on your own.

Compare the two graphs below.

How are they similar? How are they different?

Graph A

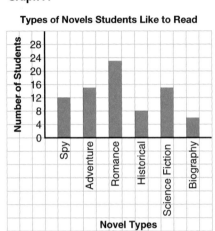

Graph B

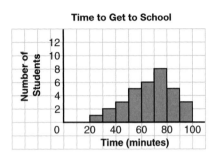

Reflect & Share

Compare your results with those of a classmate.
Why is there a space between the bars in Graph A
but not in Graph B?
Why do you think each graph is drawn the way it is?

Connect

A bar graph is used to represent measurements, or numbers, of
different items. Then, the items are compared.
Adjacent bars are separated by a space.
The length or height of each bar represents a number.

When we have a large amount of data where the numbers can be
arranged in numerical order, then grouped, we use a **histogram** to
graph the data.

In a histogram, the data are usually grouped into intervals with equal widths. Each data value belongs in exactly 1 interval.
The height of each bar represents the *frequency* or number of pieces of data in that interval.
There is no space between the bars because the data are continuous. That is, the end of one interval is the beginning of the next interval.

Example

Lucas recorded the heights, in centimetres, of all the Grade 8 students in his school. Here are his results:
147, 178, 161, 153, 130, 139, 159, 162, 151, 150, 133, 162, 147, 170, 153, 160, 174, 148, 155, 157, 163, 155, 138, 149, 152, 142, 163, 160, 155, 158, 181, 164, 158, 147, 164, 166, 177, 178, 162, 171, 169, 134, 180, 175, 140, 161, 149, 150, 187, 173, 156, 166, 164, 152, 183, 159, 137, 144, 145, 164

a) Display the data using a stem-and-leaf plot.
b) Make a frequency table.
c) What can you tell from the stem-and-leaf plot that you cannot tell from the frequency table?
d) Use the frequency table to draw a histogram.

Solution

a) In the stem-and-leaf plot, the hundreds and tens digits are the stems and the units digits are the leaves. Write the numbers in numerical order. Include repeated numbers.

Heights of Students in Centimetres

Stem	Leaf
13	0 3 4 7 8 9
14	0 2 4 5 7 7 7 8 9 9
15	0 0 1 2 2 3 3 5 5 5 6 7 8 8 9 9
16	0 0 1 1 2 2 2 3 3 4 4 4 4 6 6 9
17	0 1 3 4 5 7 8 8
18	0 1 3 7

Interval	Frequency
130–139	6
140–149	10
150–159	16
160–169	16
170–179	8
180–189	4

b) The least number is 130.
The greatest number is 187.
The range is: $187 - 130 = 57$
A suitable interval to show a range of about 50 is 10.
Since the least number is 130,
the first interval in the frequency table is 130–139.
The next interval is 140–149, and so on, to 180–189.
Use the stem-and-leaf plot to make the frequency table at the left.

c) You can identify each piece of data from the stem-and-leaf plot. In the frequency table, you know only how many pieces of data are in each interval. You do not know the values of the data.

d) The horizontal axis represents the heights.
Label it in intervals of 10, from 130 to 190.
The vertical axis represents the frequency of each interval.
Draw vertical bars whose lengths are the frequencies.
Write a title for the graph. Label each axis.

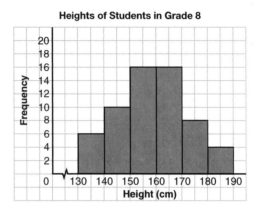

Practice

1. Draw the most appropriate graph to display each set of data. Explain your choice.

a)

Photo Package Price ($)	0–4	5–9	10–14	15–19	20–24	25–29	30–34	35–39	40–44
Frequency	6	13	24	29	35	46	25	21	12

b)

Type of Flower	Tulip	Carnation	Rose	Daffodil	Orchid	Lilac	Lily
Number Ordered	129	230	115	98	67	85	145

2. a) What information does this table provide?

Hours Per Week Students Participate in Sports
Elementary School

Hours Per Week	% Girls	% Boys
0–2	12	9
3–5	50	40
6–8	21	23
9–11	10	11
12–14	7	17

b) Draw appropriate graphs to display the data.

c) What can you infer from the table or the graphs?

3. For each set of data, would you use a bar graph or a histogram to display the data? Explain your choice.

a) Ravi recorded how many pairs of each shoe type were sold in his shoe store in April.

Sandals: 109; Dress Shoes: 46; Soccer Cleats: 65;
Athletic Shoes: 89; Rubber Boots: 77

b) Adriana recorded the times, in seconds, taken by 60 female swimmers to swim the 100-m backstroke.

120, 135, 98, 102, 87, 96, 145, 99, 77, 106, 113, 124, 126, 84, 95, 102, 128, 137, 111, 130, 122, 151, 117, 108, 129, 134, 133, 153, 148, 132, 126, 117, 119, 86, 133, 157, 140, 122, 107, 110, 139, 141, 140, 100, 155, 109, 114, 123, 133, 152, 135, 144, 133, 151, 128, 130, 145, 150, 111, 140

4. Assessment Focus Leah recorded the number of ice cream bars sold at the community centre each day in the month of July:

101, 112, 125, 96, 132, 125, 116, 97, 124, 136, 123, 119, 78, 105, 118, 130, 87, 108, 114, 99, 126, 86, 94, 117, 125, 107, 122, 119, 114, 105, 93

a) Display the data using a stem-and-leaf plot.

b) Find the median and the mode.
What does each measure tell you about the data?

c) Use the stem-and-leaf plot to make a frequency table.

d) What can you tell from the stem-and-leaf plot that you cannot tell from the frequency table?

e) Use the frequency table to draw a histogram.

f) What can you infer about the ice cream sales in July?

5. For each set of data below:
   i) Draw a histogram.
   ii) What do you know from the graph that you do not know from looking at the data?
   iii) Can the graphs be used to estimate the median? The mode? Explain your reasoning.

a) Data were collected on the prices, in dollars, of 44 different styles of jeans:
25, 59, 45, 120, 105, 100, 78, 45, 37, 49, 27, 19, 48, 39, 40, 55, 89, 95, 65, 40, 50, 33, 59, 62, 80, 74, 150, 78, 43, 35, 130, 75, 69, 105, 115, 110, 120, 80, 49, 40, 60, 109, 89, 72

b) Data were collected on the thickness, in millimetres, of 50 randomly chosen books from the library:
6, 15, 35, 12, 76, 80, 34, 22, 15, 17, 35, 40, 70, 25, 11, 45, 36, 28, 17, 20, 55, 63, 39, 47, 52, 60, 77, 81, 100, 39, 40, 75, 33, 92, 18, 22, 17, 30, 13, 44, 63, 28, 20, 31, 40, 34, 45, 40, 45, 38

Take It Further

6. Four hundred Grade 8 students in the Ottawa area participated in a math contest. The results are shown in the table. Elias wanted to compare his school results, recorded below, to the city wide results.
67, 78, 93, 56, 68, 64, 98, 70, 59, 64, 83, 72, 75, 88, 69, 78, 76, 29, 55, 69, 77, 83, 88, 76, 48, 71, 60, 82, 80, 90, 55, 74, 73, 81, 58, 66, 81, 31, 47, 78, 99, 66, 64, 72, 80, 94, 75, 68

Percent of Students	Range of Scores
8	91 to 100
21	81 to 90
29	71 to 80
22	61 to 70
9	51 to 60
7	41 to 50
3	31 to 40
1	21 to 30
0	11 to 20
0	1 to 10

a) What type of graph would you use to display the city results? Justify your choice.
b) What type of graph would you use to display Elias' school results? Justify your choice.
c) Draw the graphs you chose in parts a and b.
d) Did Elias' school do better than all the other schools? How do you know?

Reflect

When would you group data before displaying them on a graph?
When would you *not* group data?
Include examples in your explanation.

Using *Fathom* to Draw Histograms and Investigate Outliers

Focus | Draw a histogram using *Fathom*.

You will use *Fathom* to draw a histogram, change an interval width, and find the mean and median of the data.

Here are the heights, in centimetres, of all the Grade 8 students who were on the school track team: 164, 131, 172, 137, 175, 168, 146, 176, 175, 173, 155, 170, 172, 160, 168, 178, 174, 184, 189

To use *Fathom* to draw a histogram for these data, follow these steps:

1. Open *Fathom*. From the File menu, select New.

2. To enter the title:
Click the New Collection icon , then click the screen.
Double-click Collection 1.
Type **Heights of Grade 8 Students on the Track Team**, and click OK.

3. To enter the data:
Click the New Case Table icon , then click the screen.
Click <new>; type **Height**, then press Enter.
Input the data under the heading Height. Press Enter each time.

4. To graph the data:
Click the New Graph icon .
Click the screen. Two axes appear.
Click the column heading, *Height*, and drag it to the horizontal axis.
Select *Histogram* from the drop-down menu in the top right.
A histogram is created.

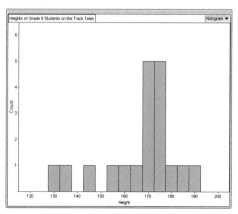

To view the heights that make up one bar, click on a bar.
It turns red.
In the table, the heights that correspond to that bar turn blue.

Fathom automatically selects a width for each bar.
To see the width, double-click the graph.
The text window for the graph appears.
You will see:

> Information about this graph:
> Histogram: Bin width: **5.0000** starting at: **127.50**
>   The **height** axis is horizontal from **115.00** to **205.00**
>   The **Count** axis is vertical from 0 to **6.5000**

The "Bin width" represents the width of the bars.
The "starting at" tells the starting value.

➤ To change the Bin width to 10, double-click the blue text
that follows *Bin width*: and enter 10.
To change the starting number to 130, double-click the
blue text that follows *starting at*: and enter 130.
To change the Count axis, double-click the blue text that currently
shows 6.5 and enter 10.
The histogram changes.

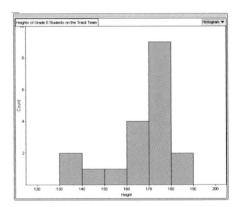

➤ To find the mean height, double-click the *Collection Box* to open
the *Inspector* for the collection.
Select the *Measures tab*. Rename <new> to Mean. Press Enter.
Right-click the *Formula* column for the *Mean* measure,
and select *Edit Formula*.
Double-click *Functions*, double-click *Statistical*, double-click *One
Attribute*, and then double-click *mean*. Type **Height** in the brackets.
The word *mean* turns red and the word *Height* turns blue.
Press Enter. The mean is shown under *Value*.

➤ To find the median height, double-click the *Collection Box* to open the *Inspector* for the collection.
Select the *Measures* tab. Rename <new> to Median. Press Enter.
Right-click the *Formula* column for the *Median* measure, and select *Edit Formula*. Double-click *Functions*, double-click *Statistical*, double-click *One Attribute*, and then double-click *median*.
Type **Height** in the brackets.
The word *median* turns red and the word *Height* turns blue.
Press Enter. The median is shown below the mean, under *Value*.

➤ To find the effects of outliers on the mean and median, input these heights into the table: 125, 127
The histogram, the mean, and the median change to reflect the new heights.
How do these new heights affect the mean and median? Explain.

➤ Replace the heights added above with the heights 198 and 200.
How do these new heights affect the mean and median? Explain.

1. Use *Fathom* to draw a histogram for these data.
They are the donations, in dollars, that were made to a Toy Wish Fund.
5, 2, 3, 9, 10, 5, 2, 8, 7, 15, 14, 17, 28, 30, 16, 19, 4, 7, 9, 11, 25, 30, 32, 15, 27, 18, 9, 10, 16, 22, 34, 19, 25, 18, 20, 17, 9, 10, 15, 35

a) What are the bin widths created by *Fathom*?
Do these make sense? Explain.

b) Use *Fathom* to find the mean and median.

c) Add some outliers to your table.
State the values you added.
How do the new mean and median compare to their original values? Explain.

2. Repeat question 1 for these data.
They are the purchases, in dollars, made by customers at a grocery store.
55.40, 48.26, 28.31, 14.12, 88.90, 34.45, 51.02, 71.87, 105.12, 10.19, 74.44, 29.05, 43.56, 90.66, 23.00, 60.52, 43.17, 28.49, 67.03, 16.18, 76.05, 45.68, 22.76, 36.73, 39.92, 112.48, 81.21, 56.73, 47.19, 34.45

Explore

Work with a partner.
In a chart on the board, record the colour of your partner's eyes.
When you have the data for the whole class,
draw a circle graph to display the data.

Reflect & Share

Compare your graph with that of another pair of classmates.
Which is the most frequent eye colour in your class?
Suppose you drew a circle graph for the eye colours
of all the students in the school.
How would the graph compare to the graph you drew? Explain.

Connect

A circle graph is used to graph parts of one whole.
Each piece of data is written as a fraction of the whole.
This fraction is then used to find the sector angle in the circle graph.
Each fraction is also written as a percent.
The method is illustrated in the following example.

Example

All the students in two Grade 8 classes were asked how they get
to school each day. Here are the results:
11 rode their bikes, 13 walked, 26 came by bus,
and 15 were driven by car.
Construct a circle graph to illustrate the data.

Solution

For each type of transport, write the number of students as a
fraction of 65, the total number of students:
Bike: $\frac{11}{65}$; walk: $\frac{13}{65}$; bus: $\frac{26}{65}$; car: $\frac{15}{65}$
To calculate each sector angle, multiply each fraction by 360°.
To calculate each percent, multiply each fraction by 100%.
The results are shown in the following table.

How Students Get to School	Number of Students	Sector Angle (Using Fractions)	Percent
Bike	11	$\frac{11}{65} \times 360° \doteq 61°$	$\frac{11}{65} \times 100\% \doteq 17\%$
Walk	13	$\frac{13}{65} \times 360° = 72°$	$\frac{13}{65} \times 100\% = 20\%$
Bus	26	$\frac{26}{65} \times 360° = 144°$	$\frac{26}{65} \times 100\% = 40\%$
Car	15	$\frac{15}{65} \times 360° \doteq 83°$	$\frac{15}{65} \times 100\% \doteq 23\%$
Total	65	360°	100%

Round each sector angle to the nearest degree. Round each percent to the nearest whole number.

To graph the data, draw a large circle. Draw one radius.
Use a protractor to measure the greatest sector angle, 144°.
Draw the radius to make a sector with this angle.
Continue to make a sector for each of the other angles.
Label each sector with its percent.

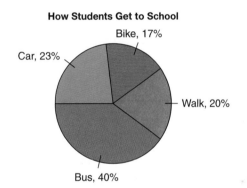

How Students Get to School

Bike, 17%
Car, 23%
Walk, 20%
Bus, 40%

Practice

1. Liam asked students at Silver Lake Summer Camp, "Where are you from?" He recorded these results: Ottawa 60, Toronto 90, Belleville 30, Hamilton 36, Kingston 45, Oshawa 20, Burlington 24, Oakville 10, Napanee 30, St. Catharines 15
   a) Display the data in a circle graph.
   b) What can you tell from the graph that you cannot easily tell from the data?

2. a) Which of the sets of data on the next page could be represented by a circle graph? Explain.
   b) For each set of data that is not best represented by a circle graph, which graph is most appropriate? Explain.

c) Display each set of data using the type of graph you chose.

i) Adam recorded the eye colours of all Grades 7 and 8 students in his school.

Eye Colour	Brown	Green	Blue	Other
Students	60	20	30	10

ii) Celine collected data on the average number of hours per week that Grade 8 girls and boys in her school played video games.

Number of Boys as a Percent	10	25	40	15	10
Number of Girls as a Percent	20	35	30	10	5
Hours Per Week Playing Video Games	0–2	3–5	6–8	9–11	12–14

iii) The table shows the area, in square kilometres, of some Canadian National Parks.

National Park	Banff	Nahanni	Pukaskwa	Prince Albert	Jasper
Area (km²)	6641	4766	1878	3874	10 878

iv) Tobias and Fabiana work evenings at Mamma Maria's restaurant. They recorded their tips for one week.

Day	Mon.	Tues.	Wed.	Thurs.	Fri.	Sat.	Sun.
Tips Tobias Earned ($)	20	12	28	15	32	35	38
Tips Fabiana Earned ($)	15	20	22	30	25	40	28

3. The school library budget to buy new books is $5000. The librarian has the following information about the types of books students borrowed in one month.

Type of Book Borrowed	Number of Students
History	125
Science	90
Biography	65
Geography	52
Fiction	110
Reference	88
French	70

a) Display the data in a circle graph.

b) How much money should be spent on each type of book? Explain.

4. **Assessment Focus** This table shows the method of transport used by U.S. residents entering Canada in 2002.

United States Residents Entering Canada in 2002

Method of Transport	Number
Automobile	33 424 000
Plane	4 224 000
Train	121 000
Bus	1 582 000
Boat	886 000
Other	641 000

Source: Statistics Canada, Canadian Statistics, CANSIM table 427-0001. Last modified: 2005-05-18.
http://www40.statcan.ca/l01/cst01/arts34.htm   Date extracted: April 11, 2005

a) How many U.S. residents visited Canada in 2002?
b) What percent of U.S. residents entered Canada by boat?
c) What fraction of U.S. residents entered Canada by plane?
d) Display the data in a circle graph.
e) What else do you know from the table or circle graph?
   Write as much as you can.

5. Here is the nutritional information on the wrapper of a roasted almond granola bar.

Protein	2.0 g
Fat	4.7 g
Sugar	6.4 g
Starch	7.3 g
Dietary Fibre	1.3 g

Use a circle graph to display these data.

Reflect

Which type of data is best represented by a circle graph?
Support your answer with an example.

Identifying Key Verbs in Math Problems

Many math problems are questions or puzzles presented without a context.
The question may ask you to do something with math information.
It may include a *key verb* to tell you what to do.
Here are some key verbs and their meanings.

Calculate, evaluate:	Find the answer to a question that uses any or all of the operations: add, subtract, multiply, and divide
Classify:	Sort objects into groups according to a rule. Tell the rule.
Compare:	Tell what is the same and what is different.
Construct:	Make a model. Draw accurately.
Describe, explain, represent:	Use numbers, words, pictures, diagrams, charts, and/or models to tell about your work and show your thinking.
Draw:	Create a picture or sketch.
Estimate:	Make a reasoned guess about what you know.
Justify:	Give reasons and evidence to show your answer makes sense.
Model:	Show with objects and/or pictures.
Predict:	Tell what you think will happen, based on what you know.
Relate:	Show and explain a connection between ideas, objects, drawings, numbers, and events.
Simplify:	Make simpler but equal.
Solve:	Develop a solution to a problem.
Show your work:	Record important calculations, numbers, words, pictures, diagrams, graphs, charts, and/or models. Include the steps you used to get your answer so that someone else could repeat it.

 **Check**

Identify the key verb, then solve each problem.

1. These figures can be used to construct different solids.
You can use more than one of each figure.

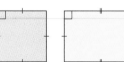

  a) List the different solids that could be made.
  b) Construct one solid. Draw the others.

c) The square has an area of 4 square units.
The rectangle has an area of 6 square units.
Four circles just fit inside the square.
One triangle is one-half the rectangle.
Another triangle is one-half the square.
Estimate the surface area of each solid
that is made from these figures.

2. Here are some data from Statistics Canada for 2001.

Province or Territory	Median age (years)	Percent of population in each age group		
		0–19	65+	20–64
Newfoundland and Labrador	38.4	25.0	12.3	62.7
Prince Edward Island	37.7	27.3	13.7	59.0
Nova Scotia	38.8	25.0	13.9	61.1
New Brunswick	38.6	24.8	13.6	61.7
Quebec	38.8	24.2	13.3	62.5
Ontario	37.2	26.3	12.9	60.8
Manitoba	36.8	28.1	14.0	58.0
Saskatchewan	36.7	29.2	15.1	55.8
Alberta	35.0	28.3	10.4	61.4
British Columbia	38.4	25.0	13.6	61.4
Yukon Territory	36.1	29.0	6.0	64.9
Northwest Territories	30.1	35.0	4.4	60.7
Nunavut	22.1	46.5	2.2	51.2

Source: Statistics Canada, *2001 Census of Canada, analysis series*. Catalogue no. 96F0030XIE2001002. Last modified 2005-03-08.
http://www12.statcan.ca/english/census01/Products/Analytic/companion/age/ewt1.cfm
Date extracted: April 11, 2005

a) For each province, relate the median age to the age groups.
How do they compare? Explain.

b) Which different graphs could you draw to represent these data?

c) Suppose you need to decide where to spend health care dollars
for an aging population. Justify where most money should be
spent.

d) Where might you increase money spent on education?
Justify your answer.

e) Write your own problem about these data. Solve your problem.

Unit Review

Review any lesson with

online tutorial

What Do I Need to Know?

A *census* collects data from all the people in a population.

A *sample* collects data from some of the people in a population.

A *double-bar graph* or a *double-line graph* displays two sets of data on the same grid.

A *histogram* displays data that can be grouped into intervals. There is no space between the bars in a histogram.

The *mean, median,* and *mode* are the three *measures of central tendency.*

An *outlier* is a number in a set of data that is much greater than or much less than most of the numbers in the data set.

What Should I Be Able to Do?

For extra practice, go to page 492.

LESSON

5.1 **1.** For each situation, explain why data are collected from a sample and not a census.
  a) To test the quality of batteries
  b) To find the number of hours a light bulb will burn
  c) To find the percent of Grade 8 students who know the multiplication tables

2. In each case, identify the collected data as from a sample or a census. Justify your answer.
  a) To find if the Grade 8 class should play soccer or basketball, each student in the class voted.

  b) To find the percent of teenagers in Ontario who work part-time, all high school students in Hamilton were surveyed.

5.2 **3.** Rob researched accident statistics in his community. He learned that, in the past year, 20 people were injured in car accidents and 5 people were injured in motorcycle accidents. Rob inferred, "It is safer to ride a motorcycle than it is to drive a car." Is Rob's inference valid? Explain.

5.2
5.3

4. Odakota recorded the number of Grades 7 and 8 students born in each season.

Number of Students Born in Each Season

Season	Summer	Autumn	Winter	Spring
Grade 7	11	10	7	5
Grade 8	4	9	14	6

a) Which types of graphs could be used to display these data? Explain your choices.
b) Display the data using one of your choices from part a.
c) Write a convincing argument based on the graph. Justify your argument.

5. a) Display the data in the table using a suitable graph. Justify your choice.

Average Monthly Rainfall (mm)

Month	Halifax	Yellowknife
January	153	13
February	134	11
March	128	12
April	115	10
May	106	17
June	90	17
July	94	34
August	111	44
September	94	31
October	134	35
November	153	25
December	180	18

b) What trends does the graph show? Explain.
c) What inferences can you make from the graph? Explain.

6. This table shows the numbers of boys and girls who attended the school football games.

Game	1	2	3	4	5	6	7
Girls	35	67	71	69	56	63	65
Boys	52	58	60	78	67	61	63

a) Display the data using a suitable graph. Explain your choice.
b) Glenna looked at the data and inferred, "More girls than boys went to the games." Is Glenna's inference valid? If your answer is yes, explain how Glenna might have made the inference. If your answer is no, what inference would you make from the data? Explain.

5.4 **7.** The table shows the annual incomes of the employees of *Computers For You*.

Position	Salary ($)
1 owner	108 000
1 senior manager	81 500
2 managers	72 000
4 engineers	67 000
2 researchers	55 500
4 assistants	45 000
1 clerk	24 000

a) Find the measures of central tendency for these salaries.
b) Which measure best represents the salaries? Explain.
c) What inferences can you make from these data?

d) What are the outliers? How do these affect the measures of central tendency? Explain.

8. Is each conclusion true or false? Explain.

a) The mean test score was 68%. Therefore, one-half the class scored above 68%.

b) The mode colour of eyes in James' class is brown. Therefore, most of the students have brown eyes.

c) A random sample of 100 people had a mean income of $35 000. Therefore, a random sample of 200 people would have a mean income of $70 000.

5.5 **9.** The data show the times, in minutes and seconds, for swimmers to swim a 400-m freestyle race.
3:28, 2:56, 4:25, 3:42, 5:33, 3:57, 3:45, 4:29, 5:03, 4:55, 3:58, 2:55, 4:17, 3:29, 4:31, 3:30, 4:12, 4:21, 4:53, 5:06, 4:47, 3:50, 4:28, 4:09, 5:01, 3:46, 4:27, 4:51, 5:12, 4:58

a) Convert the data to seconds. Display the data using a stem-and-leaf plot.

b) State the range, median, and mode of the data.

c) Use the stem-and-leaf plot to set up a frequency table with intervals.

d) How does the frequency table differ from the stem-and-leaf plot?

e) Use the frequency table to draw a histogram.

5.2
5.6 **10.** Ester surveyed 120 people in her neighbourhood to find out what time people usually wake up on a weekday. She recorded her results in this table.

What Time People Wake Up

Time	Number of People
Before 5 a.m.	3
5 a.m. to 5:59 a.m.	25
6 a.m. to 6:59 a.m.	44
7 a.m. to 7:59 a.m.	31
8 a.m. to 8:59 a.m.	11
After 9 a.m.	6

a) Display the data in a circle graph.

b) What percent of people wake up after 6:59 a.m?

c) In which time interval do approximately 20% of the people wake up?

d) What inferences could you make from these data? Explain.

11. A sample survey was conducted to find how old car owners were when they got their first car.

Age	16–20	21–25	26–30	31–35	36–40
Number of First Time Car Owners	12	28	16	9	3

a) What are two possible ways to display these data?

b) Display the data using both methods from part a.

c) In a group of 1000 car owners, about how many of them were over 30 when they got their first car? Explain.

Practice Test

242, 225, 296, 352,
305, 260, 313, 220,
255, 236, 304, 220,
195, 215, 292, 281,
277, 310, 272, 219,
252, 311, 263, 252,
207, 283, 222, 245,
229, 278, 237, 400,
229, 201, 324, 335,
299, 356, 410, 348,
192, 355, 342, 324,
358, 369, 293, 347,
308, 317, 301, 220,
268, 415, 288

1. The lengths of the songs, in seconds, from a "Let's Dance"
CD collection set are shown at the left.
a) Use an appropriate graph to display the data. Explain your choice.
b) Describe any trends the graph shows.
c) An advertisement for the CD said, "The average song is no less than
5 minutes long." Is this true? Give reasons for your answer.
d) Make a convincing argument based on the graph.
Explain the reason for your argument.

2. The mean mark on a quiz was 7.
The median mark was 6 and the mode was 5.
Fifteen students wrote the quiz.
The lowest mark was 1 and the highest mark was 10.
a) Write a possible set of marks.
b) One student, who was absent, wrote the quiz later. Her mark was 20.
How does this mark affect the mean, the median, and the mode?
Explain.

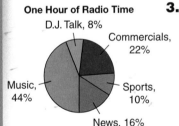

One Hour of Radio Time
D.J. Talk, 8%
Commercials, 22%
Music, 44%
Sports, 10%
News, 16%

3. A radio station broadcasts 24 h a day, 7 days a week.
The circle graph shows how one hour of a radio station's
air-time is used. Suppose the radio station decided it wanted
to allow more time for music. How do you think this could
be achieved? Draw a new circle graph to show your answer.
Explain your graph.

4. A random sample of 1200 Canadians were asked:
"Overall, do you think Canada would benefit or lose out from
having a common currency with the United States?"
Here are the results:

Greatly Benefit	Somewhat Benefit	No Impact	Somewhat Lose Out	Greatly Lose Out
17%	27%	8%	22%	20%

A student claimed, "There are as many Canadians who would like
a common currency with the U.S. as those who do not."
Did the student interpret the results correctly? Explain.

What would you like to know about people in your community?
You will collect primary data on a topic that interests you.
You will design and conduct a survey.
You will display and analyse the data.

Work in a group.

Part 1

Choose a topic that interests you.
You may pick from this list or select one of your own.
- Careers that interest students
- Personal statistics such as age, height, heart rate, shoe size, and so on
- Which languages people speak
- Which professional sports people watch
- Which electronic games students play
- Which community activities people participate in
- How students spend their money

Design a survey. Think about:
- What information do you want to find out about the population?
- What data do you need to collect to get that information?
- Is the survey question clear, precise, and unbiased?

Test your survey on other classmates. Make changes if necessary.

Decide on the population you will sample. Be specific.
Decide on the sample size and the method you will use
to pick your sample.

Conduct the survey. Collect data from males and females.
Record the results.

Part 2

Organize the data. Think about:
- Can a table be used to organize the data?
- What is the most suitable type of graph to display the data?
- What other types of graphs could be used to display the data?

Analyse the results.
Make inferences and convincing arguments based on your analysis.
Review your method of collecting data.
Were there any flaws or problems with the survey, or the way the data were collected?

Part 3

Write a report.
Present the results of your survey to your class.
Describe how you analysed the data and made inferences and arguments.
How do you know if your survey results are reliable and valid?

Check List

Your work should include:

✓ how you designed your survey and planned your data collection

✓ correctly constructed tables and graphs

✓ inferences and conclusions, based on reasonable arguments

✓ a clear explanation and analysis of the procedures you used

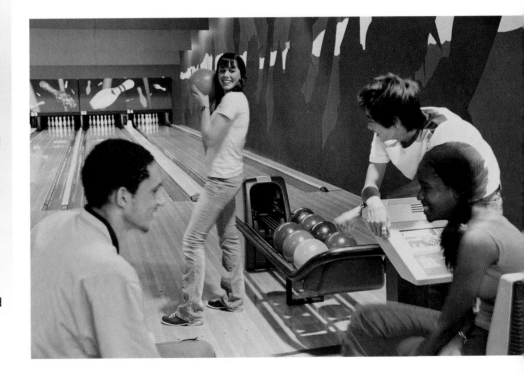

Reflect on the Unit

Describe the different graphs that can be used to display primary data.
How do you use tables and graphs to make inferences and convincing arguments?
Include examples in your explanation.

UNIT

6

Circles

- What is a circle?
- Where do you see circles?
- What do you know about a circle?
- What might be useful to know about a circle?

What You'll Learn

- Measure the radius, diameter, and circumference of a circle.
- Investigate and explain the relationships among the radius, diameter, and circumference of a circle.
- Develop formulas to find the circumference and area of a circle.
- Estimate the measures for a circle, then use the formulas to calculate these measures.
- Draw a circle, given its radius, area, or circumference.
- Develop formulas to find the surface area and volume of a cylinder.

Why It's Important

Measuring circles is an extension of measuring polygons in earlier grades.

236

Key Words

- diameter
- radius
- circumference
- irrational number

Skills You'll Need

Rounding Measurements

1 cm = 10 mm
and, 1.6 cm = 16 mm
When a measurement in centimetres has 1 decimal place,
the measurement is given to the nearest millimetre.

1 cm

1.6 cm

1 m = 100 cm
1.3 m = 130 cm
1.37 m = 137 cm ← **When a measurement in metres has 2 decimal places, the measurement is given to the nearest centimetre.**

1 m = 1000 mm
1.4 m = 1400 mm
1.43 m = 1430 mm
1.437 m = 1437 mm ← **When a measurement in metres has 3 decimal places, the measurement is given to the nearest millimetre.**

Example

Write each measurement to the nearest millimetre.
a) 25.2 mm **b)** 3.58 cm

Solution

a) 25.2 mm
Since there are 2 tenths, round down.
25.2 mm is 25 mm to the nearest millimetre.

b) 3.58 cm
To round to the nearest millimetre, round to 1 decimal place.
Since there are 8 hundredths, round up.
3.58 cm is 3.6 cm to the nearest millimetre.

✓ Check

1. **a)** Round to the nearest metre.
   i) 4.38 m **ii)** 57.298 m **iii)** 158.5 cm

   b) Round to the nearest millimetre and nearest centimetre.
   i) 47.2 mm **ii)** 47.235 cm **iii)** 1.0579 m

Focus Measure radius and diameter, and discover their relationship.

Which attribute do these objects share?

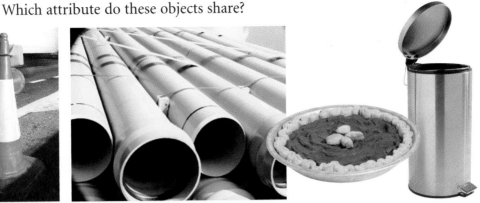

Work with a partner.
You will need circular objects, a compass, ruler, and scissors.

➤ Use a compass. Draw a large circle.
Use a ruler.
Draw a line segment that joins two points on the circle.
Measure the line segment. Label the line segment with its length.
Draw and measure other segments that join two points on the circle.
Find the longest segment in the circle.
Repeat the activity for other circles.

➤ Trace a circular object.
Find a way to locate the centre of the circle.
Measure the distance from the centre to the circle.
Measure the distance across the circle, through its centre.
Record the measurements in a table.
Repeat the activity with other circular objects.
What pattern do you see in your results?

Reflect & Share

Compare your results with those of another pair of classmates.
Where is the longest segment in any circle?
What relationship did you find between the distance across a circle through its centre, and the distance from the centre to the circle?

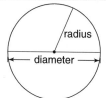

A circle is a closed curve.
All points on a circle are the same distance
from the centre of the circle.
This distance is the **radius** of the circle.

The distance across a circle through
its centre is the **diameter** of the circle.
The radius is one-half the diameter.
Let r represent the radius, and d the diameter.
Then, $r = \frac{1}{2}d$, or $r = \frac{d}{2}$
The diameter is two times the length of the radius.
That is, $d = 2r$

Example

Draw a line segment.
Construct a circle for which this segment is:

a) the radius   b) the diameter

Solution

a) Draw a line segment.
   Place the compass point at one end.
   Place the pencil point at the other end.
   Draw a circle.

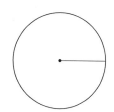

b) Measure the line segment in part a.
   Draw a congruent segment.
   Mark its midpoint.
   Place the compass point at the midpoint.
   Place the pencil point at one end of the segment.
   Draw a circle.

Practice

1. Draw a circle with radius 6 cm.
   What is the diameter of the circle? Explain.

2. Draw a circle with diameter 8 cm.
   What is the radius of the circle? Explain.

The word *radii* is the plural of *radius*.

3. a) How many radii does a circle have?
   b) How many diameters does a circle have?

4. A circle has diameter 3.8 cm. What is the radius?

5. A circle has radius 7.5 cm. What is the diameter?

6. A circular tabletop is to be cut from a rectangular piece of wood that measures 1.20 m by 1.80 m. What is the radius of the largest tabletop that could be cut? Justify your answer. Include a sketch.

7. A glass has a circular base with radius 3.5 cm. A rectangular tray has dimensions 40 cm by 25 cm. How many glasses will fit on the tray? What assumptions do you make?

8. a) Draw a circle.
   Draw a diameter. Label it AB.
   Choose a point P on the circle.
   Join AP and PB. Measure ∠APB.
   b) Choose another point Q on the circle.
   Join AQ and QB. Measure ∠AQB.
   c) Repeat parts a and b for a different circle.
   What do you notice?

9. Assessment Focus Your teacher will give you a large copy of this logo. Find the radius and diameter of each circle in this logo. Show your work.

Take It Further

10. A circular area of grass needs watering. A sprinkler is to be placed at the centre of the circle. Explain how you would locate the centre of the circle. Include a diagram in your explanation.

Number Strategies

The mean of 3 numbers is 30. The least number is 26. What might the other 2 numbers be? Find as many answers as you can.

This is the logo for the Aboriginal Health Department of the Vancouver Island Health Authority.

Reflect

How are the diameter and radius of a circle related?
When you know the diameter, how can you find the radius?
When you know the radius, how can you find the diameter?
Include examples in your explanation.

Focus | Develop and use the formula for the circumference of a circle.

You know the relationship between the radius and diameter of a circle.
You will now investigate how these two measures are related to
the distance around a circle.

Explore

Work with a partner.
You will need 6 circular objects, dental floss, scissors, and a ruler.

Choose an object.
Use dental floss to measure the distance around it.
Measure the radius and diameter of the object.
Record these measures in a table.

Object	Distance Around (cm)	Radius (cm)	Diameter (cm)

Repeat the activity for the remaining objects.
What patterns can you see in the table?
How is the diameter related to the distance around?
How is the radius related to the distance around?
For each object, calculate:
- distance around ÷ diameter
- distance around ÷ radius

What do you notice?

Reflect & Share

Compare your results with those of another pair of classmates.
Work together to write a formula for the distance around a circle,
when you know its diameter.

Connect

The distance around a circle is its **circumference**.
For any circle, the circumference, C, divided by the diameter, d, is $\frac{C}{d}$,
which is constant with value approximately 3.

So, the circumference is approximately 3 times the diameter.
And, the circumference C divided by the radius, r, is $\frac{C}{r}$, which is approximately 6.
So, the circumference is approximately 6 times the radius.

For any circle, the ratio $\frac{C}{d} = \pi$

You will learn about other irrational numbers in Unit 8.

We use the symbol π because the value of $\frac{C}{d}$ is an **irrational number**; that is, π represents a decimal that never repeats and never terminates.

The symbol π is a Greek letter that we read as "pi."

$\pi \doteq 3.14$

So, the circumference is π multiplied by d.

The circumference of a circle is also the perimeter of the circle.

We write: $C = \pi d$

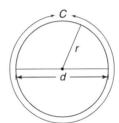

Since $d = 2r$, the circumference is also π multiplied by $2r$.
We write: $C = \pi \times 2r$, or $C = 2\pi r$
When we know the radius or diameter of a circle, we can use one of the formulas above to find the circumference of the circle.

Example 1

The face of a toonie has a radius of 1.4 cm.
a) What is the diameter of the face?
b) Estimate the circumference of the face.
c) Calculate the circumference.
   Give the answer to the nearest millimetre.

1.4 cm

Solution

a) The diameter $d = 2r$, where r is the radius.
   Substitute: $r = 1.4$
   $d = 2 \times 1.4$
   $\quad = 2.8$
   The diameter is 2.8 cm.

The circumference is a length, so its units are units of length such as centimetres, metres, or millimetres.

b) The circumference is approximately 3 times the diameter:
   $3 \times 2.8 \text{ cm} \doteq 3 \times 3 \text{ cm}$
   $\qquad\qquad\qquad = 9 \text{ cm}$
   The circumference is approximately 9 cm.

c) *Method 1*

The circumference is: $C = \pi d$

Substitute: $d = 2.8$

$C = \pi \times 2.8$

Key in: $\boxed{\pi}$ $\boxed{\times}$ 2.8 $\boxed{\text{ENTER} =}$

to display 8.79645943

So, $C \doteq 8.796$

$\doteq 8.8$

Method 2

The circumference is: $C = 2\pi r$

Substitute: $r = 1.4$

$C = 2 \times \pi \times 1.4$

Key in: 2 $\boxed{\times}$ $\boxed{\pi}$ $\boxed{\times}$ 1.4 $\boxed{\text{ENTER} =}$

to display 8.79645943

So, $C \doteq 8.796$

$\doteq 8.8$

The circumference is 8.8 cm, to the nearest millimetre.

Use a calculator. If your calculator does not have a π key, use 3.14 instead.

When we know the circumference,
we can use a formula to find the diameter.
Use the formula $C = \pi d$. To isolate d, divide each side by π.

$$\frac{C}{\pi} = \frac{\pi d}{\pi}$$

$$\frac{C}{\pi} = d$$

So, $d = \dfrac{C}{\pi}$

Example 2

A circular pond has circumference 12 m.

a) Estimate the lengths of the diameter and radius of the pond.

b) Calculate the diameter and radius.

Give the answers to the nearest centimetre.

Solution

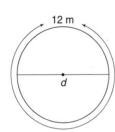
12 m
d

a) Since the circumference is approximately 3 times the diameter, then the diameter is about $\frac{1}{3}$ the circumference.
One-third of 12 m is 4 m. So, the diameter is about 4 m.
The radius is $\frac{1}{2}$ the diameter. One-half of 4 m is 2 m.
So, the radius of the pond is about 2 m.

b) The diameter is: $d = \dfrac{C}{\pi}$

Substitute: $C = 12$

$d = \dfrac{12}{\pi}$

$\doteq 3.8197$

Use a calculator.
Do not clear the calculator.

The radius is $\frac{1}{2}$ the diameter.
Divide the number in the calculator display by 2.

$r \doteq 1.9099$

The diameter is 3.82 m to the nearest centimetre.
The radius is 1.91 m to the nearest centimetre.

1. Estimate the circumference of each circle.

a)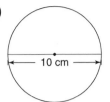
10 cm

b)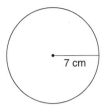
7 cm

c)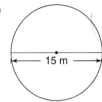
15 m

2. Calculate the circumference of each circle in question 1.
   Give the answers to 1 decimal place.

3. Estimate the diameter and radius of each circle.

a)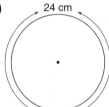
24 cm

b)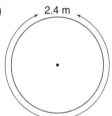
2.4 m

c)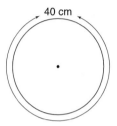
40 cm

4. Calculate the diameter and radius of each circle in question 3.
   Give the answers to the nearest millimetre.

5. When you estimate the circumference, you use 3 instead of π.
   Is the estimated value greater than or less than the calculated
   value? Explain.

6. A circular garden has diameter 2.4 m.
   a) The garden is to be enclosed with plastic edging.
      How much edging is needed?
   b) The edging costs $4.53/m.
      What is the cost to edge the garden?

Math Link

Science
The orbit of Earth around the sun is approximately a circle.
The radius of the orbit is about 1.5×10^8 km.
How could you calculate the circumference of the orbit?
The orbit of a communication satellite around Earth is approximately a circle.
The satellite is about 35 800 km above Earth's surface.
What would you need to know to be able to calculate the radius of the orbit?

Number Strategies

A rectangular plot of land has area 256 m².

List the possible whole-number dimensions of the plot.

7. **Assessment Focus** A bicycle tire has a stone stuck in it. The radius of the tire is 46 cm. Every time the wheel turns, the stone hits the ground.
   a) How far will the bicycle travel between the stone hitting the ground for the first time and the second time?
   b) How many times will the stone hit the ground when the bicycle travels 1 km?

   Show your work.

8. A carpenter is making a circular tabletop with circumference 4.5 m. What is the radius of the tabletop in centimetres?

9. Can you draw a circle with circumference 33 cm? If you can, draw the circle and explain how you know its circumference is correct. If you cannot, explain why it is not possible.

10. a) What if you double the diameter of a circle. What happens to the circumference?
    b) What if you triple the diameter of a circle. What happens to the circumference?

Take It Further

11. Suppose a metal ring could be placed around Earth at the equator.
    a) The radius of Earth is 6378.1 km. How long is the metal ring?
    b) Suppose the length of the metal ring is increased by 1 km. Would you be able to crawl under the ring, walk under the ring, or drive in a school bus under the ring? Explain how you know.

Reflect

When you know the circumference of a circle, how can you calculate its radius and diameter? Include an example in your explanation.

Explore

Work in a group of 4.
You will need scissors, glue, a compass, and protractor.

➤ Each of you draws a circle with radius 8 cm.
   Each of you chooses one of these:

   - 4 congruent sectors
   - 8 congruent sectors
   - 10 congruent sectors
   - 12 congruent sectors

   The angle at the centre of the circle is 360°. Divide this angle by the number of sectors. The quotient is the sector angle.

➤ Use a protractor to divide your circle into the number of sectors you chose.
   Cut out the sectors.
   Arrange the sectors to approximate a parallelogram.

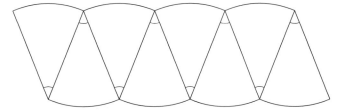

The area of a parallelogram is base × height.

Glue the sectors on paper.
Calculate the area of the parallelogram.
Estimate the area of the circle.

Reflect & Share

Compare your answer for the area of the circle with those of your group members.
Which area do you think is closest to the area of the circle? Explain.
How could you improve your estimate for the area?

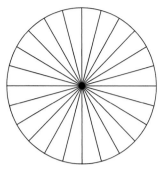

A circle, with radius 10 cm, was cut into 24 congruent sectors. The sectors were then arranged to form a parallelogram.

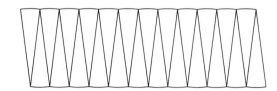

The more congruent sectors we use, the closer the area of the parallelogram is to the area of the circle.

If the number of sectors is large enough, the parallelogram is almost a rectangle.

The sum of the two longer sides of the rectangle is equal to the circumference, C.

$C = 2\pi r$
$\quad = 2\pi \times 10 \text{ cm}$
$\quad = 20\pi \text{ cm}$

So, each longer side is: $\frac{20\pi \text{ cm}}{2} = 10\pi \text{ cm}$

Each of the two shorter sides is equal to the radius, 10 cm.

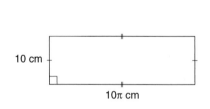 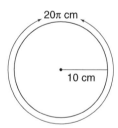

The area of the rectangle is: $10\pi \text{ cm} \times 10 \text{ cm} = 10^2\pi \text{ cm}^2$
$= 100\pi \text{ cm}^2$

So, the area of the circle with radius 10 cm is $100\pi \text{ cm}^2$.
We can apply this idea to a circle with radius r.

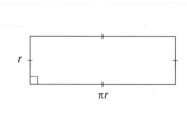 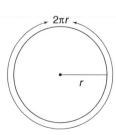

So, the area, A, of a circle with radius r is: $\pi r \times r = \pi r^2$

So, $A = \pi r^2$

We can use this formula to find the area of any circle when we know its radius.

Example

The face of a dime has diameter 1.8 cm.

a) Estimate the area of the face of the dime.

b) Calculate the area. Give the answer to 2 decimal places.

Solution

The diameter of the face of a dime is 1.8 cm.

So, its radius is:

$\frac{1.8 \text{ cm}}{2} = 0.9$ cm

a) The area of the face of the dime is about $3 \times r^2$.

$$r \doteq 1$$
$$\text{So, } r^2 = 1$$
$$\text{and } 3 \times r^2 = 3 \times 1$$
$$= 3$$

The area of the face of the dime is approximately 3 cm².

b) Use the formula: $A = \pi r^2$

Substitute: $r = 0.9$

$A = \pi \times 0.9^2$

Use a calculator.

Key in: $\boxed{\pi}$ $\boxed{\times}$ 0.9 $\boxed{x^2}$ $\boxed{\text{ENTER} \atop =}$ to display 2.544690049

$A \doteq 2.544\ 69$

The area of the face of the dime is 2.54 cm² to 2 decimal places.

If your calculator does not have an x^2 key, key in 0.9 × 0.9 instead of 0.9².

Since 1 mm = 0.1 cm

Then 1 mm² = 1 mm × 1 mm

$\qquad = 0.1$ cm × 0.1 cm

$\qquad = 0.01$ cm²

This illustrates that when an area in square centimetres has 2 decimal places, the area is given to the nearest square millimetre. In the *Example*, the area 2.54 cm² is written to the nearest square millimetre.

1. Estimate the area of each circle.

a)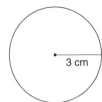
3 cm

b)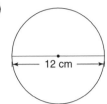
12 cm

c)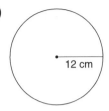
12 cm

2. Calculate the area of each circle in question 1.
Give the answers to the nearest square millimetre.

3. a) Use the results of questions 1 and 2.
What if you double the radius of a circle.
What happens to its area?
b) What do you think happens to the area of a circle
if you triple its radius?
Justify your answers.

4. **Assessment Focus** Use 0.5-cm grid paper.
Draw a circle with radius 5 cm.
Draw a square outside the circle that just encloses the circle.
Draw a square inside the circle so that its vertices lie on the circle.
a) How can you use the areas of the two squares to estimate
the area of the circle?
b) Check your estimate in part a by calculating the area
of the circle.
c) Repeat the activity for circles with different radii.
Record your results.
Show your work.

5. In the biathlon, athletes shoot at targets.
Each target is 50 m from the athlete.
Find the area of each target.
a) The target for the athlete who is standing is a
circle with diameter 11.5 cm.
b) The target for the athlete who is prone is a circle
with diameter 4.5 cm.
Give the answers to the nearest square centimetre.

1 cm

0.01 m

6. a) A square has side length 1 cm.

    i) What is the area of the square in square centimetres?

    ii) What is the area of the square in square metres?

    iii) Use the results of parts i and ii to write 1 cm² in square metres.

b) A calculator display shows the area of a circle as
7.068583471 m².
What is this area rounded to the nearest square centimetre?

7. In curling, the target area is a bull's eye series of 4 concentric circles.

Concentric circles have the same centre.

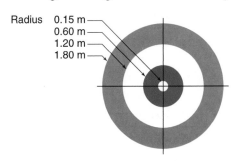

Radius 0.15 m
          0.60 m
          1.20 m
          1.80 m

a) Calculate the area of the smallest circle.
Write the area to the nearest square centimetre.

b) When a smaller circle overlaps a larger circle, a ring is formed.
Calculate the area of each ring on the target area to the
nearest square centimetre.

8. The bottom of a swimming pool is a circle with circumference 31.4 m.
What is the area of the bottom of the pool?
Give the answer to the nearest square metre.

Take It Further

9. A large pizza has diameter 35 cm.
Two large pizzas cost $19.99.
A medium pizza has diameter 30 cm.
Three medium pizzas cost $24.99.
Which is the better deal: 2 large pizzas or 3 medium pizzas?
Justify your answer.

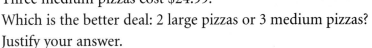

Reflect

When you know the radius of a circle,
how can you calculate its area?
Include an example in your explanation.

Mid-Unit Review

LESSON

6.1 **1.** A circle has radius 3.6 cm. What is its diameter?

2. A circle has diameter 3.6 cm. What is its radius?

3. a) Draw a large circle.
Label its centre C.
Mark points P, Q, and R on the circle.
Join P, Q, and R to form △PQR.
Join QC and RC. These line segments form 2 angles at C.
Measure ∠QPR and the smaller ∠QCR.
How are these angles related?

b) Repeat part a for a different circle. Is the relationship in part a still true? Explain.

6.2 **4.** The face of a penny has radius 9.5 mm.
a) Estimate the circumference of the penny.
b) Calculate the circumference. Give the answer to the nearest tenth of a millimetre.

5. An auger is used to drill a hole in the ice, for ice fishing.
The diameter of the hole is 25 cm. What is the circumference of the hole?

6. Explain how you could calculate the circumference of a paper plate.

7. There is a clock on the Peace Tower in Ottawa. The circumference of the clock face is approximately 15.02 m.
a) Estimate the diameter and radius of the clock face.
b) Calculate the diameter and radius of the clock face to the nearest centimetre.

8. a) How is the circumference of a circle with radius 9 cm related to the circumference of a circle with diameter 9 cm?
b) Draw both circles in part a.

6.3 **9.** The radius of a circular tray is 14.4 cm. What is its area to the nearest square millimetre?

10. The diameter of a circle is 58 m. What is its area to the nearest square centimetre?

11. A circular table has radius 56 cm. A tablecloth covers the table. The edge of the cloth is 10 cm below the tabletop. What is the area of the tablecloth?

12. a) How is the area of a circle with radius 6 cm related to the area of a circle with diameter 6 cm?
b) Draw both circles in part a. Do the diagrams justify your answer in part a? Explain.

6.4 Volume of a Cylinder

Focus Develop and use the formula for the volume of a cylinder.

A cylinder is formed when a circle is translated through the air so that the circle is always parallel to its original position.

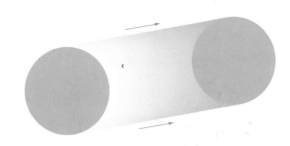

How does this relate to the triangular prism in *Unit 3*, page 112?

Explore

Work with a partner.
You will need cylindrical objects and a ruler.
Choose a cylindrical object.
Calculate its volume.
How did you use the diameter and the radius in your calculations?
How did you use π?

Reflect & Share

Share your method for calculating the volume with another pair of classmates.
Work together to write a formula for the volume of a cylinder.
Use any of diameter, radius, height, and π in your formula.

Connect

A cylinder is a prism.
The volume of a prism is base area × height.
A can of baked beans is cylindrical.
Its diameter is 7.4 cm. Its height is 10.5 cm.
To find the volume of the can, first find the area of its base.
The base is a circle, with diameter 7.4 cm.
So, its radius is: $\frac{7.4 \text{ cm}}{2} = 3.7$ cm
The base area: $A = \pi r^2$
Substitute: $r = 3.7$
$A = \pi (3.7)^2$
The height of the can is 10.5 cm.

So, its volume is: V = base area × height

$$= \pi(3.7)^2 \times 10.5$$

Use a calculator.

Key in: $\boxed{\pi}$ $\boxed{\times}$ 3.7 $\boxed{x^2}$ $\boxed{\times}$ 10.5 $\boxed{\text{ENTER}=}$ to display 451.588236

$V \doteq 451.6$

The volume of the can of baked beans is about 452 cm³.

We can apply this idea to
write a formula for the volume
of any cylinder.
Its radius is r.
So, its base area is πr^2.
Its height is h.

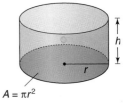

$A = \pi r^2$

So, its volume is: V = base area × height

$$= \pi r^2 \times h$$

$$= \pi r^2 h$$

So, a formula for the volume of a cylinder is $V = \pi r^2 h$,
where r is the radius of its base, and h its height.

Example

The base of a juice can is a circle
with diameter 6.8 cm.
The height of the can is 12.2 cm.
What is the volume of the can?

6.8 cm

12.2 cm

Solution

The radius of the base is: $\frac{6.8 \text{ cm}}{2} = 3.4$ cm
Use the formula for the
volume of a cylinder:
$V = \pi r^2 h$
Substitute: $r = 3.4$ and $h = 12.2$

Use a calculator.

$V = \pi(3.4)^2 \times 12.2$
Key in: $\boxed{\pi}$ $\boxed{\times}$ 3.4 $\boxed{x^2}$ $\boxed{\times}$ 12.2 $\boxed{\text{ENTER}=}$ to display 443.0650951
$V \doteq 443.07$
The volume of the can is about 443 cm³.

Capacity is measured in litres or millilitres.
Since 1 cm³ = 1 mL, the capacity of the can in the *Example* is
about 443 mL.

Give each volume to the nearest cubic unit.

1. Calculate the volume of each cylinder.

a) 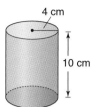 4 cm

10 cm

b) 15 mm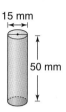

50 mm

c) 2.9 m

12.4 m

2. A candle mould is cylindrical. Its radius is 5 cm and its height is 20 cm. What is the capacity of the mould?

3. Frozen apple juice comes in cylindrical cans. A can is 12 cm high and has radius 3.5 cm.
a) What is the capacity of the can?
b) Apple juice expands when it freezes. The can is filled to 95% of its volume. What is the volume of apple juice in the can?

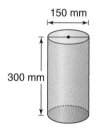
150 mm

300 mm

4. A core sample of earth is cylindrical. The length of the core is 300 mm. Its diameter is 150 mm. Calculate the volume of earth in cubic millimetres and cubic centimetres.

5. **Assessment Focus** A concrete column in a parkade is cylindrical. The column is 10 m high and has diameter 3.5 m.
a) What is the volume of concrete in one column?
b) There are 127 columns in the parkade. What is the total volume of concrete?
c) What if the concrete in part a is made into a cube. What would the dimensions of the cube be?

Reflect

How is the volume of a cylinder related to the volume of a triangular prism?
How are these volumes different?

Explaining Solutions

To explain a solution using only words helps to communicate why you did what you did.
When you explain, show the numbers or models you used.

Here are two solutions to this problem:
Six students wrote a math test.
Their mean mark was 68%.
Another student wrote the test and scored 89%.
What is the mean mark for the 7 students?

Reading/ Viewing

Talking/ Listening

Writing/ Representing

Talking/ Listening

Writing/ Representing

Jol explains:

I took the mean mark and multiplied it by 6, the number of students. This gave me the total marks for all of them. Next, I added the mark of the seventh student to get the total marks for everyone. I then divided the total marks by 7, the total number of students who took the test. This gave me the mean mark for the 7 students.

Jol writes:

$$\frac{(\text{original mean}) \times (\text{number of students}) + (\text{7th mark})}{\text{total number of students}} = \text{new mean}$$

$$\frac{68 \times 6 + 89}{7} = \frac{408 + 89}{7} = \frac{497}{7} = 71$$

The mean mark for the 7 students is 71%.

Veronica explains:

I subtracted the mean from the mark of the 7th student. This told me how many marks above the mean the 7th student had. I shared these marks equally among the 7 students. Then I added the extra marks to the original mean to get the new mean.

Veronica writes:

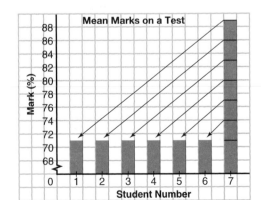

$$89 - 68 = 21$$
$$21 \text{ extra marks} \div 7$$
$$= 3 \text{ extra marks each}$$
$$68 + 3 = 71$$
The mean for 7 students is 71%.

 **Check**

Write your solution to each problem using words, numbers, pictures, and/or models. Explain your solution using only words.

Strategies

- Make a table.
- Use a model.
- Draw a diagram.
- Solve a simpler problem.
- Work backward.
- Guess and check.
- Make an organized list.
- Use a pattern.
- Draw a graph.
- Use logical reasoning.

1. A large piece of construction paper is 0.01 mm thick. It is cut in half and one piece is placed on the other to make a pile. These two pieces are cut in half and all four pieces are placed in a pile. The process continues. After the pieces have been cut and piled 10 times, what is the height of the pile in centimetres?

2. A magician opened a show at the mall. On Day 1, 50 people attended the show. On Day 2, there were 78 people. On Day 3, there were 106 people. This pattern continues. When will there be at least 200 people in the audience?

3. This pentagon has each vertex connected with every other. How many triangles are in the pentagon?

4. A ball is dropped from a height of 2.30 m. After each bounce, it reaches 40% of its height before the bounce.
   a) What height does it reach after each of the first four bounces?
   b) After how many bounces does the ball reach a height of about 1 cm?

Lake	Area (km²)
Superior	82 100
Michigan	57 800
Huron	59 600
Erie	25 700
Ontario	18 960

5. The Great Lakes are some of the largest freshwater lakes in the world. The areas are listed in the chart at the left. About what percent of the total area is each lake?

6. How many breaths have you taken in your lifetime?

7. A bag contains ten marbles. One marble is red, two are blue, three are white, and four are black. A marble is picked at random. What is the probability of picking:
   a) red?
   b) not white?
   c) neither black nor red?
   d) either black or white?

6.5 Surface Area of a Cylinder

Focus Develop and use the formula for the surface area of a cylinder.

Explore

Work with a partner.
You will need a cardboard tube, scissors, and sticky tape.
Cut out two circles to fit the ends of the tube.
Tape a circle to each end of the tube.
You now have a cylinder.
Find a way to calculate the surface area of the cylinder.

Reflect & Share

Share your method for finding the surface area with that of another pair of classmates.
Work together to write a formula for the surface area of any cylinder.

Connect

A cylinder has height 10 cm and radius 4 cm.
To find the surface area of the cylinder, we think of its net.
The bases of the cylinder are 2 congruent circles.
The curved surface of the cylinder is a rectangle.

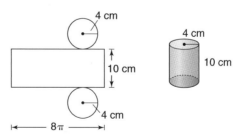

The height of the rectangle is equal to the height of the cylinder.
The base of the rectangle is equal to the circumference of
the base of the cylinder, which is: $2\pi(4 \text{ cm}) = 8\pi$ cm
So, the surface area of the cylinder is:
SA = area of 2 congruent circles + area of rectangle
$$= 2(\pi(4)^2) + 10 \times 8\pi$$

Use a calculator.

$$= 32\pi + 80\pi$$

Key in: 32 $\boxed{\times}$ $\boxed{\pi}$ $\boxed{+}$ 80 $\boxed{\times}$ $\boxed{\pi}$ $\boxed{\text{ENTER}\atop =}$ to display 351.8583772

$SA \doteq 351.858$

The surface area of the cylinder is approximately 352 cm².

To find a formula for the surface area of any cylinder, we can apply this idea to a cylinder with height h and radius r.
Sketch the cylinder and its net.

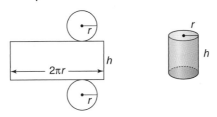

On the net:
The height of the rectangle is h.
The base of the rectangle is the circumference of
the base of the cylinder: $2\pi r$
The surface area of the cylinder is:
$SA = 2(\pi r^2) + 2\pi r(h)$
$SA = 2\pi r^2 + 2\pi rh$

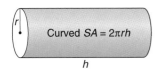

Curved $SA = 2\pi rh$

If a cylinder is like a cardboard tube, and has no circular bases, its surface area is the curved surface only: curved $SA = 2\pi rh$

Example

A manufacturer produces a can with height 7 cm and diameter 5 cm. What is the surface area of the can, to the nearest square millimetre?

Solution

The radius of the can:
$r = \frac{5 \text{ cm}}{2} = 2.5$ cm
Use the formula for the
surface area of a cylinder.
$SA = 2\pi r^2 + 2\pi rh$
Substitute: $r = 2.5$ and $h = 7$
$SA = 2\pi(2.5)^2 + 2\pi(2.5)(7)$

Use a calculator.

Key in: 2 $\boxed{\times}$ $\boxed{\pi}$ $\boxed{\times}$ 2.5 $\boxed{x^2}$ $\boxed{+}$ 2 $\boxed{\times}$ $\boxed{\pi}$ $\boxed{\times}$ 2.5 $\boxed{\times}$ 7 $\boxed{\text{ENTER}\atop =}$

to display 149.225651

$SA \doteq 149.2257$

The surface area of the can is 149.23 cm² to the nearest
square millimetre.

Give each area to the nearest square unit, unless stated otherwise.

1. Calculate the curved surface area of each tube.

a)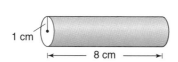

1 cm

8 cm

b)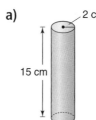

10 cm

3 cm

c)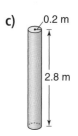

2 m

20 m

2. Calculate the surface area of each cylinder.

a) 2 cm

15 cm

b) 25 mm

230 mm

c) 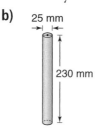 0.2 m

2.8 m

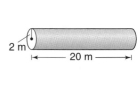

3. A cylindrical tank has diameter 3.8 m and length 12.7 m.
What is the surface area of the tank?

4. Cylindrical paper dryers are used in pulp and paper mills.
One dryer has diameter 1.5 m and length 2.5 m.
What is the area of the curved surface of this dryer?

5. A wooden toy kit has different painted solids.
One solid is a cylinder with diameter 2 cm and height 14 cm.
a) What is the surface area of the cylinder?
b) One can of paint covers 40 m².
How many cylinders can be painted with one coat of paint?

6. Assessment Focus A soup can has diameter 6.6 cm.
The label on the can is 8.8 cm high.
There is a 1-cm overlap on the label.
What is the area of the label?

Reflect

How is the formula for the surface area of a cylinder related to the
net of the cylinder? Include a diagram in your explanation.

Unit Review

Review any lesson with

etext
online tutorial

What Do I Need to Know?

☑ **Measurements in a Circle**

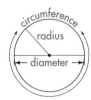

The distance from the centre to the circle is the *radius*.
The distance across the circle, through the centre, is the *diameter*.
The distance around the circle is the *circumference*.

☑ **Circle Formulas**

In a circle, let the radius be r, the diameter d,
the circumference C, and the area A.

Then $d = 2r$

$r = \frac{1}{2}d$

$C = \pi d$, or $C = 2\pi r$

$d = \frac{C}{\pi}$

$A = \pi r^2$

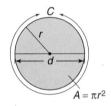

$A = \pi r^2$

π is an irrational number that is approximately 3.14.

☑ **Cylinder Formulas**

Let the height of a cylinder be h and its radius r.
The volume of a cylinder is: $V = \pi r^2 h$
The curved surface area of a cylinder is: Curved $SA = 2\pi rh$
The total surface area of a cylinder is: $SA = 2\pi r^2 + 2\pi rh$

What Should I Be Able to Do?

For extra practice, go to page 493.

LESSON

6.1

1. a) Mark two points on a piece of paper. Join the points. Use this line segment as the radius of a circle. Draw the circle. What is its diameter?

b) Mark two points. Join the points. Use this line segment as the diameter of a circle. Draw the circle. What is its radius?

2. Trace a large circular object. Explain how to find the radius and diameter of the circle you have drawn.

6.2

3. The diameter of a circle is 43 mm. What is its circumference? Give the answer to the nearest millimetre and to the nearest centimetre.

4. The circumference of a large crater is 219.91 m. What is the radius of the crater to the nearest centimetre?

6.3

5. The radius of a circle is 12 m. What is its area? Give the answer to the nearest square centimetre.

6. The diameter of a circular mirror is 28.5 cm. What is the area of the mirror? Give the answer to the nearest square millimetre.

6.2
6.3

7. Choose a radius. Draw a circle. What if you halve the radius.

a) What happens to the circumference?

b) What happens to the area? Explain.

8. A goat is tied to an 8-m rope in a field.

a) What area of the field can the goat graze?

b) What is the circumference of the area in part a?

6.4

9. A can of creamed corn has a label that indicates its capacity is 398 mL. The height of the can is 10.5 cm. The diameter of the can is 7.2 cm.

a) Calculate the capacity of the can in millilitres.

b) Give a reason why the answer in part a is different from the capacity on the label.

6.5

10. A piece of sculpture comprises 3 cylindrical columns. Each column has diameter 1.2 m. The heights of the columns are 3 m, 4 m, and 5 m. The surfaces of the cylinders are to be painted. Calculate the area to be painted. (The base each column sits on will *not* be painted.)

Practice Test

1. Draw a circle. Measure its radius.
   Calculate its diameter, circumference, and area.

2. Calculate the circumference and area of each circle.
   Give each answer to the nearest unit or square unit.

   a)

   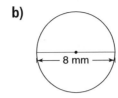

   15 cm

   b)

   8 mm

   c)

   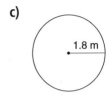

   1.8 m

3. Arleen has 50 m of plastic edging.
   She uses the edging to enclose a circular garden.
   a) What is the circumference of the garden? Explain.
   b) What is the radius of the garden?
   c) What is the area of the garden?
   d) Arleen fills the garden with topsoil to a depth of 15 cm.
   What is the volume of topsoil in the garden?

4. What if you are asked to draw a circle with circumference 100 cm.
   a) You have a compass. Explain why it is not possible for you to
   draw a circle with this circumference.
   b) Draw a circle whose circumference is as close to 100 cm
   as possible.
   i) What is this circumference?
   ii) What is the radius of this circle?

5. Which has the greater volume?
   • a piece of paper rolled into a cylinder lengthwise, or
   • the same piece of paper rolled into a cylinder widthwise
   Justify your answer. Include diagrams in your answer.

Unit Problem Circle Designs

Have you noticed how many designs use circles?

You will use circles to create a design.
Your design *must* include:

- a circle with area approximately 80 cm^2
- a circle with circumference approximately 50 cm

Your design may include:

- lines
- curves
- more than the two required circles

Here are some questions to consider:

- Will I use colour?
- What is the scale of my design?
- Which tools will I use?

Your design is to be in one of these forms:

- logo for a company
- wallpaper
- sign for a shop
- poster for an art show
- quilt sample
- sundial
- greetings card
- another idea (approved by your teacher)

Your finished project must include:

- the design
- any calculations you made
- a written description of the design and how you drew it

Check List

Your work should show:

✓ how you estimated and calculated area and circumference

✓ correct measurements and calculations

✓ a diagram of your design

✓ a clear explanation of your design, with correct use of mathematical language

Reflect on the Unit

How are circles and cylinders related?

Write a paragraph to explain what you learned about circles and cylinders.

UNIT
7

Geometry

Logos, badges, signs, and banners are designed to attract attention. Many designs use geometric concepts.

What geometric concepts are in these figures?

What You'll Learn

- Identify and describe angles formed by intersecting and parallel lines.
- Identify the relationships among the angles of a triangle.
- Construct line segments and angles.
- Create and solve problems involving lines, angles, and triangles.

Why It's Important

A knowledge of the geometry of lines, angles, and triangles is required in careers such as trades (carpentry, plumbing, welding), engineering, interior design, science, and architecture.

266

Key Words

- complementary angles
- complement
- supplementary angles
- supplement
- opposite angles
- perpendicular
- transversal
- alternate angles
- corresponding angles
- interior angles
- bisect
- bisector
- perpendicular bisector
- angle bisector
- circumcentre
- circumcircle

Skills You'll Need

Using a Protractor to Measure Angles

To measure an angle:

Place the base line of a protractor along one arm of the angle, with the centre of the base line at the vertex of the angle.

Read the angle measure from the scale that has its 0 on the arm of the angle.

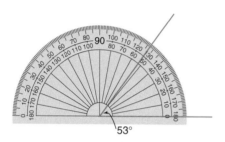

This acute angle measures 53°.

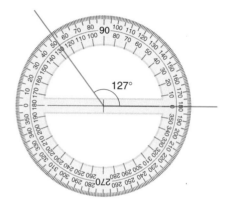

This obtuse angle measures 127°.

 **Check**

1. Draw a large irregular quadrilateral. Label it ABCD.
Measure each angle.
Label each angle with its measure.
Find the sum of the measures of the angles.

Describing Transformations

Here is an example of 3 transformations of Figure A.

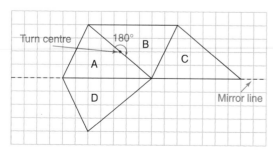

- Figure B is the image of Figure A after a 180° rotation about the midpoint of the side they share. This midpoint is the turn centre.
- Figure C is the image of Figure A after a translation of 7 units right.
- Figure D is the image of Figure A after a reflection in a horizontal line through the side Figure A and Figure D share. The reflection line is the mirror line.

2. In the drawing on page 268, which transformation relates Figure C to Figure B?

3. Look at this pattern.

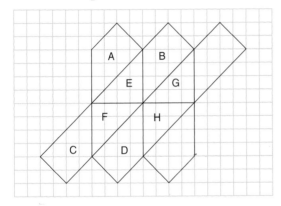

Describe the transformation for which:
a) Figure B is the image of Figure A.
b) Figure C is the image of Figure A.
c) Figure E is the image of Figure G.
d) Figure H is the image of Figure G.
e) Figure D is the image of Figure B.

Using a Ruler and Compass for Constructions

To construct △PQR with QR = 8 cm, PQ = 6 cm, and PR = 5 cm:
Sketch the triangle first.

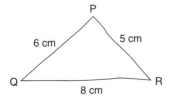

Use a ruler to draw QR = 8 cm.
With compass and pencil points 6 cm apart, place the compass point on Q.
Draw an arc above QR.
With compass and pencil points 5 cm apart, place the compass point on R.
Draw an arc above QR. Make sure the arcs intersect.
Label the point of intersection P.
Join PQ and PR. Label each side with its length.

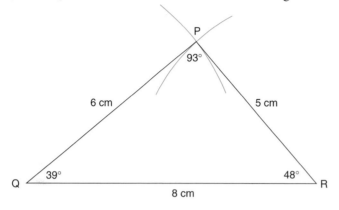

The arcs could have been drawn below QR.

The angles are measured with a protractor. They are accurate to the nearest degree.
Recall how to use 3 letters to label an angle:

$\angle PQR = 39°$
$\angle PRQ = 48°$
$\angle RPQ = 93°$

✓ Check

4. For each triangle below:
   a) Draw the triangle, then measure all the angles.
   b) Name the triangle.
   c) How many different ways can you classify the triangle?
      i) △ABC with AB = 6 cm, BC = 7 cm, and AC = 10 cm
      ii) △DEF with DE = 4 cm, EF = 5 cm, and DF = 8 cm
      iii) △GHJ with GH = 5 cm, HJ = 7 cm, and GJ = 5 cm
      iv) △KMN with KM = 6 cm, MN = 6 cm, and KN = 6 cm

5. Suppose you are given the lengths of 3 line segments.
   How can you tell if they can form a triangle?

6. A scalene triangle has side lengths 6, x, and 16.
   The side lengths are listed from shortest to longest.
   What are the possible whole-number values for x?

First Nations people use a drying rack to
dry fish and animal hides.
The drying rack in this picture is used
in a Grade 2 classroom to dry artwork.

Each end support of the drying rack
forms 4 angles.
What do you notice about these angles?
What is the greatest possible value of each angle?
What is the least possible value of each angle?

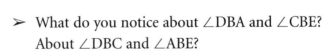

Explore

Work on your own.
You will need a circular
protractor and a large
copy of this diagram.

Measure all the angles.
Label each angle with its measure.

➤ Add these angles.
∠DBA + ∠DBC
∠DBC + ∠CBE
∠DBA + ∠ABF
What do you notice? Explain your results.

➤ What do you notice about ∠DBA and ∠CBE?
About ∠DBC and ∠ABE?

Reflect & Share

Compare your results with those of a classmate.
Suppose line AC were rotated about B.
How would the results change? Explain.
Suppose line FB were rotated about B.
How would the results change? Explain.

| One complete turn is 360°. | So, one-half of one complete turn is $\frac{1}{2}$ of 360° = 180°. | And, one-quarter of one complete turn is $\frac{1}{4}$ of 360° = 90°. |

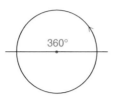 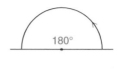 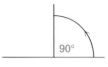

➤ When a 90° angle is divided into smaller angles, the sum of the smaller angles is 90°.

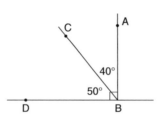

$\angle ABD = 90°$

$\angle ABC = 40°$

$\angle CBD = 50°$

$\angle ABC + \angle CBD = 40° + 50°$

$= 90°$

When two angles have a sum of 90°, one angle is the **complement** of the other.
In the diagram above, $\angle ABC$ is the complement of $\angle CBD$; and $\angle CBD$ is the complement of $\angle ABC$.
Two angles whose sum is 90° are **complementary**.
The angles do *not* have to share an arm.

➤ When a 180° angle is divided into smaller angles, the sum of the smaller angles is 180°.

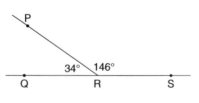

$\angle QRS = 180°$

$\angle PRQ = 34°$

$\angle PRS = 146°$

$\angle PRQ + \angle PRS = 34° + 146°$

$= 180°$

An angle of 180° is described as the angle on a straight line or a straight angle.

When two angles have a sum of 180°, one angle is the **supplement** of the other.
In the diagram above, $\angle PRQ$ is the supplement of $\angle PRS$; and $\angle PRS$ is the supplement of $\angle PRQ$.

Two angles whose sum is 180° are **supplementary**.
The angles do *not* have to share an arm.

➤ Extend PR to T.
Then $\angle PRT = 180°$
So, $\angle SRT = \angle PRT - \angle PRS$
$\angle SRT = 180° - 146°$
$= 34°$
So, $\angle PRQ = \angle SRT$
These are **opposite angles**.

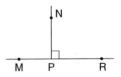

Similarly, $\angle PRS = \angle QRT$
These are opposite angles.

This property applies to any pair of intersecting lines.
So, for any intersecting lines, the opposite angles are equal.

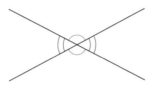

When two supplementary angles are equal,
each angle is 90°.
$\angle MPN = 90°$ and $\angle NPR = 90°$
We say that NP is perpendicular to MR.

Example

Find the measure of each angle. Justify your answers.

a) $\angle ACE$ **b)** $\angle BCE$ **c)** $\angle FCD$

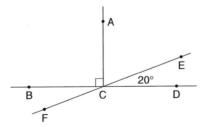

Solution

a) Since $\angle BCD$ is a straight angle,
$\angle BCA$ and $\angle DCA$ are supplementary.
$\angle BCA = 90°$
So, $\angle DCA = 180° - 90°$
$= 90°$

Since $\angle DCA = 90°$,
$\angle ACE$ and $\angle DCE$ are complementary.
$\angle DCE = 20°$
So, $\angle ACE = 90° - 20°$
$\qquad = 70°$

b) $\angle BCE = \angle BCA + \angle ACE$
We know that $\angle BCA = 90°$ and $\angle ACE = 70°$.
So, $\angle BCE = 90° + 70°$
$\qquad = 160°$

Which other ways could you use to find these measures?

c) Since BD and FE intersect, then $\angle FCD$ and $\angle BCE$ are opposite angles.
So, $\angle FCD = \angle BCE$
$\qquad = 160°$

Practice

1. In the diagram, below left, what is the measure of each angle?
a) $\angle AEC$ **b)** $\angle DEB$ **c)** $\angle CEB$ **d)** $\angle CED$

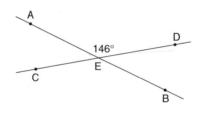

2. In the diagram, above right, name:
a) the angle opposite $\angle PWQ$
b) the complement of $\angle VWT$
c) two angles supplementary to $\angle QWR$
d) the supplement of $\angle SWR$

3. $\angle RQS = 56°$ and $\angle PQR$ is a right angle.
Find the measure of each angle.
Justify your answers.
a) $\angle PQT$
b) $\angle SQP$
c) $\angle RQT$

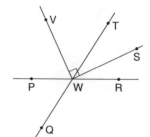
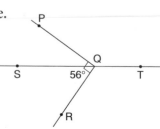

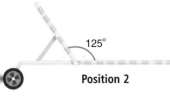

4. The back of a garden lounge chair has 3 possible positions.
  a) In position 1, the obtuse angle is 100°. What is the acute angle?
  b) In position 2, the obtuse angle is 125°. What is the acute angle?
  c) In position 3, the acute angle is 28°. What is the obtuse angle?

125°

Position 2

5. a) Name two 53° angles.
  b) Name two 127° angles.
  c) Name a 143° angle.

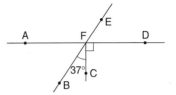

6. Assessment Focus
  a) Sketch opposite angles that are supplementary. How many possible angles are there?
  b) Sketch opposite angles that are complementary. How many possible angles are there?
  c) Sketch two supplementary angles that could not be opposite angles. How many possible angles are there?
  d) Sketch two complementary angles that could not be opposite angles. How many possible angles are there?

7. Find the measures of ∠EBD and ∠DBC. Explain your reasoning.

Remember that matching arcs in a diagram show equal angles.

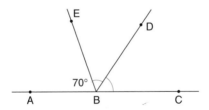

Take It Further

8. On a cross-country trail, Karen runs round the course shown. Through which angle does she turn at each point A, B, C, and D? Explain. What is the total angle Karen turns through? Show how you know.

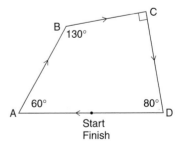

Reflect

Draw a diagram that contains at least one pair of complementary angles and one pair of supplementary angles. Explain how you know the angles are complementary or supplementary.

Focus Use technology to investigate intersecting lines.

1. Open *The Geometer's Sketchpad.*

2. From the **Toolbox**, choose 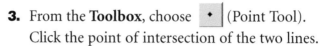 (Straightedge Tool).
   Move to a point on the screen,
   then click and drag to draw a
   line segment.
   Click and drag to form a second
   line segment that intersects
   the first segment.

3. From the **Toolbox**, choose ⚫ (Point Tool).
   Click the point of intersection of the two lines.

4. From the **Toolbox**, choose **A** (Text Tool).
   Click the five points to assign labels A, B, C, D, E.
   E labels the point of intersection of segments AB and CD.

5. From the **Toolbox**, choose ➤ (Selection Arrow Tool).
   Click a blank area of the screen to deselect all objects.
   Click C, E, B in order.
   From the **Measure** menu, choose **Angle**.
   Click and drag the angle measure into ∠CEB.

6. Repeat *Step 5* to find the measure of ∠AED.

7. From the **Toolbox**, choose ➤.
   Click a blank area of the screen to deselect all objects.
   Click and drag point B.
   What do you notice about the measures of opposite angles
   ∠CEB and ∠AED?

8. Repeat *Step 5* to find the measure of ∠BED.

9. From the **Measure** menu, choose **Calculate**.
   Click m∠CEB.
   Click ➕ on the calculator screen.
   Click m∠BED.
   Click OK on the calculator screen.

m∠CEB means
the measure of ∠CEB.

10. What do you notice about m∠CEB + m∠BED?

11. From the **Toolbox**, choose ↖▸.
Click a blank area of the screen to deselect all objects.
Click and drag point C.
What do you notice about the sum of the measures of
supplementary angles ∠CEB and ∠BED?

12. From the **File** menu, choose **New Sketch** to get a new screen.

13. From the **Toolbox**, choose ◢▸.
Click and drag anywhere on the screen to draw a line segment.

14. From the **Toolbox**, choose ◦▕.
Click the screen to create a point not on the line segment.

15. From the **Toolbox**, choose ↖▸.
Click a blank area of the screen to deselect all objects.
Click the point and the line segment.

16. From the **Construct** menu, choose **Perpendicular Line**.

17. From the **Toolbox**, choose ◢▸.
Draw a line segment with one endpoint at the intersection of
the two segments.

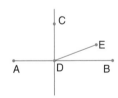

18. From the **Toolbox**, choose **A**.
Click the five points in order to assign labels A, B, C, D, E.

19. Repeat *Step 5* to find the measures of ∠CDE and ∠EDB.

20. Repeat *Steps 9* and *10* for ∠CDE and ∠EDB.

21. From the **Toolbox**, choose ↖▸.
Click a blank area of the screen to deselect all objects.
Click and drag point E, ensuring that DE always lies
between CD and DB.

22. What do you notice about the sum of the measures of
complementary angles ∠CDE and ∠EDB?

7.2 Angles in a Triangle

Explore

Work with a partner.
You will need tracing paper, scissors, and protractor.
Draw a right triangle, an acute triangle, and an obtuse triangle.
Trace each triangle so you have 6 triangles.

➤ Cut out one right triangle. Cut off its angles.
Place the vertices of the three angles together, so adjacent sides touch.
What do you notice?
Repeat the activity with an acute triangle and an obtuse triangle.
What do you notice?
What can you say about the sum of the angles in the triangles you drew?

➤ Use the original triangles.
Measure each angle with a protractor and label it with its measure.
Find the sum of the angles in each triangle.
Does this confirm your results from cutting off the angles? Explain.

Label the vertices so you can identify them.

Reflect & Share

Compare your results with those of another pair of classmates.
What can you say about the sum of the angles in each of the 6 triangles?

Connect

We use a variable to represent an angle measure. It is easier to describe and indicate equal angles.

We can show that the sum of the angles in a triangle is the same for all triangles.
In any rectangle, a diagonal divides the rectangle into two congruent triangles.

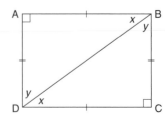

So, △ABD and △CDB are congruent because corresponding sides are equal.

Since the triangles are congruent, corresponding angles are equal.

That is, $\angle DAB = \angle BCD = 90°$

$\qquad \angle ABD = \angle CDB = x$

$\qquad \angle ADB = \angle CBD = y$

When there is only one angle at a vertex, we can use a single letter to name the angle.

$\angle ABC = 90°$ because it is the angle in a rectangle.

So, $x + y = 90°$

In △ABD, $\angle A + \angle ABD + \angle ADB = 90° + x + y$

$\qquad\qquad\qquad\qquad\qquad\qquad = 90° + 90°$

$\qquad\qquad\qquad\qquad\qquad\qquad = 180°$

Note that the sum of the acute angles in a right triangle is 90°.

Similarly, in △CDB, the sum of the angles is 180°.

We can divide any triangle into 2 right triangles by drawing a perpendicular from the greatest angle to the base.

In △JKN, draw the perpendicular from K to JN at M.

We use variables to label the acute angles.

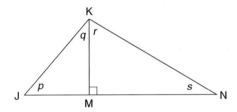

So, $\angle JMK = \angle NMK = 90°$

In △JKM, $p + q = 90°$,

because the sum of the acute angles in a right triangle is 90°.

In △KMN, $r + s = 90°$,

because the sum of the acute angles in a right triangle is 90°.

So, in △JKN,

$\angle J + \angle JKN + \angle N = p + q + r + s$

$\qquad\qquad\qquad\qquad\qquad = (p + q) + (r + s)$

$\qquad\qquad\qquad\qquad\qquad = 90° + 90°$

$\qquad\qquad\qquad\qquad\qquad = 180°$

So, the sum of the angles in any triangle is 180°.

We can use this property to find angles in triangles and other figures.

Example 1

Find the measure of ∠C in △ABC.

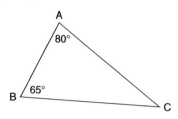

Solution

The sum of the angles is 180°.

$$\angle A + \angle B + \angle C = 180°$$

Substitute for ∠A and ∠B.

$$80° + 65° + \angle C = 180° \qquad \text{Add the angles.}$$

$$145° + \angle C = 180°$$

We can also solve this equation by subtracting 145° from each side.

To find the number to be added to 145 to make 180, subtract:

$$\angle C = 180° - 145°$$
$$= 35°$$

Example 2

Find the measures of ∠R and ∠P in △PQR.

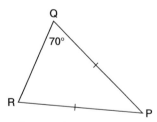

Solution

Since PQ = PR, the triangle is isosceles.
This means ∠R = ∠Q.
Since ∠Q = 70°, then ∠R = 70°
The sum of the angles is 180°.

$$\angle P + \angle Q + \angle R = 180°$$
$$\angle P + 70° + 70° = 180°$$
$$\angle P + 140° = 180°$$
$$\angle P = 180° - 140°$$
$$= 40°$$

Practice

Use the *Glossary* to find the meanings of terms you have forgotten.

1. Draw a large obtuse triangle on dot paper.
Measure each angle.
Find the sum of the measures of the angles.

2. What are the measures of ∠K and ∠M in △KJM?
Justify your answers.

3. a) In △ABC, find the measures of ∠A, ∠B, and ∠C without measuring. Explain your work.
b) Check by measuring with a protractor.

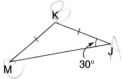

4. In △ABC, below left, find the measures of ∠ACB and ∠A.

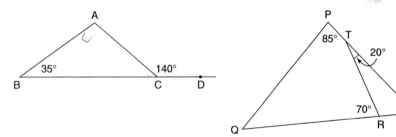

5. In △PQS, above right, find the measure of each angle.
Justify your answers.
a) ∠TRS   **b)** ∠PSQ   **c)** ∠PQS

6. Assessment Focus   For each part below, if your answer is yes, sketch as many triangles as you can and label the angles.
If your answer is no, explain.
a) Can two angles in a triangle be supplementary?
b) Can two angles in a triangle be complementary?

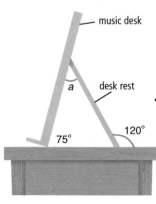

7. The music desk of a grand piano is supported by a desk rest. The side view of the music desk, desk rest, and top of the piano is a triangle.
Use the given angles to find the angle *a* that the desk rest makes with the music desk. Explain your reasoning.

8. Sketch a rectangle with diagonals that intersect at 40°.
Find the measures of all the other angles.
Explain how you know.

9. a) Sketch an isosceles triangle with one angle 130°.
  b) Draw the line of symmetry.
     Find the measures of all other angles.

10. An acute triangle has three acute angles.
  a) Can an obtuse triangle have three obtuse angles?
     Two obtuse angles? Explain.
  b) Can a right triangle have three right angles?
     Two right angles? Explain.

11. Figure ABCDE
is a regular pentagon.
Find the measure of ∠A.

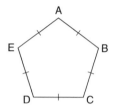

12. A truss is a framework for supporting a bridge.
There are different types of trusses.
The Pratt Truss uses right isosceles triangles.
Design your own truss using triangles.
Measure each angle. Find the sum of
the angles in each triangle.
What do you notice?

13. How many different kinds of triangles are there?
Sketch each triangle you name.

Take It Further

14. Find the measure of
∠PQR in this circle.
Point O is the centre
of the circle.
Explain your reasoning.

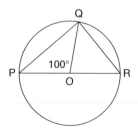

Reflect

How could you explain to a classmate who missed this
lesson why the sum of the angles in a triangle is 180°?

Using *The Geometer's Sketchpad* to Investigate Angles in a Triangle

| Focus | Use technology to investigate angles in a triangle. |

1. Open *The Geometer's Sketchpad*.

2. From the **Toolbox**, choose (Straightedge Tool).
Move to a point on the screen, then click and drag to draw a line segment.
Draw a second line segment that shares an endpoint with the first segment.
Complete a triangle by drawing a third line segment.

3. From the **Toolbox**, choose **A** (Text Tool).
Click the three points in order to label the vertices of the triangle A, B, C.

4. From the **Toolbox**, choose (Selection Arrow Tool).
Click a blank area of the screen to deselect all objects.
Click A, B, C in order.
From the **Measure** menu, choose **Angle**.
Click and drag the angle measure into ∠ABC.

5. Repeat *Step 4* to get the measures of ∠BAC and ∠ACB.

6. From the **Toolbox**, choose .
Click a blank area of the screen to deselect all objects.

7. From the **Measure** menu, choose **Calculate**.
Click m∠ABC. Click ⊞ on the calculator screen.
Click m∠BAC. Click ⊞.
Click m∠ACB. Click OK.
What do you notice about m∠ABC + m∠BAC + m∠ACB?

8. From the **Toolbox**, choose .
Click a blank area of the screen to deselect all objects.
Click and drag point A.
What do you notice about m∠ABC as you drag point A?
What do you notice about m∠BAC? m∠ACB?
What do you notice about m∠ABC + m∠BAC + m∠ACB?

This tiling pattern uses only parallelograms.

How could you create the tiling by translating a parallelogram?
How could you create the tiling by rotating a parallelogram
about the midpoint of one side?

Explore

Work with a partner.
You will need tracing paper, isometric dot paper, and a ruler.
- Draw four congruent parallelograms on dot paper.
- Label the parallelograms as shown below.
  M is the midpoint of EF.

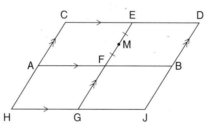

Trace the figure. Label the tracing to match the figure.

➤ Place the tracing on top of the original figure.
  Rotate the tracing 180° about point M.
  What do you notice about these pairs of angles?
  ∠CEF and ∠EFB
  ∠DEF and ∠EFA

➤ Place the tracing on top of the original figure.
  Slide the tracing along EG until E coincides with F.
  What do you notice about these pairs of angles?
  ∠DEF and ∠BFG   ∠EFA and ∠FGH
  ∠EFB and ∠FGJ   ∠CEF and ∠AFG

➤ What can you say about these pairs of angles?
  ∠DEF and ∠EFB   ∠CEF and ∠EFA

Reflect & Share

Compare your results with those of another pair of classmates.
Work with your classmates. Write as many statements as you can
about angle properties of parallel lines.

Connect

PQ and RS are parallel lines.
KN is a **transversal**.
F is the midpoint of JM.

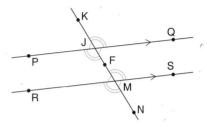

➤ After a rotation of 180° about point F,
∠PJM coincides with ∠JMS.
So, ∠PJM = ∠JMS
∠PJM and ∠JMS are between the parallel lines and are on
opposite sides of the transversal.
They are called **alternate angles**.
Similarly, ∠QJM and ∠JMR are alternate angles.
So, ∠QJM = ∠JMR

➤ After a translation along line KN, ∠QJM coincides with ∠SMN.
So, ∠QJM = ∠SMN
Look at the parallel lines and the transversal.
∠QJM and ∠SMN are in the same relative position.
That is, each angle is below a parallel line and
to the right of the transversal.
These angles are called **corresponding angles**.
Similarly, ∠KJQ and ∠JMS are corresponding angles.
Each angle is above a parallel line and to the right of
the transversal.
∠KJQ = ∠JMS
And, ∠KJP and ∠JMR are corresponding angles.
Each angle is above a parallel line and to the left of
the transversal.
∠KJP = ∠JMR

The 4th pair of corresponding angles is ∠PJM and ∠RMN.
Each angle is below a parallel line and to the left
of the transversal.
∠PJM = ∠RMN

➢ ∠QJM is the supplement of ∠KJQ because
∠QJM + ∠KJQ = 180°.
But, ∠KJQ = ∠JMS
So, ∠QJM is the supplement of ∠JMS.
That is, ∠QJM + ∠JMS = 180°
∠QJM and ∠JMS are between the parallel lines and are on the
same side of the transversal. They are called **interior angles**.
Similarly, ∠PJM and ∠JMR are interior angles.
That is, ∠PJM + ∠JMR = 180°

Example

Two transversals intersect parallel lines.
Find the measure of each angle.
Justify each answer.

a) ∠HBC

b) ∠AFC

c) ∠AGB

d) ∠BGF

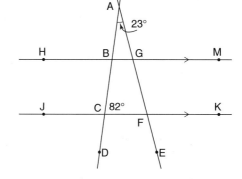

Solution

a) ∠HBC and ∠BCF are alternate angles.
So, ∠HBC = ∠BCF = 82°
∠HBC = 82°

b) In △AFC, the sum of the angles is 180°.
So, ∠A + ∠ACF + ∠AFC = 180°
23° + 82° + ∠AFC = 180°
105° + ∠AFC = 180°
∠AFC = 180° − 105°
= 75°

c) ∠AGB and ∠AFC are corresponding angles.

So, ∠AGB = ∠AFC = 75°

∠AGB = 75°

∠GFC is the same angle as ∠AFC, which is 75°.

d) ∠BGF and ∠GFC are interior angles.

So, ∠BGF + ∠GFC = 180°

∠BGF + 75° = 180°

∠BGF = 180° − 75°

= 105°

Practice

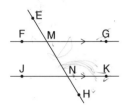

1. Use lined paper. Draw a larger copy of this diagram. Measure and mark all the angles.

  a) Name the corresponding angles. Are they equal?

  b) Name the alternate angles. Are they equal?

  c) Name the interior angles. Are they supplementary?

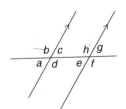

2. In this diagram, each angle is labelled with a variable. Use the variables to name:

  a) two pairs of corresponding angles

  b) two pairs of alternate angles

  c) two pairs of interior angles

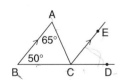

3. Look at this diagram.

  a) Name two parallel line segments.

  b) Name two transversals.

  c) Name two corresponding angles.

  d) Name two alternate angles.

  e) Find the measures of ∠ECD, ∠ACE, and ∠BCA.

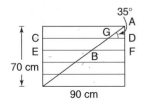

4. A gate is made of 5 parallel boards, with a cross piece for support. The plans for the gate are shown.

  a) What is the measure of ∠AGC? ∠CGB?

  b) What is the measure of ∠ABE? ∠ABF? ∠DGB? Justify each answer.

5. In this diagram, find the measures of ∠SPQ, ∠TPR, and ∠QPR.
Explain how you know.

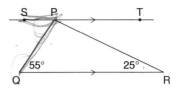

6. Are the pairs of angles below equal, complementary, or supplementary?
Explain how you know.
Where possible, justify each answer more than one way.
   a) ∠SRY and ∠RTK   **b)** ∠SQR and ∠SVT
   c) ∠RTK and ∠QRT   **d)** ∠YRT and ∠RTV
   e) ∠QRT and ∠RTV   **f)** ∠RTK and ∠VTW
   g) ∠SRQ and ∠QSR   **h)** ∠SRY and ∠SRQ

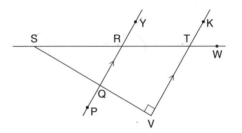

Mental Math

Which number does not belong in the pattern below?

Which number should go in its place?
Explain your reasoning.

1, 4, 9, 16, 25, 36, 48, 64, …

7. **Assessment Focus**
   a) Find the measure of each unmarked angle in this diagram.
   b) Use the properties of angles between parallel lines to justify that the sum of the angles in a triangle is 180°.

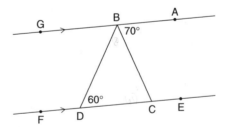

8. Find the measures of ∠KGH, ∠KGF, ∠HGJ, and ∠GHJ.

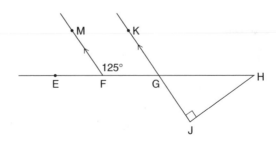

9. In trapezoid PQRS, PQ is parallel to SR, and diagonal PR is drawn. SP = SR, ∠SRP = 50°, and ∠PRQ = 85° Find the measures of ∠RQP, ∠SPR, and ∠PSR. Justify your answers.

10. a) Sketch parallelogram ABCD with ∠A = 51°.
  b) How can you use what you know about parallel line segments and a transversal to find the measures of the other 3 angles in the parallelogram? Explain your work.
  c) When is a quadrilateral a parallelogram? Explain.

11. Suppose there are 2 line segments that look parallel. How could you tell if they are parallel?

12. Draw 2 line segments that are not parallel, and a transversal. Measure the alternate, corresponding, and interior angles. What do you notice?

13. Find the value of x, and the measures of ∠ADC and ∠CAD.

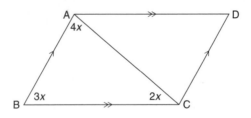

Your World

Make a list of where you see parallel lines and transversals in your community, around the house, or on a construction site. Sketch diagrams to illustrate your list.

Reflect

Draw two parallel lines and a transversal. Label the angles. Explain how you remember which are corresponding angles, which are alternate angles, and which are interior angles.

Using *The Geometer's Sketchpad* to Investigate Parallel Lines and Transversals

Focus Use technology to investigate parallel lines and transversals.

1. Open *The Geometer's Sketchpad*.

2. From the **Toolbox**, choose (Straightedge Tool).
   Move to a point on the screen, then click and drag to draw a line segment.

3. From the **Toolbox**, choose (Point Tool).
   Click the screen to create a point *not* on the line segment.

4. From the **Toolbox**, choose (Selection Arrow Tool).
   Click a blank area of the screen to deselect all objects.
   Click the point and the line segment.
   From the **Construct** menu, choose **Parallel Line**. A line is drawn through the point parallel to the line segment.

5. From the **Toolbox**, choose .
   Draw a transversal that intersects the line and line segment.

6. From the **Toolbox**, choose .
   Click each point of intersection.
   Click the line to mark a point,
   so there is a point on each side
   of the transversal.

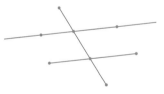

7. From the **Toolbox**, choose **A** (Text Tool).
   Click the points, in order, to assign labels A to H.

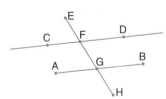

Alternate Angles between Parallel Lines

8. From the **Toolbox**, choose .
   Click a blank area of the screen to deselect all objects.
   Click D, F, G in order. From the **Measure** menu, choose **Angle**.
   Click and drag the angle measure into ∠DFG.

9. Repeat *Step 8* for ∠FGA.

10. From the **Toolbox**, choose [cursor icon].
Click a blank area of the screen to deselect all objects.
Click and drag point E.
What do you notice about the measures of alternate angles ∠DFG and ∠FGA as you drag point E?

11. Repeat *Steps 8* to *10* for ∠CFG and ∠FGB.

Corresponding Angles between Parallel Lines

12. From the **File** menu, choose **New Sketch**.

13. Repeat *Steps 2* to *7* to create a new set of parallel lines cut by a transversal.

14. Repeat *Steps 8* to *10* for: ∠DFG and ∠BGH; ∠CFG and ∠AGH; ∠EFD and ∠FGB; ∠EFC and ∠FGA
What do you notice about the pairs of corresponding angles?

Interior Angles between Parallel Lines

15. From the **File** menu, choose **New Sketch**. Repeat *Step 13*.

16. From the **Toolbox**, choose [cursor icon].
Use the **Measure** menu to measure ∠DFG and ∠FGB.

17. Use the **Measure** menu to calculate the sum of ∠DFG and ∠FGB.
What do you notice about ∠DFG + ∠FGB?

18. From the **Toolbox**, choose [cursor icon].
Click a blank area of the screen to deselect all objects.
Click and drag point E so that point F is always between point C and point D.
What do you notice about the sum of the measures of interior angles ∠DFG and ∠FGB?

19. Repeat *Steps 16* to *18* for ∠CFG and ∠FGA.

Angles between Non-Parallel Lines

20. Change *Steps 1* to *19*, as necessary, to find the angle measures between a transversal and 2 non-parallel lines.
What do you notice?

Mid-Unit Review

LESSON

7.1
7.2

1. Find the measure of each angle.
Explain how you know.
a) ∠JKM
b) ∠NJM
c) ∠JNM

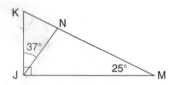

2. Find the measure of each angle.
Justify your answers.
a) ∠CBA
b) ∠CAB
c) ∠ACB

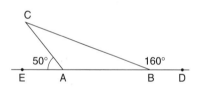

3. A triangle is drawn with angle
measures that are whole numbers.
a) What is the greatest possible
angle measure in an
obtuse triangle?
b) What is the greatest possible
angle measure in an
acute triangle?

7.3 **4.** Find the measure of each angle in
the following diagram.
Explain how you know.
a) ∠BEF   b) ∠ADE
c) ∠BAD   d) ∠GDE

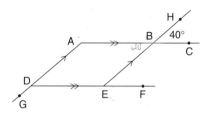

7.1
7.2
7.3

5. Find the measure of each angle.
Justify your answers.
a) ∠SQT
b) ∠QRW
c) ∠PRV

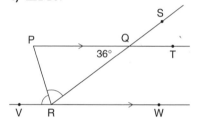

6. Sketch this diagram.
Calculate the measure of each
unmarked angle.
Write the measures on your
diagram.

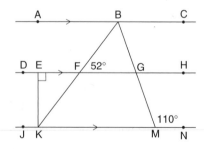

Focus Use a variety of methods to construct bisectors of line segments and angles.

Recall that a rhombus has all sides equal and opposite angles equal.

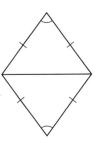

Each diagonal cuts the rhombus into 2 congruent isosceles triangles.
How do you know the triangles are isosceles?
How do you know the triangles are congruent?

You will investigate ways to cut line segments and angles into 2 equal parts.

Explore

Work with a partner.
You may need rulers, protractors, tracing paper, plain paper, and Miras.
Use any method or tools.

- Draw a line segment on plain paper.
  Draw a line perpendicular to the line segment that cuts the line segment in half.
- Draw an angle on plain paper.
  Draw the line that cuts the angle in half.

Reflect & Share

Compare your results and strategies with those of another pair of classmates.
Could you use your strategy to cut your classmates' line segment in half? Explain.
Could you use your classmates' strategy to cut your angle in half? Explain.

Connect

➤ When you draw a line to divide a line segment into two equal parts, you **bisect** the segment. The line you drew is a **bisector** of the segment.

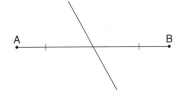

When the bisector is drawn at right angles to the segment, the line is the **perpendicular bisector** of the segment.

➤ Each diagonal of a rhombus is a line of symmetry.
The diagonals intersect at right angles.
The diagonals bisect each other.
So, each diagonal is the perpendicular bisector of the other diagonal.

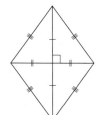

We can use these properties of a rhombus to construct the perpendicular bisector of a line segment.
Think of the line segment as a diagonal of a rhombus.
As we construct the rhombus, we also construct the perpendicular bisector of the segment.

Use a ruler and compass.

A ————————— B

• Draw any line segment AB.

• Set the compass so the distance between the compass and pencil points is greater than one-half the length of AB.

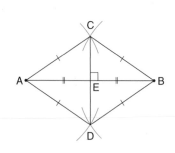

A ————————— B

• Place the compass point on A.
Draw an arc above and below the segment.
Do not change the distance between the compass and pencil points. Place the compass point on B. Draw an arc above and below the segment to intersect the first 2 arcs you drew.

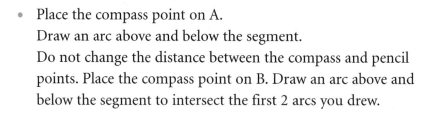

• Label the points C and D where the arcs intersect.
Join the points to form rhombus ACBD. Draw the diagonal CD. The diagonals intersect at E. CD is the perpendicular bisector of AB. That is, AE = EB and ∠AEC = ∠CEB = 90°
To check that the perpendicular bisector has been drawn correctly, measure the two parts of the segment to check they are equal, or measure the angles to check each is 90°.

Note that any point on the perpendicular bisector of a line segment is the same distance from the endpoints of the segment.
For example, AC = BC and AD = BD

➢ When you divide an angle into two equal parts, you **bisect** the angle.
We can use the properties of a rhombus to construct the bisector of an angle.
Think of the angle as one angle of a rhombus.

BJ and BK are 2 sides of the rhombus; BJ = BK.

- Draw ∠B as one angle of a rhombus. With compass point on B, draw an arc that intersects one arm at K and the other arm at J.

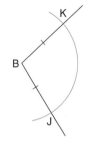

- Do not change the distance between the compass and pencil points. Place the compass point on K. Draw an arc between the arms of the angle. Place the compass point on J. Draw an arc to intersect the second arc you drew. Label the point M where the arcs intersect.

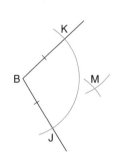

JM and MK are the other 2 sides of the rhombus.

BM is a diagonal of the rhombus.

- Join KM and MJ to form rhombus BKMJ. Draw a line through BM. This line is the **angle bisector** of ∠KBJ. That is, ∠KBM = ∠MBJ

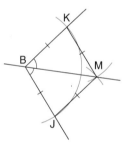

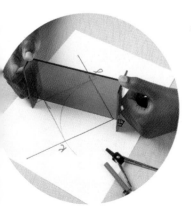

To check that the bisector of ∠B has been drawn correctly, we can:
- Measure the two angles formed by the bisector. They should be equal.
- Fold the angle so the bisector is the fold line. Arm JB should coincide with arm KB.
- Place a Mira along BM. The reflection image of arm JB should coincide with arm KB, and vice versa.

Example

Draw any obtuse angle.
Use a ruler and compass to bisect the angle.
Measure the angles to check.

Solution

Draw obtuse ∠B.
With compass point on B, draw an arc
to intersect the arms of the angle.
Label the points of intersection F and G.
Place the compass point on F. Draw an
arc between the arms of the angle.
Do not adjust the compass. Place the compass point on G.
Draw an arc to intersect the second arc.
Label the point H where the arcs intersect.
Draw a line through BH.
This line is the bisector of ∠B.

Use a protractor to check.
∠FBG = 126°
∠FBH = 63° and ∠GBH = 63°
∠FBH + ∠GBH = 63° + 63°
$$= 126°$$
$$= ∠FBG$$

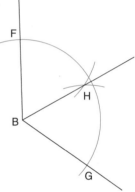

Practice

Show all construction lines.

1. a) Draw line segment AB.
    Use a ruler and compass to draw its perpendicular bisector.
   b) Choose three different points on the bisector.
    Measure the distance to each point from A and from B.
    What do you notice? Explain.

2. Find out what happens if you try to draw the perpendicular
   bisector of a segment when the distance between the compass
   and pencil points is:
   a) equal to one-half the length of the segment
   b) less than one-half the length of the segment

3. You have used Miras and paper folding to bisect an angle. What is the advantage of using a ruler and compass?

4. Use a ruler and compass.
   a) Draw acute ∠PQR. Bisect the angle.
   b) Draw obtuse ∠GEF. Bisect the angle.

Use the *Glossary* if you have forgotten what a reflex angle is.

5. Draw a reflex angle.
   a) How many different methods can you find to bisect this angle?
   b) Describe each method.
      Check that each bisector you draw is correct.

6. a) Draw line segment HJ.
      Draw the perpendicular bisector of HJ.
   b) Bisect each right angle in part a.
   c) How many angle bisectors did you need to draw in part b? Explain.

7. Draw a large △ABC. Cut it out.
   Fold the triangle so B and C coincide. Open the triangle.
   Fold it so A and B coincide. Open the triangle.
   Fold it so A and C coincide. Open the triangle.
   Use a ruler to draw a line along each crease.
   a) Measure the angles each crease makes with one side.
      What do you notice?
   b) Label point K where the creases meet.
      Measure KA, KB, and KC.
      What do you notice?
   c) What have you constructed by folding?

8. Draw a large △PQR.
   Construct the perpendicular bisector of each side.
   Label point C where the bisectors meet.
   Draw the circle with centre C and radius CP.

9. a) How could you use the construction in question 8 to draw a circle through any 3 points that do not lie on a line?
   b) Mark 3 points as far apart as possible.
      Draw a circle through the points.
      Describe your construction.

Number Strategies

Jean received these test marks in geography:

72, 84, 88, 76, 64, 84

Find the mean, median, and mode.

"Circum" is Latin for "around." So, the circumcircle is the circle that goes around a triangle.

The point at which the perpendicular bisectors of the sides of a triangle intersect is called the circumcentre.

10. Assessment Focus

Use a ruler and compass.

a) Draw a large isosceles △ABC, with AB = AC.

b) Bisect ∠A.

c) Draw the perpendicular bisector of BC.

d) What do you notice about the bisectors?

e) Will the result in part d be true for:

    i) all isosceles triangles?

    ii) all equilateral triangles?

    iii) all scalene triangles?

Explain your answers. Show your work.

Take It Further

11. Use a ruler and protractor to draw quadrilateral ABCD.
AB = 7 cm, AD = 5 cm, ∠DAB = 105°, ∠ABC = 100°, and ∠ADC = 80°
Join BD. Construct the circumcircle of △ABD.
Does the circle pass through C?
Do you think this would happen for all quadrilaterals? Investigate.

Science

The circumcentre of an equilateral triangle is also the balancing point of the triangle. For example, you can balance the triangle on the tip of a pencil using the circumcentre. Gravity acts on all parts of a structure. Gravity pulls a structure downward through a point called the **centre of gravity**. The centre of gravity of an equilateral triangle is its circumcentre.

circumcentre

Reflect

How many bisectors can an angle have?

How many bisectors can a line segment have?

How many perpendicular bisectors can a line segment have?

Draw a diagram to illustrate each answer.

Focus | Use a variety of methods to construct angles.

You have used a protractor to construct an angle of a given measure.
What if you do not have a protractor.
How can you construct angles?

Explore

Work with a partner.
You will need a ruler, compass, Mira, and tracing paper.

➤ Use your knowledge of equilateral triangles and angle bisectors to construct a 30° angle.
➤ Use your knowledge of perpendicular bisectors and angle bisectors to construct a 45° angle.

Explain the steps you took to construct each angle.

Reflect & Share

Compare strategies with another pair of classmates.
How many different strategies can you use to construct a 30° angle?
A 45° angle?

Connect

You can construct a 60° angle by drawing two sides of an equilateral triangle.
You can construct a 90° angle by drawing the perpendicular bisector of a line segment.
You can use these constructions to draw other angles.

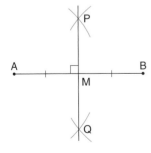

➤ Here is one method to draw a 135° angle.
135° = 90° + 45°
So, draw a 90° angle and a 45° angle at a common vertex.
Draw line segment AB.
Use a ruler and compass to draw the perpendicular bisector of AB.

$\angle AMP = \angle PMB = 90°$
Use a ruler and compass to bisect $\angle PMB$.
Label point R on the angle bisector.

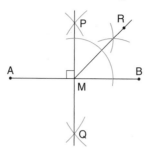

Then, $\angle PMR = \angle RMB = \frac{1}{2}$ of $90°$

So, $\angle PMR = 45°$

And, $\angle AMR = \angle AMP + \angle PMR$
$= 90° + 45°$
$= 135°$

➤ We can use the properties of a rhombus
to construct a 90° angle.
Use a ruler and compass.
Draw a line. Mark a point C away from the line.
C is on one arm of the 90° angle. The other arm is on the line.
Place the compass point on C.
Draw an arc that intersects the line
at two points, D and E.
DE is one diagonal of the rhombus.
CD and CE are two sides of the rhombus.

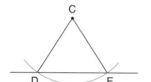

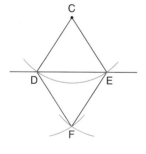

Do not change the distance between the compass and
pencil points.
Place the compass point on D. Draw an arc below DE.
Place the compass point on E.
Draw an arc below DE to intersect the previous arc.
Label point F where the arcs intersect.
Then, DF and EF are the other two sides of the rhombus.

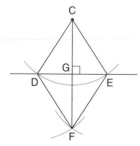

Draw CF, the other diagonal of the rhombus.
Label point G where the diagonals intersect.
Since the diagonals of a rhombus intersect at right angles,
$\angle CGE = 90°$.

The following *Example* explains how to draw a 120° angle.

Example

Use only a ruler and compass.

a) Construct △DBC, with ∠B = 120°.

b) Explain how you know ∠B = 120°.

Solution

a) Draw line segment AC. Mark point B on AC.
Make the distance between the compass and pencil points equal to AB.
Place the compass point on B. Draw an arc.
Do not change the distance between the compass and pencil points. Place the compass point on A.
Draw an arc to intersect the first arc.
Label point D where the arcs intersect. Join DB. Join DC.

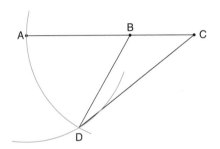

Measure to check that the angle is correct.

Then, ∠DBC = 120°

b) Sketch the construction.
Join AD.
Since the distance between the compass and pencil points is equal to AB,
then AB = BD = AD.
So, △ABD is equilateral.
And, ∠A = ∠ABD = ∠ADB = 60°
∠ABD and ∠DBC are supplementary angles.
So, ∠ABD + ∠DBC = 180°
60° + ∠DBC = 180°
∠DBC = 180° − 60°
= 120°

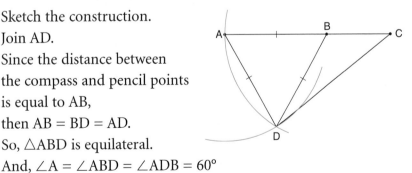

Use only a ruler and compass in questions 1 to 3.

1. Construct a large equilateral △PQR.
Bisect ∠P to construct a 30° angle.

2. Construct an isosceles right △DEF, with ∠D = 90°.
   a) What is the measure of ∠E? ∠F? How do you know?
   b) How can you construct a 45° angle?

3. a) Use your knowledge of constructing a 90° angle and a
      30° angle to construct a 120° angle.
   b) Compare your method with the method in the *Example*.
      Which is easier? Explain.

4. (Assessment Focus)   Use any methods or tools.
   a) Construct an isosceles triangle with one 30° angle.
   b) Construct an isosceles triangle with two 30° angles.
   How many different ways can you construct each triangle?
   Include a construction with each method.

5. Construct a 300° angle.
   Can you do this in more than one way? Explain.

6. Describe one way to construct a 240° angle.

7. Construct parallelogram ABCD with:
   ∠A = ∠C = 150°, ∠B = ∠D = 30°, AD = BC = 3 cm,
   AB = CD = 5 cm

8. Triangle ABC has angle measures related this way:
   ∠A is 3 times ∠C, and ∠B is twice ∠C.
   a) What is the measure of ∠A? ∠B? ∠C?
   b) Use only a ruler and compass. Construct △ABC.

Mental Math

Kenji completed a
bicycle race in
2 h and 23 min.

He finished at 4:15 p.m.

At what time did the
race start?

Reflect

List all the angles between 0° and 180° you can construct
using only a ruler and compass.
Choose one angle from the list. Construct the angle.
Describe how you did it.

Creating and Solving Geometry Problems

Focus | Create and solve problems involving angle measurement.

Explore

Work with a partner.

➤ You will need a copy of this diagram.

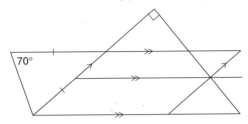

Find the measures of all the angles.

➤ Make up a geometry problem that involves a triangle and parallel lines.
Sketch your diagram.
Remember to mark equal side lengths, if appropriate.
Find the measures of all the angles.
On a copy of your diagram, write the fewest angle measures a person would need to find all the measures.
Trade diagrams with another pair of classmates.
Find the measures of the angles in your classmates' diagram.

Reflect & Share

Compare your answers with those of your classmates.
Were you able to find all the measures? Explain.
Were your classmates able to find all the measures on your diagram?
If not, find out why.

Connect

Look back through the unit to review any concepts you are unsure about.

To solve problems that involve angle measures associated with triangles and parallel lines, you might need to use any or all of the geometry concepts you learned in this unit.
We often use variables to describe angle measures.
In each case, the variable represents one angle measure.

Example 1

Find the values of
w, x, y, and z
in this diagram.

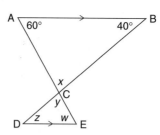

Solution

Each variable represents one angle measure.
In △ABC, the sum of the angles is 180°.
So, ∠A + ∠B + ∠ACB = 180°

$$60° + 40° + x = 180°$$
$$100° + x = 180°$$

Solve this equation by inspection.

> **Think:** What do we add to 100° to get 180°?

$$x = 80°$$

x and y are measures of opposite angles.
So, $y = x = 80°$
$$y = 80°$$

Transversal AE intersects parallel line segments AB and DE.
So, alternate angles are equal.
∠DEC = ∠BAC
$$w = 60°$$

What other method could we use to find the value of *z*?

Transversal DB intersects parallel line segments AB and DE.
So, alternate angles are equal.
∠CDE = ∠ABC
$$z = 40°$$

Example 2

Find the values of
q, w, and y
in this diagram.

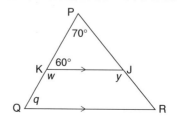

Solution

Transversal PQ intersects parallel
line segments KJ and QR.
So, corresponding angles are equal.
$\angle KQR = \angle PKJ$
$q = 60°$

Angles that form a straight line are supplementary.
So, $\angle PKJ + \angle JKQ = 180°$
$60° + w = 180°$
$w = 120°$

What other method could we use to find the value of w?

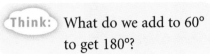 **Think:** What do we add to 60° to get 180°?

We cannot find the value of y directly.
First, we need to find the measure of $\angle PJK$.
In $\triangle PKJ$, the sum of the angles is 180°.
So, $\angle P + \angle PKJ + \angle PJK = 180°$
$70° + 60° + \angle PJK = 180°$
$130° + \angle PJK = 180°$
$\angle PJK = 50°$
Angles that form a straight line are supplementary.
So, $\angle PJK + \angle KJR = 180°$
$50° + y = 180°$
$y = 130°$

Practice

1. Look at this diagram.
   a) Name a pair of
   parallel line segments.
   b) Name two transversals.
   c) Name two corresponding
   angles.
   d) Name two alternate angles.
   e) Name two complementary angles.
   f) Find the values of x, y, and z.

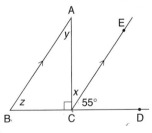

2. a) Write an equation for the sum of the angles in △ABC.
  b) Solve the equation to find the value of x.

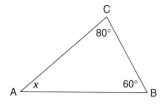

3. Look at the diagram below.
  a) Name two isosceles triangles.
  b) Name two complementary angles.
  c) Find the values of x and y.

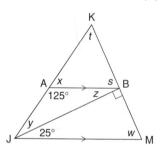

4. In the diagram below, find the values of x, y, z, w, t, and s. Show all your work. Justify your answers.

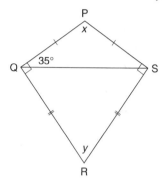

5. Assessment Focus Use a copy of this diagram, or draw your own diagram.
  Measure and label one or more angles, the fewest you need to know to calculate the remaining measures.
  Label each unmarked angle with a variable.
  Find the value of each variable. Justify your answers. Show your work.

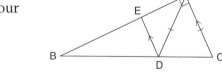

6. Point O is the centre of the circle.
Find the values of *x*, *y*, *z*, and *w*.
Explain how you know.

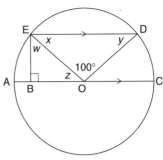

7. Use a ruler and lined paper.
Draw a diagram that has corresponding angles,
alternate angles, and interior angles between parallel lines.
Use a protractor to measure and label one angle.
Calculate the measures of all the other angles.
Justify your answers.

8. Explain why ∠BCA and ∠ECD are complementary.

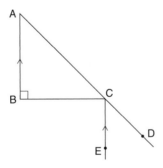

Take It Further

9. In this diagram, is AD parallel to BC?
Explain.

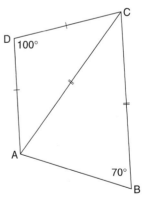

How does the use of variables help solve some
geometry problems?
Explain using examples.

Reasonable Solutions and Concluding Statements

Katja invites 8 friends to her birthday party, which takes place in 4 days. Katja's age in years will be 2 less than twice the number of people she invites. How old will Katja be on her birthday? Which solution is reasonable?

a) 12   **b)** 14   **c)** 16   **d)** 18

Show why 14 is the reasonable solution.

To be sure you have a reasonable solution:

1. Identify the features of the problem (context, math information, problem statement) to help understand the problem.
   Think of what might be a reasonable answer.

2. Use words, numbers, pictures, models, graphic organizers, and/or symbols to solve the problem. Record how you solved it.

3. Think back on the problem:
   • Check if the solution makes sense with the *context*.
   • Check that the *math information* was used appropriately.
   • Check that the solution answers the *problem*.

4. Write a concluding statement to answer the problem.

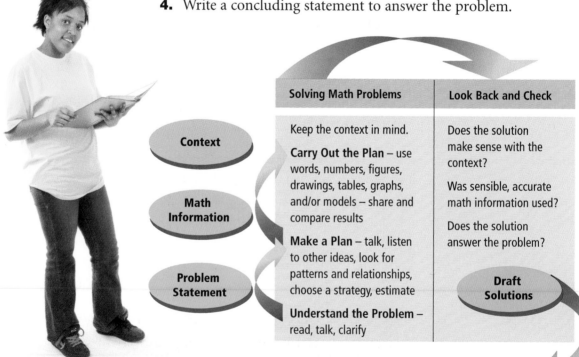

It is common to use part of the problem statement in the concluding statement.

How (old) (will Katja be) (on her birthday?)

(Katja will be) (14 years old) (on her birthday.)

Some Dos and Don'ts for Problem Solving

- Do keep asking yourself, "Does this make sense?"
- Don't work without writing anything down.
- Do look for patterns.
- Don't stay with a strategy when it is not working.

✓ Check

Show how you might use math information from the problem to find each answer. Which solutions are reasonable? Why? Write a concluding statement for each word problem.

1. You need enough cake to serve 18 people. A 20-cm square cake serves four people. How many 30-cm square cakes are needed to provide equal servings for 18 people?
   a) 2 **b)** 4 **c)** 10 **d)** 12

2. Exactly five diagonals can be drawn from one vertex of a convex polygon. How many sides does the polygon have?
   a) 4 **b)** 5 **c)** 6 **d)** 8

3. Mr. Davis wants to encourage his son Mike to save money. Every month, when Mike puts $3 into a savings account, Mr. Davis doubles the amount of money in the account. Mike now has $90 in the account. How many months ago did Mike start to save?
   a) 2 **b)** 3 **c)** 4 **d)** 5

4. One hundred sixty-three students are going on a field trip. One bus holds 60 students. How many buses are needed?
   a) 2 **b)** 2.72 **c)** 3 **d)** 4

Robotics Engineer

The robots from novels, movies, and television shows that look and act like humans exist only in science fiction. Scientists *are* developing machines with certain human or animal capabilities, and lesser robots — such as automated arms like Canadarm — have been in use for years.

Imagine a human task: screwing a light bulb into a socket. Sounds simple, doesn't it? To design a robotic machine that can do this again and again without error is a major feat of engineering. The robot must have mobility: it has to grasp the bulb, pull it from a bin, line it up with the socket, turn the bulb into the socket, move up as the bulb goes in, and release it when done. It also needs agility: a light bulb is delicate — if the robot holds it too tightly or tries to force it into place when it isn't aligned, the bulb will break.

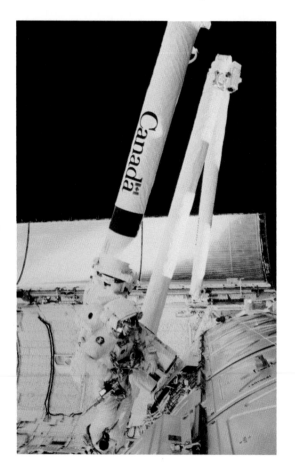

The mobility design of a robot is an exercise in geometry. How many joints should this robotic arm have? How long should each segment be? Through which angle must the arm bend and how much does it need to rotate? The robotics engineer calculates the best design for the robot's intended tasks. One advantage of robots over simple industrial machines is that a robot can be programmed to do different tasks. Today it might screw light bulbs into sockets; next week it might screw small bolts into pre-drilled holes. The tasks are similar, but the geometry has changed and the programming needs to change as well.

Providing robots with mobility and agility is one of the biggest challenges for today's robotics engineers.

Unit Review

What Do I Need to Know?

☑ **Intersecting Lines**

When two lines intersect,
opposite angles are equal.
∠MPN = ∠KPQ and ∠MPK = ∠NPQ

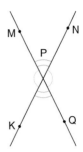

☑ **Supplementary Angles**

Angles that form a straight line are *supplementary*.
Supplementary angles have a sum of 180°.
∠KPM + ∠MPN = 180°

☑ **Complementary Angles**

Angles that form a right angle are *complementary*.
Complementary angles have a sum of 90°.
∠QRT + ∠TRS = 90°

☑ **Angles in a Triangle**

The *sum of the angles* in a triangle is 180°.

- ∠A + ∠B + ∠C = 180°
- An *equilateral triangle* has all sides equal
  and all angles equal.
  Each angle in an equilateral triangle is 60°.
- An *isosceles triangle* has two sides equal
  and two angles equal.

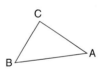

☑ **Perpendicular Bisector**

The *perpendicular bisector* of a line segment
divides the segment into two equal parts.
Line CD is the perpendicular bisector
of segment AB.
AE = EB
∠AEC = ∠CEB = 90°

✓ Bisector of an Angle

The *bisector of an angle* divides the angle
into two equal angles.
Line QS is the bisector of ∠PQR.
∠PQS = ∠SQR

✓ Parallel Lines and a Transversal

DE is a transversal that intersects parallel lines
AC and FH.

- Alternate angles are equal.

    ∠CBG = ∠BGF and ∠ABG = ∠BGH

- Corresponding angles are equal.

    ∠DBC = ∠BGH

    ∠DBA = ∠BGF

    ∠CBG = ∠HGE

    ∠ABG = ∠FGE

- Interior angles are supplementary.

    ∠CBG + ∠BGH = 180°

    ∠ABG + ∠BGF = 180°

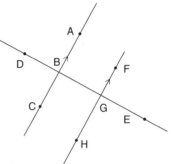

What Should I Be Able to Do?

For extra practice, go to page 494.

LESSON

7.1 **1.** Look at the diagram below. Name:

   a) an angle supplementary
   to ∠BFC

   b) an angle complementary to
   ∠EFD

   c) an angle opposite ∠AFB

   d) an angle supplementary
   to ∠EFC

   e) an angle opposite ∠BFC

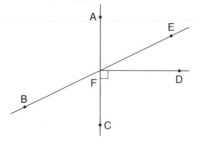

2. In the diagram in question 1,
the measure of ∠EFD is 25°.
Find the measure of each other
angle. Explain how you know.

   a) ∠AFE

   b) ∠AFB

   c) ∠BFC

3. An angle measures 34°.
What is the measure of:

   a) its complementary angle?

   b) its supplementary angle?

   c) its opposite angle?

4. Look at this diagram.

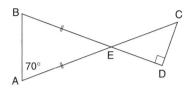

a) Name two pairs of opposite angles.

b) Name a right triangle.

c) Name an isosceles triangle.

d) Find the measures of ∠ABE, ∠BEA, ∠CED, and ∠ECD. Justify your answers.

5. Suppose you know one angle of an isosceles triangle. How can you find the measures of the other two angles? Use examples in your explanation.

6. A metronome is used by people practising musical instruments to help them keep time.
The pendulum can be adjusted to tick at different speeds. At one particular instant, the pendulum forms △ABC, with ∠A = 75° and ∠B = 62°.

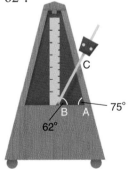

a) Find the measure of ∠C.

b) At another instant, ∠B is 79°. Draw the triangle formed. Find the new measure of ∠C.

7. A right triangle is drawn with angle measures that are whole numbers of degrees.

a) What is the greatest possible measure of an acute angle in the triangle? Explain.

b) What is the least possible measure of an acute angle in the triangle? Explain.

7.3 **8.** Look at this diagram.

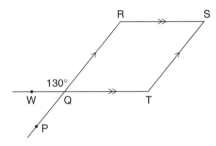

a) Name two pairs of parallel line segments.

b) Name four pairs of interior angles.

c) Find the measures of ∠RQT, ∠QRS, ∠RST, and ∠STQ. Justify your answers.

d) Look at the results of part c. What do they tell you about opposite angles in a parallelogram?

7.4 **9. a)** Draw line segment AB. Fold the paper to construct the perpendicular bisector.

b) Draw line segment CD. Use a Mira to construct the perpendicular bisector.

c) Draw line segment EF. Use a ruler and compass to construct the perpendicular bisector.

d) Which of the three methods is most accurate? Justify your answer.

7.4 **10. a)** Draw acute ∠BAC. Fold the paper to construct the angle bisector.

b) Draw right ∠DEF. Use a Mira to construct the angle bisector.

c) Draw obtuse ∠GHJ. Use a ruler and compass to construct the angle bisector.

d) Which of the three methods is most accurate? Justify your answer.

7.5 **11.** Use only a ruler and compass. Construct each triangle.

a) △ABC with ∠A = 90°, ∠B = 60°, and ∠C = 30°

b) △DEF with ∠D = 120°, ∠E = 45°, and ∠F = 15°

7.6 **12.** Look at this diagram.

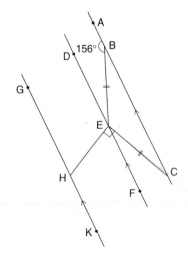

a) Name 3 parallel lines.

b) Name an isosceles triangle.

c) Find the measure of each angle. Give reasons for your answers.

   i) ∠BCE   **ii)** ∠CEF

   iii) ∠FEH   **iv)** ∠EHG

   v) ∠HED   **vi)** ∠HEB

13. There are two buildings. A worker is standing on top of the shorter building. She looks up to the top of the taller building.

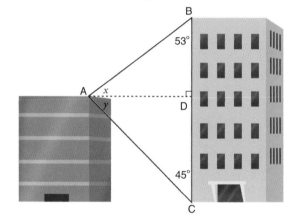

a) What is the angle of elevation, x?

b) The worker then looks down to the base of the taller building. What is the angle of depression, y?

c) What is the measure of ∠BAC?

d) What is the sum of the angles in △ABC? Are your answers to parts a and b correct? How do you know?

Practice Test

1. **a)** Which angle is both a complementary and a supplementary angle? Explain.

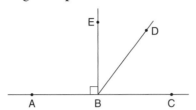

   b) Suppose the measure of ∠EBD is 38°.
   Explain how to find the measures of ∠DBC and ∠ABD.

2. In the diagram below left, find the values of *v*, *w*, *x*, and *y*.
   Justify your answers.

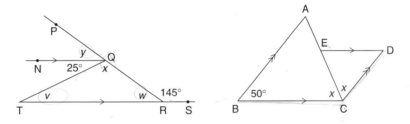

3. In the diagram above right, find the measures of
   ∠DCB, ∠DCE, and ∠CAB. Give reasons.
   What can you say about △ABC and △CDE? Explain.

4. **a)** Draw a large acute △ABC.
   Construct the bisectors of the angles.
   Use a different method for each bisector.
   b) Draw a large obtuse △DEF.
   Construct the perpendicular bisector of each side.
   Use a different method for each bisector.
   Describe each method used.

5. In this diagram, which angles can
   you *not* find the measures of?
   Explain what you would need to
   know to help you find the measures.

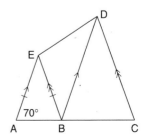

You have to design one of a series of mathematical banners to be hung on posts outside the school.

Your design will be created on 4 large pieces of chart paper.
It has to include:

- geometric figures, including triangles
- parallel lines with transversals
- geometric constructions

Work in a group of 4.
Brainstorm a theme and design ideas for your banner.
Sketch your banner.
Each person is responsible for one piece of the banner.
Make sure the pattern or design continues across a seam.
Create your banner.

On a sketch of the banner, show all construction lines.

Alternatively, use *The Geometer's Sketchpad* to illustrate each construction on your banner.

Write about your banner.

Explain your choice of design or pattern, and how it relates to the geometric concepts in this unit.

Check List

Your work should show:

✓ a detailed sketch including construction lines

✓ a description of the figures and angles in your part of the banner

✓ your understanding of geometric language and ideas

✓ the construction methods you used

Reflect on the Unit

Summarize what you have learned about angles formed by intersecting and parallel lines, and angles of a triangle. Use words and diagrams in your summary.

Pack It Up!

Work with a partner.

Imagine that you work for a packaging company.
You have a sheet of thin card that measures 28 cm by 43 cm.
You use the card to make a triangular prism with the
greatest volume. The triangular faces of the prism
are right isosceles triangles.

As you complete this *Investigation*, include all your work in a
report that you will hand in.

Materials:
- sheets of paper measuring 28 cm by 43 cm
- piece of thin card measuring 28 cm by 43 cm
- centimetre ruler
- 0.5-cm grid paper
- scissors
- tape

Part 1

Here is one prism and its net.

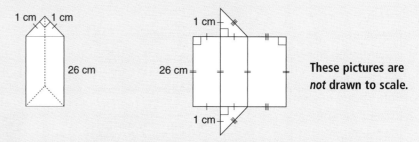

These pictures are *not* drawn to scale.

Each triangular face of this prism has base and height 1 cm.

Increase the base and height of the triangular faces by 1 cm
each time.
Calculate the length of the prism and its volume.
Copy and complete this table.

Triangular Face			Length of Prism (cm)	Volume of Prism (cm³)
Base (cm)	Height (cm)	Area (cm²)		
1	1			
2	2			

Continue to increase the base and height of the triangular face.
Use a sheet of paper to draw each net if you need to.
The length of the prism decreases each time, so the net always fits on the piece of paper.

➤ When do you know that the table is complete?
➤ What patterns do you see in the table? Explain.
➤ From the table, what is the greatest volume of the prism? What are its dimensions?

Part 2

Use 0.5-cm grid paper.
Draw a graph of *Volume* against *Base of triangular face*.
Join the points.

➤ What inferences can you make from the graph?
➤ How can you find the greatest volume from the graph?

Part 3

Use a piece of thin card that measures 28 cm by 43 cm. Construct the net for your prism.
Cut out, then fold and tape the net to make the prism.

Take It Further

Does the prism with the greatest volume also have the greatest surface area? Write about what you find out.

Square Roots and Pythagoras

Some of the greatest builders are also great mathematicians. Geometry is their specialty. Look at the architecture on these pages.
What aspects of geometry do you see?

In this unit, you will develop strategies to measure distances that cannot be described exactly, using whole numbers, fractions, or decimals.

What You'll Learn

- Relate the area of a square to the length of its side.
- Understand that the square root of a non-perfect square is approximate.
- Estimate and calculate the square root of a whole number.
- Draw a circle, given its area.
- Investigate and apply the Pythagorean Theorem.

Why It's Important

The Pythagorean Theorem enables us to measure distances that would be impossible to measure using only a ruler. It enables a construction worker to make a square corner without using a protractor.

Key Words

- square number
- perfect square
- square root
- irrational number
- leg
- hypotenuse
- Pythagorean Theorem
- Pythagorean triple

321

Skills You'll Need

Areas of a Square and a Triangle

Area is the amount of surface a figure covers. It is measured in square units.

Example 1

Find the area of each figure.

a)

5 cm

b)

5 cm

4 cm

Solution

a) The figure is a square.

The area of a square is: $A = s^2$

Substitute: $s = 5$

$A = 5^2$

$\quad = 5 \times 5$

$\quad = 25$

The area is 25 cm².

b) The figure is a triangle.

The area of a triangle is: $A = \frac{1}{2}bh$

Substitute: $b = 4$ and $h = 5$

$A = \frac{1}{2}(4 \times 5)$

$\quad = \frac{1}{2}(20)$

$\quad = 10$

The area is 10 cm².

✔ Check

1. Find the area of each figure.

a)

6.5 cm

b)

3 cm

2 cm

c)

3 cm   3 cm

d)

4.5 cm

2.5 cm

Square Numbers

When we multiply a number by itself, we square the number.
We can use exponents to write a **square number**.
4^2 means: $4 \times 4 = 16$
We say, "Four squared is sixteen."
16 is a square number, or a **perfect square**.
One way to model a square number is to draw a square
whose area is equal to the square number.

Example 2

Show that 49 is a square number. Use symbols, words, and a diagram.

Solution

With symbols: $49 = 7 \times 7 = 7^2$
With words: "Seven squared is forty-nine."

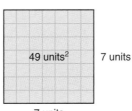

49 units²   7 units

7 units

✓ Check

2. Show that 36 is a square number. Use a diagram, symbols, and words.

3. Write each number in exponent form.
   a) 25   **b)** 81   **c)** 64   **d)** 169

4. List the first 15 square numbers.

5. Here are the first 3 triangular numbers.

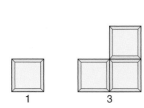 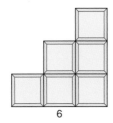

1   3   6

   a) Write the next 3 triangular numbers.
   b) Add consecutive triangular numbers. What do you notice? Explain.

Square Roots

Squaring and finding a **square root** are inverse operations.

For example, $7^2 = 49$ and $\sqrt{49} = 7$

We can model a square root with a diagram.

The *area* of a square shows the *square number*.

The *side length* of the square shows a *square root* of the square number.

We say, "A square root of 36 is 6."

We write: $\sqrt{36} = 6$

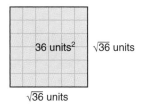

36 units²   √36 units

√36 units

Example 3

Find a square root of 64.

Solution

Method 1

Think of a number that, when multiplied by itself, produces 64.

$$8 \times 8 = 64$$

So, $\sqrt{64} = 8$

Method 2

Visualize a square with an area of 64 units².

Find its side length.

$$64 = 8 \times 8$$

So, $\sqrt{64} = 8$

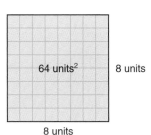

64 units²   8 units

8 units

✓ Check

6. Find each square root.

  a) $\sqrt{1}$ **b)** $\sqrt{25}$ **c)** $\sqrt{81}$ **d)** $\sqrt{9}$

  e) $\sqrt{16}$ **f)** $\sqrt{100}$ **g)** $\sqrt{121}$ **h)** $\sqrt{225}$

Constructing and Measuring Squares

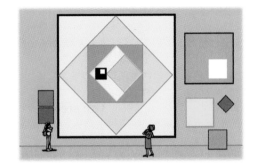

Focus | Use the area of a square to find the length of a line segment.

All squares are *similar*.
They come in many sizes, but always have the same shape.
How many ways can you describe a square?

Explore

Work with a partner.
You will need 1-cm grid paper.
Copy the squares below.
Without using a ruler, find the area and side length of each square.

What other squares can you draw on a 4 by 4 grid?
Find the area and side length of each square.
Write all your measurements in a table.

Reflect & Share

How many squares did you draw?
Describe any patterns in your measurements.
How did you find the area and side length of each square?
How did you write the side lengths of squares C and D?

Connect

We can use the properties of a square to find its area or side length.

Area of a square = length × length
$$= (\text{length})^2$$

When the side length is l, the area is l^2.
When the area is A, the side length is $\sqrt{A}$.

Example 1

A square has side length 10 cm.
What is the area of the square?

Solution

Area = (length)² or $A = l^2$
$$A = 10^2$$
$$= 100$$
The area is 100 cm².

Example 2

A square has area 81 cm².
What is the side length of the square?

Solution

Length = $\sqrt{\text{Area}}$ or $l = \sqrt{A}$
$$l = \sqrt{81}$$
$$= 9$$
The side length is 9 cm.

We can calculate the length of any segment on a grid by thinking of it as the side length of a square.

To find the length of the line segment AB:

Construct a square on the segment.
Find the area of the square.
Then, the length of the segment is the square root of the area.

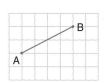

To construct a square on segment AB:
Rotate segment AB 90° counterclockwise about A, to get segment AC.
Rotate segment AC 90° counterclockwise about C, to get segment CD.
Rotate segment CD 90° counterclockwise about D, to get segment DB.

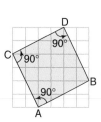

Here are two methods to find the area of the square.

Method 1

Draw an enclosing square and subtract areas.
Draw square EFGH along grid lines
so each vertex of ABDC lies on
one side of the enclosing square.

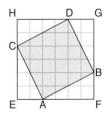

The area of EFGH = 6^2 units^2

$\qquad\qquad\quad = 36$ units^2

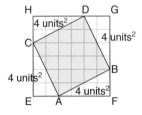

The triangles formed by the enclosing square are congruent.
Each triangle has area: $\frac{1}{2}(4)(2)$ units^2 = 4 units^2
So, the 4 triangles have area 4×4 units^2 = 16 units^2

The area of ABDC = Area of EFGH − Area of triangles

$\qquad\qquad\qquad = 36$ units^2 − 16 units^2

$\qquad\qquad\qquad = 20$ units^2

So, the side length of ABDC is: AB = $\sqrt{20}$ units

Method 2

Cut the square into smaller figures, then rearrange.
Cut and move two triangles to form a figure
with side lengths along grid lines.
Count squares to find the area.
The area of the new figure is 20 units^2.
So, the area of square ABDC = 20 units^2
And, the side length of the square, AB = $\sqrt{20}$ units

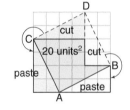

Since 20 is not a square number,
we cannot write $\sqrt{20}$ as a whole number.
Later in this unit, you will learn how to find an approximate value for
$\sqrt{20}$ as a decimal.

Practice

1. Simplify.

a) 3^2 b) $\sqrt{1}$ c) 4^2 d) $\sqrt{64}$

e) 7^2 f) $\sqrt{144}$ g) 10^2 h) $\sqrt{169}$

i) 6^2 j) $\sqrt{121}$ k) 12^2 l) $\sqrt{625}$

2. Copy each square on grid paper. Find its area.
Then write the side length of the square.

a) 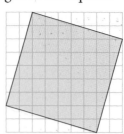 b)   c)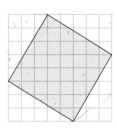

3. The area A of a square is given. Find its side length.
Which side lengths are whole numbers?

a) $A = 36$ cm² **b)** $A = 49$ m² **c)** $A = 95$ cm² **d)** $A = 108$ m²

4. Copy each segment on grid paper.
Draw a square on each segment.
Find the area of the square and the length of the segment.

a)   b)

c)   d)

5. The Great Pyramid at Giza is the largest pyramid in the world.
The area of its square base is about 52 441 m².
What is the length of each side of the base?

6. **Assessment Focus**
On square dot paper, draw a square with an area of 2 units².
Write to explain how you know the square does have this area.

Take It Further

7. Suppose you know the length of the diagonal of a square.
How can you find the side length of the square? Explain.

Reflect

How are square roots related to exponents?
How is the area of a square related to its side length?
How can we use this relationship to find the length of
a line segment? Include an example in your explanation.

Estimating Square Roots

Focus | Develop strategies for estimating a square root.

You know that the square root of a given number is a number which, when multiplied by itself, results in the given number; for example, $\sqrt{121} = \sqrt{11 \times 11}$
$$= 11$$

You also know that the square root of a number is the side length of a square with area that is equal to the number. For example, $\sqrt{9} = 3$

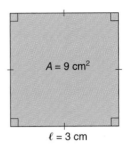

$A = 9\ cm^2$

$\ell = 3\ cm$

Explore

Work with a partner.
Use a copy of the number line below.
Place each square root on the number line to show its approximate value as a decimal: $\sqrt{2}, \sqrt{5}, \sqrt{9}, \sqrt{18}, \sqrt{24}$
Use grid paper if it helps.

0   1   2   3   4   5

Reflect & Share

Compare your answers with those of another pair of classmates.
What strategies did you use to estimate the square roots?
How could you use a calculator to check your square roots?

Here is one way to estimate the value of $\sqrt{20}$:

Find the square number closest to 20, but greater than 20.
The number is 25.
On grid paper, draw a square with area 25.
Its side length is: $\sqrt{25} = 5$
Find the square number closest to 20, but less than 20.
The number is 16.
Draw a square with area 16.
Its side length is: $\sqrt{16} = 4$
Draw the squares so they overlap.

A square with area 20 lies between these two squares.
Its side length is $\sqrt{20}$.
20 is between 16 and 25, but closer to 16.
$\sqrt{20}$ is between $\sqrt{16}$ and $\sqrt{25}$, but closer to $\sqrt{16}$.
So, $\sqrt{20}$ is between 4 and 5, but closer to 4.
An estimate of $\sqrt{20}$ is 4.4.

The *Example* illustrates another method to estimate $\sqrt{20}$.

Example

Use a number line and a calculator to estimate $\sqrt{20}$.

Solution

Think of the perfect squares closest to $\sqrt{20}$.

$\sqrt{20}$ is between 4 and 5, but closer to 4.
With a calculator, use guess and check to refine the estimate.
Try 4.4: $4.4 \times 4.4 = 19.36$ (too small)
Try 4.5: $4.5 \times 4.5 = 20.25$ (too large)
Try 4.45: $4.45 \times 4.45 = 19.8025$ (too small)
Try 4.46: $4.46 \times 4.46 = 19.8916$ (too small)
Try 4.47: $4.47 \times 4.47 = 19.9809$ (very close)
A close estimate of $\sqrt{20}$ is 4.47.

1. Copy this diagram on grid paper. Then estimate the value of $\sqrt{7}$.

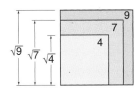

2. Use the number line below.
   a) Which placements are good estimates of the square roots? Explain your reasoning.
   b) Use the number line to estimate the value of each square root that is incorrectly placed.

3. Which two consecutive whole numbers is each square root between? How do you know?
   a) $\sqrt{5}$ b) $\sqrt{11}$ c) $\sqrt{57}$ d) $\sqrt{38}$ e) $\sqrt{171}$

4. Write five square roots whose values are between 9 and 10. Explain your strategy.

5. Is each statement true or false? Explain.
   a) $\sqrt{17}$ is between 16 and 18.
   b) $\sqrt{5} + \sqrt{5}$ is greater than $\sqrt{10}$.
   c) $\sqrt{131}$ is between 11 and 12.

6. Use guess and check to estimate the value of each square root. Record each trial.
   a) $\sqrt{23}$ b) $\sqrt{13}$ c) $\sqrt{78}$ d) $\sqrt{135}$ e) $\sqrt{62}$

To round a length in centimetres to the nearest millimetre, round to the nearest tenth.

7. Find the approximate side length of the square with each area. Give your answer to the nearest millimetre.
   a) 92 cm^2 b) 430 m^2 c) 150 cm^2 d) 29 m^2

8. A square garden has an area of 138 m^2.
   a) What are the approximate dimensions of the garden?
   b) About how much fencing would be needed to go around the garden?

9. Assessment Focus A student uses a 1-m square canvas for her painting. After framing, she wants her artwork to have an area twice the area of the canvas.
What are the dimensions of the square frame?
Show your work.

10. Most classrooms are rectangles.
Measure the dimensions of your classroom.
Calculate its area.
What if your classroom was a square.
What would its dimensions be?

Take It Further

11. A square carpet covers 75% of the area of a floor.
The floor is 8 m by 8 m.

8 m

8 m

a) What are the dimensions of the carpet?

b) What area of the floor is not covered by carpet?

12. Is the product of two perfect squares sometimes a perfect square? Always a perfect square?
Investigate to find out. Write about your findings.

A palindrome is a number that reads the same forward and backward.

13. a) Find the square root of each palindrome.

    i) $\sqrt{121}$ **ii)** $\sqrt{12\,321}$

    iii) $\sqrt{1\,234\,321}$ **iv)** $\sqrt{123\,454\,321}$

b) Continue the pattern.
Write the next 4 palindromes and their square roots.

Reflect

How can you find the perimeter of a square if you know its area?
What is your favourite method for estimating a square root of a number that is not a perfect square? Explain your choice.

Fitting In

HOW TO PLAY THE GAME:

Your teacher will give you 3 sheets of game cards. Cut out the 54 cards.

1. Place the 1, 5, and 9 cards on the table.
   Spread them out so there is room for several cards between them.
   Shuffle the remaining cards.
   Give each player six cards.

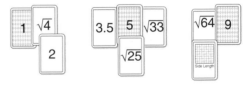

2. All cards laid on the table must be arranged from least to greatest. Take turns to place a card so it touches another card on the table.
   - It can be placed to the right of the card if its value is greater.
   - It can be placed to the left of the card if its value is less.
   - It can be placed on top of the card if its value is equal.
   - However, it cannot be placed between two cards that are already touching.

YOU WILL NEED

1 set of FITTING IN game cards; scissors

NUMBER OF PLAYERS

2 to 4

GOAL OF THE GAME

To get the lowest score

What other games could you play with these cards?
Try out your ideas.

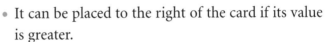

In this example, the $\sqrt{16}$ card cannot be placed because the 3.5 and the 5 cards are touching.
The player cannot play that card in this round.

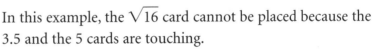

3. Place as many of your cards as you can. When no player can place any more cards, the round is over.
   Your score is the number of cards left in your hand.
   At the end of five rounds, the player with the lowest score wins.

Investigating Square Roots with a Calculator

Focus | Use a calculator to investigate square roots.

 ➤ We can use a calculator to calculate a square root.
To find a square root of 16:
On a calculator,
press: $\boxed{\sqrt{}}$ 16 $\boxed{)}$ $\boxed{\text{ENTER} \atop =}$ to display 4
A square root of 16 is 4.

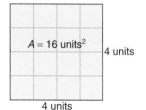

4 units

A = 16 units²

4 units

Check by multiplying.
Press: 4 $\boxed{\times}$ 4 $\boxed{\text{ENTER} \atop =}$ to display 16

If you use a different calculator,
what keystrokes do you use to find square roots?

➤ Many square roots are not whole numbers.
To find a square root of 20:
On a calculator,
press: $\boxed{\sqrt{}}$ 20 $\boxed{)}$ $\boxed{\text{ENTER} \atop =}$ to display 4.472135955
A square root of 20 is approximately 4.5.

Area = 20 cm²

√20 cm

➤ We investigate what happens when we check our answer.
Compare using a scientific calculator
with a 4-function calculator.
On a 4-function calculator, press: 20 $\boxed{\sqrt{}}$ $\boxed{\times}$ 20 $\boxed{\sqrt{}}$ $\boxed{=}$
What do you see in the display?
On a scientific calculator,
press: $\boxed{\sqrt{}}$ 20 $\boxed{)}$ $\boxed{\times}$ $\boxed{\sqrt{}}$ 20 $\boxed{)}$ $\boxed{\text{ENTER} \atop =}$
What do you see in the display?
Which display is accurate? How do you know?

➤ Check what happens when you enter
4.472135955 × 4.472135955 into both calculators.
What if you multiplied using pencil and paper.
Would you expect a whole number or a decimal? Explain.

Recall from *Unit 6* that π is another irrational number.

➤ $\sqrt{20}$ cannot be described exactly by a decimal.
The decimal for $\sqrt{20}$ never repeats and never terminates.
A number like $\sqrt{20}$ is called an **irrational number**.

When we know the area of a circle, we can use square roots to calculate its radius.

The area of a circle, A, is about $3r^2$.
So, r^2 is about $\frac{1}{3}$ of A, or $\frac{A}{3}$.
Similarly, the area $A = \pi r^2$; so $r^2 = \frac{A}{\pi}$
To find r, we take the square root.

So, $r = \sqrt{\dfrac{A}{\pi}}$
We can use this formula to calculate the radius of a circle when we know its area.

A circular rug has area 11.6 m².
To calculate the radius of the rug, use: $r = \sqrt{\dfrac{A}{\pi}}$
Substitute: $A = 11.6$

$r = \sqrt{\dfrac{11.6}{\pi}}$

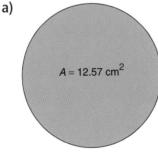
$A = 11.6 \text{ m}^2$
r

Use a calculator.
Key in: $\sqrt{\ }$ 11.6 $\div$ π $)$ ENTER =
to display 1.92156048

$r \doteq 1.92$
The radius of the rug is about 1.92 m, to the nearest centimetre.

✓ Check

1. Calculate the radius and diameter of each circle.
Give the answers to 1 decimal place.

a)

$A = 12.57 \text{ cm}^2$

b)

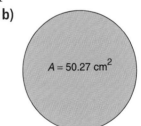
$A = 50.27 \text{ cm}^2$

c)

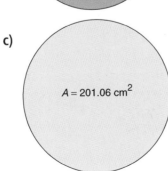
$A = 201.06 \text{ cm}^2$

d)

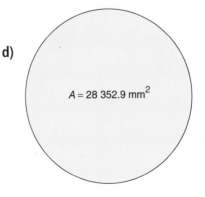
$A = 28\,352.9 \text{ mm}^2$

2. Draw each circle in question 1.

Mid-Unit Review

LESSON

8.1 **1.** Copy each square onto
1-cm grid paper.
  i) Find the area of each square.
  ii) Write the side length of each
square as a square root.
  iii) Which areas can be written
using exponents? Explain.
  a)   **b)**

  c)

2. a) The area of a square is 24 cm².
What is its side length?
  b) The side length of a square is
9 cm. What is its area?
  c) Explain the relationship
between square roots and
square numbers.
Use diagrams, symbols,
and words.

8.1
8.2 **3.** Copy this square onto
1-cm grid paper.

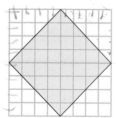

  a) What is the area of the square?

b) Write the side length of the
square as a square root.
c) Estimate the side length to the
nearest millimetre.

8.2 **4.** Between which two consecutive
whole numbers does each square
root lie? How do you know?
  a) $\sqrt{3}$ **b)** $\sqrt{65}$
  c) $\sqrt{57}$ **d)** $\sqrt{30}$

5. What is a square root of 100?
Use this fact to predict the square
root of each number.
Use a calculator to check.
  a) 900   **b)** 2500
  c) 400   **d)** 8100
  e) 10 000   **f)** 1 000 000

6. a) Draw a circle with area 113 cm².
  b) Does the circle in part a have an
area of exactly 113 cm²?
How do you know?

7. The opening of the fresh air intake
pipe for a furnace is circular.
Its area is 550 cm².
What are the radius and diameter of
the pipe? Give the answers to the
nearest millimetre.

8. The top of a circular concrete
footing has an area of 4050 cm².
What is the radius of the circle?
Give the answer to the nearest
millimetre.

8.3 The Pythagorean Relationship

Focus Discover a relationship among the side lengths of a right triangle.

In *Lesson 8.1*, you learned how to use the properties of a square to find the length of a line segment.

We will now use the properties of a right triangle to find the length of a line segment. A right triangle has two **legs** that form the right angle. The third side of the right triangle is called the **hypotenuse**.

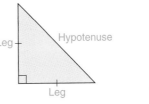

Isosceles right triangle   Scalene right triangle

Explore

Work on your own.
You will need grid paper.

➤ Copy segment AB.
  Find the length of the segment by drawing a square on it.
➤ Copy segment AB again.
  Draw a right triangle that has segment AB as its hypotenuse.
  Draw a square on each side.
  Find the area and side length of each square.
➤ Draw 3 different right triangles, with a square on each side.
  Find the area and side length of each square.
  Record your results in a table.

Use the corner of a sheet of paper or a protractor to check that the angles in the square are right angles.

	Area of Square on Leg 1	Length of Leg 1	Area of Square on Leg 2	Length of Leg 2	Area of Square on Hypotenuse	Length of Hypotenuse
Triangle 1						
Triangle 2						
Triangle 3						
Triangle 4						

Reflect & Share

Compare your results with those of another classmate.
What relationship do you see among the areas of the squares on the sides of a right triangle? How could this relationship help you find the length of a side of a right triangle?

Here is a right triangle, with a square drawn on each side.

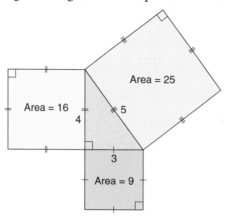

Area = 25

Area = 16

5

4

3

Area = 9

Later in this unit, we will use *The Geometer's Sketchpad* to verify that this relationship is true for all right triangles.

The area of the square on the hypotenuse is 25.
The areas of the squares on the legs are 9 and 16.
Notice that: $25 = 9 + 16$
This relationship is true for all right triangles:
The area of the square on the hypotenuse is equal to the sum of the areas of the squares on the legs.
This relationship is called the **Pythagorean Theorem**.
This theorem is named for the Greek mathematician, Pythagoras, who first wrote about it.

We can use this relationship to find the length of any side of a right triangle, when we know the lengths of the other two sides.

Example

Find the length of the unmarked side in each right triangle.
Give the lengths to the nearest millimetre.

a)

4 cm   4 cm

b)

10 cm

5 cm

Solution

a) The unmarked side is the hypotenuse. Label it h.

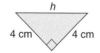

h

4 cm   4 cm

b) The unmarked side is a leg. Label it l.

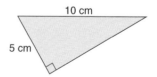

10 cm

5 cm

ℓ

The area of the square on the
hypotenuse is h^2.
The area of the squares on
the legs are 4^2 and 4^2.
So, $h^2 = 4^2 + 4^2$
Use the order of operations.
Square, then add.

$$h^2 = 16 + 16$$
$$h^2 = 32$$

The area of the square
on the hypotenuse is 32.
So, the side length of
the square is: $h = \sqrt{32}$
Use a calculator.
$h \doteq 5.6569$
So, the hypotenuse is
approximately 5.7 cm.

The area of the square on the
hypotenuse is 10^2.
The areas of the squares on
the legs are l^2 and 5^2.
So, $10^2 = l^2 + 5^2$
Square each number.
$100 = l^2 + 25$
To solve this equation,
subtract 25 from
each side.
$100 - 25 = l^2 + 25 - 25$
$$75 = l^2$$
The area of the square
on the leg is 75.
So, the side length of
the square is: $l = \sqrt{75}$
$l \doteq 8.66025$
So, the leg is approximately 8.7 cm.

Practice

1. The area of the square on each side of a triangle is given.
Is the triangle a right triangle? How do you know?

a)
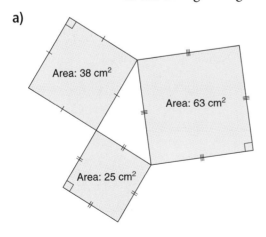

b)
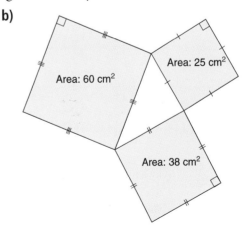

Write the first
6 multiples of
each number:
4, 9, 11, 12

Find the lowest
common multiple of
these numbers.

2. Find the length of the hypotenuse in each right triangle.

a)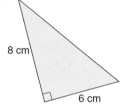

b)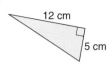

c)

d)

3. Find the length of the unmarked leg in each right triangle.

a)

b)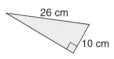

c)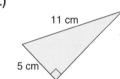

d)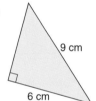

4. Find the length of the unmarked side in each right triangle.

a)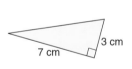

b)

c)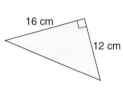

Math Link

History
"Numbers Rule the Universe!" That was the belief held
by a group of mathematicians called the Brotherhood
of Pythagoreans. Their power and influence became so
strong that fearful politicians forced them to disband.
Nevertheless, they continued to meet in secret and
teach Pythagoras' ideas.

The 3 whole-number side lengths of a right triangle are called a *Pythagorean triple*.

5. Look at the answers to questions 1 to 4, and at the triangles in *Connect*. Identify the triangles that have all 3 side lengths that are whole numbers.
  a) List the lengths of the legs and the hypotenuse for each of these triangles. Try to arrange the measures to show patterns.
  b) What patterns do you see? Explain the patterns.
  c) Extend the patterns. Explain your strategy.

6. Mei Lin uses a ruler and compass to construct a triangle with side lengths 3 cm, 5 cm, and 7 cm. Before Mei Lin constructs the triangle, how can she tell if the triangle will be a right triangle? Explain.

7. **Assessment Focus**
The hypotenuse of a right triangle is $\sqrt{18}$ units.
What are the lengths of the legs of the triangle?
How many different answers can you find?
Sketch a triangle for each answer. Explain your strategies.

8. On grid paper, draw a line segment with each length. Explain how you did it.
  a) $\sqrt{5}$ **b)** $\sqrt{10}$ **c)** $\sqrt{13}$ **d)** $\sqrt{17}$

Take It Further

9. Use grid paper.
Draw a right triangle with a hypotenuse with each length.
  a) $\sqrt{20}$ units   **b)** $\sqrt{89}$ units   **c)** $\sqrt{52}$ units

10. a) Sketch a right triangle with side lengths: 3 cm, 4 cm, 5 cm
  b) Imagine that each side is a diameter of a semicircle. Sketch a semicircle on each side.
  c) Calculate the area of each semicircle you drew. What do you notice? Explain.

Reflect

When you know the side lengths of a triangle, how can you tell if it is a right triangle? Use examples in your explanation.

Using *The Geometer's Sketchpad* to Verify the Pythagorean Theorem

Focus   Use a computer to investigate the Pythagorean relationship.

1. Open *The Geometer's Sketchpad.*
2. From the **Graph** menu, select **Show Grid**.
3. From the **Graph** menu, select **Snap Points**.

To construct right △ABC:

4. From the **Toolbox**, choose ⊡.
   Construct points at $(0, 0)$, $(0, 4)$, and $(3, 0)$.

   From the **Toolbox**, choose **A**.
   Click the 3 points in the order listed above to label them A, B, and C.

5. From the **Toolbox**, choose ⊡.
   Click the 3 points to highlight them.
   From the **Construct** menu, choose **Segments**.
   You now have a right triangle.
   Click and drag any vertex, and it remains a right triangle.

6. Click points A, B, and C.
   From the **Construct** menu, choose **Triangle Interior**.
   From the **Display** menu, choose **Color**.
   From the pull-down menu, pick a colour.
   Click the triangle to deselect it.

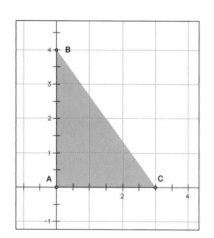

To construct a square on each side of △ABC:

7. From the **Toolbox**, choose ⊡.
   Double-click point B.
   The flash shows this is now a centre of rotation.
   Click point B, point C, and segment BC.
   From the **Transform** menu, choose **Rotate**.
   Enter 90 degrees. Click **Rotate**.

The minus sign in front of the angle measure means a clockwise rotation.

8. Click anywhere on the screen to deselect the rotated segment.
   Make sure points B and C, and segment BC, are still selected.
   Double-click point C to mark a new centre of rotation.
   From the **Transform** menu, choose **Rotate**.
   Enter −90 degrees. Click **Rotate**.

9. From the **Toolbox**, choose ⟋ .
   Join the points at the ends of the rotated segments
   to form a square on side BC.

10. From the **Toolbox**, choose **A** .
   Double-click one unlabelled vertex of the square. Type D. Click OK.
   Double-click the other unlabelled vertex. Type E. Click OK.

11. From the **Toolbox**, choose ↖ .
   Click the vertices of the square. If other points or segments are
   highlighted, deselect them by clicking them.
   From the **Construct** menu, choose **Quadrilateral Interior**.
   From the **Display** menu, choose **Color**.
   From the pull-down menu, pick a colour.

12. From the **Measure** menu, choose **Area**.
   The area of square BDEC appears.

13. Repeat *Steps 7* to *12* to construct and measure a square on each
   of the other two sides of the triangle.
   Decide the direction of rotation for each line segment;
   it could be 90 degrees or −90 degrees.
   Label the vertices F, G, and H, I.

14. Drag a vertex of the triangle and observe what happens to the
   area measurements.
   What relationship is shown?

To use *The Geometer's Sketchpad* calculator:

15. From the **Measure** menu, choose **Calculate**.
   Click the area equation for the smallest square.
   Click + .
   Click the area equation for the next smallest square.
   Click OK .

16. Drag a vertex of the triangle. How do the measurements change?
   How does *The Geometer's Sketchpad* verify the
   Pythagorean Theorem?

Use *The Geometer's Sketchpad* to investigate "what if" questions.

17. What if the triangle was not a right triangle.
   Is the relationship still true?

Communicating Solutions

The draft solution to a problem is often messy. To communicate a solution clearly, the draft solution must be tidy.

Communicating means talking, writing, drawing, or modelling to describe, explain, and justify your ideas to others. The draft solution is revised and edited for clear communication. The final solution presents only the steps needed to arrive at the answer. The steps are listed in the correct order.

Draft Solution

Represent Final Solution

Communicate Solutions – Represent, justify, and prove to others

Revise and edit draft solutions:
1. Show all the math information you *used*; that is, words, numbers, drawings, tables, graphs, and/or models.
2. Remove any information that you did not use.
3. Arrange the steps in a logical order.
4. Create a concluding statement.
5. Check the criteria to assess your communication.

Here are some criteria for good communication of math solutions:
- The solution is complete. All steps are shown. Other students can follow the steps and come to the same conclusion.
- The steps are in a logical order.
- The calculations are accurate.
- The spelling and grammar are correct.
- The math conventions are correct; for example, units, position of equal sign, labels and scales on graphs/diagrams, symbols, brackets, and so on.
- Where appropriate, more than one possible solution is described.
- The strategies are reasonable and are explained.
- Where appropriate, words, numbers, drawings, tables, graphs, and/or models are used to support the solution.
- The concluding statement fits the context and clearly answers the problem.

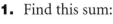

Reading and Writing in Math

✓ Check

Solve these problems. Share your work with a classmate for feedback and suggestions. Use the criteria as a guide.

1. Find this sum:
   23 + 25 + 27 + 23 + 25 + 27 + 23 +
   25 + 27 + 23 + 25 + 27 + 23 + 25 + 27
   Explain three different ways to solve
   this problem.

2. a) Six people met at a party.
      All of them exchanged handshakes.
      How many handshakes were there?
   b) How many different line segments can
      be named using the labelled points as end points? List them.

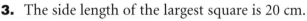

   A   B   C D   E   F

   c) How are these problems similar?

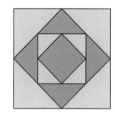

3. The side length of the largest square is 20 cm.
   a) What is the area of each purple section?
   b) What is the area of each orange section?
   Explain how you got your answers.

4. There are 400 students at a school.
   Is the following statement true?
      There will always be at least two students in the school
      whose birthdays fall on the same day of the year.
   Explain.

5. Camden has a custard recipe that needs 6 eggs, 1 cup of sugar,
   750 mL of milk, and 5 mL of vanilla. He has 4 eggs.
   He adjusts the recipe to use the 4 eggs.
   How much of each other ingredient will he need?

6. Lo Choi wants to buy a dozen doughnuts. She has a coupon.
   This week, the doughnuts are on sale for $3.99 a dozen.
   If Lo Choi uses the coupon, each doughnut is $0.35.
   Should Lo Choi use the coupon? Explain.

7. How many times in a 12-h period does the sum of the digits
   on a digital clock equal 6?

Applying the Pythagorean Theorem

Explore

Work with a partner.
Solve this problem:
A doorway is 2.0 m high and 1.0 m wide.
A square piece of plywood has side length 2.2 m.
Can the plywood fit through the door?
How do you know?
Show your work.

Reflect & Share

Compare your solution with that of another pair of classmates.
If the solutions are different, find out which is correct.
What strategies did you use to solve the problem?

Connect

Since the Pythagorean Theorem is true for all right triangles,
we can write an algebraic equation to describe it.

In the triangle at the right, the hypotenuse
has length c, and the legs have
lengths a and b.

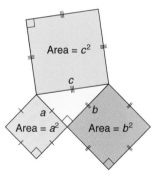

The area of the square on the
hypotenuse is $c \times c$, or c^2.

The areas of the squares on the legs
are $a \times a$ and $b \times b$, or a^2 and b^2.

So, we can say: $c^2 = a^2 + b^2$

When we use this equation,
remember that the lengths of the
legs are represented by a and b,
and the length of the hypotenuse by c.

Example 1

Find the length of each side labelled with a variable.
Give the lengths to the nearest millimetre.

a)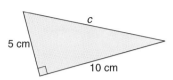

b)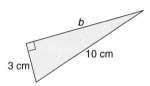

Solution

Use the Pythagorean Theorem: $c^2 = a^2 + b^2$

a) Substitute: $a = 5$ and $b = 10$

$c^2 = 5^2 + 10^2$

Square, then add.

$c^2 = 25 + 100$

$c^2 = 125$

The area of the square with side length c is 125.

So, $c = \sqrt{125}$

$c \doteq 11.180\ 34$

c is approximately 11.2 cm.

b) Substitute: $a = 3$ and $c = 10$

$10^2 = 3^2 + b^2$

Square, then add.

$100 = 9 + b^2$

Subtract 9 from each side to isolate b^2.

$100 - 9 = 9 + b^2 - 9$

$91 = b^2$

The area of the square with side length b is 91.

So, $b = \sqrt{91}$

$b \doteq 9.539\ 39$

b is approximately 9.5 cm.

Use a calculator to calculate each square root.

We can use the Pythagorean Theorem to solve problems that involve right triangles.

Example 2

A ramp has horizontal length 120 cm
and sloping length 130 cm.
How high is the ramp?

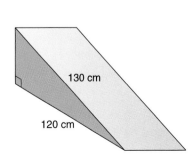

130 cm

120 cm

Solution

The height of the ramp is vertical,
so the front face of the ramp is a
right triangle.
The hypotenuse is 130 cm.

One leg is 120 cm.
The other leg is the height. Label it *a*.

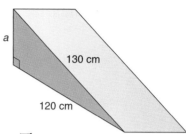

Use the Pythagorean Theorem.
$c^2 = a^2 + b^2$
Substitute: $c = 130$ and $b = 120$
$130^2 = a^2 + 120^2$ Use a calculator.
$16\,900 = a^2 + 14\,400$
Subtract 14 400 from each side to isolate a^2.
$16\,900 - 14\,400 = a^2 + 14\,400 - 14\,400$
$2500 = a^2$
The area of the square with side length *a* is 2500.
$$a = \sqrt{2500}$$
$$= 50$$
The ramp is 50 cm high.

Practice

1. Find the length of each hypotenuse labelled with a variable.

a)

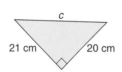

b)

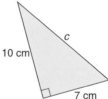

c)

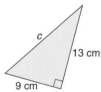

Calculator Skills

Suppose your calculator does not have a $\boxed{\sqrt{}}$ key.
How can you find $\sqrt{1089}$?

2. Find the length of each leg labelled with a variable.

a)

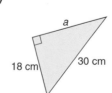

b)

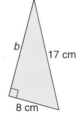

c)

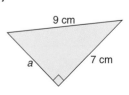

3. Find the length of each side labelled with a variable.

a)
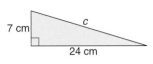

b)
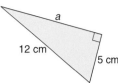

c)
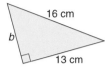

4. A 5-m ladder is leaning against a house.
It is 3 m from the base of the wall.
How high does the ladder reach?

5. Brandon constructed a right triangle with sides 10 cm and 24 cm.
   a) How long is the third side?
   b) Why are there two answers to part a?

6. Copy each diagram on grid paper.
Explain how each diagram illustrates the Pythagorean Theorem.

a)

b)

7. Alyssa has made a picture frame.
The frame is 60 cm long and 25 cm wide.
To check that the frame has square corners,
Alyssa measures a diagonal.
How long should the diagonal be?
Sketch a diagram to illustrate your answer.

8. The size of a TV set is described by the length of a diagonal
of the screen.
One TV is labelled as size 70 cm.
The screen is 40 cm high.
What is the width of the screen?
Draw a diagram to illustrate
your answer.

9. **Assessment Focus**

   Look at the grid.
   Without measuring, find
   another point that is the
   same distance from A as B is.
   Explain your strategy.
   Show your work.

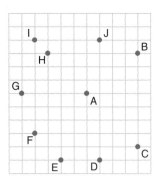

10. Joanna usually uses the sidewalk
    when she walks home from school.
    Today she is late, and so cuts
    through the field. How much
    shorter is Joanna's shortcut?

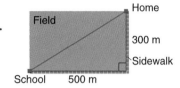

Take It Further

11. How high is the kite above the ground?

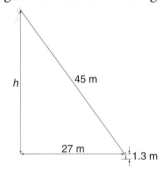

12. What is the length of the diagonal
    in this rectangular prism?

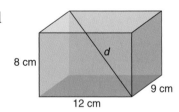

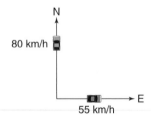

13. Two cars meet at an intersection.
    One travels north at an average speed of 80 km/h.
    The other travels east at an average speed of 55 km/h.
    How far apart are the cars after 3 h?

Reflect

When can you use the Pythagorean Theorem to solve a problem?
Use examples in your explanation.

Special Triangles

Focus | Apply the Pythagorean Theorem to isosceles and equilateral triangles.

Explore

Work with a partner.

An isosceles triangle has two equal sides.

Use this information to find the area of an isosceles triangle with side lengths 6 cm, 5 cm, and 5 cm.

Reflect & Share

Share your results with another pair of classmates.

Compare strategies.

How could you use the Pythagorean Theorem to help you find the area of the triangle?

Connect

To apply the Pythagorean Theorem to new situations, we look for right triangles within other figures.

➤ A square has four equal sides and four 90° angles.
   A diagonal creates two congruent isosceles right triangles.
   Any isosceles right triangle has two equal sides and angles of 45°, 45°, and 90°.

➤ An equilateral triangle has three equal sides and three 60° angles.
   A line of symmetry creates two congruent right triangles.
   Each congruent right triangle has angles 30°, 60°, and 90°.

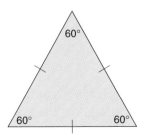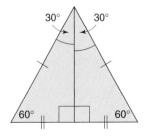

We can use the area of an equilateral triangle to find the surface area and volume of a hexagonal prism when the base is a regular hexagon.

Example

The base of a prism is a regular hexagon with side length 8 cm.
The length of the prism is 12 cm.

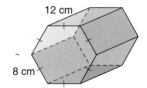

a) Find the area of the hexagonal base.
b) Find the volume of the prism.
c) Find the surface area of the prism.

Solution

a) The diagonals through the centre of a regular hexagon divide it into 6 congruent equilateral triangles.
    One of these triangles is △ABC.
    Draw the perpendicular from A to BC at D.
    AD bisects BC, so: BD = DC = 4 cm
    Label h, the height of △ABC.

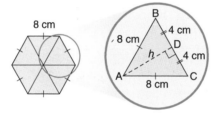

Use the Pythagorean Theorem in △ABD.
$$c^2 = a^2 + b^2$$
Substitute: $c = 8$, $a = 4$, $b = h$
$$8^2 = 4^2 + h^2$$
$$64 = 16 + h^2$$
$$64 - 16 = 16 + h^2 - 16$$
$$48 = h^2$$
So, $h = \sqrt{48}$
The height of △ABD is $\sqrt{48}$ cm.
The base of the triangle is 8 cm.
So, the area of △ABC $= \frac{1}{2} \times 8 \times \sqrt{48}$

There are 6 congruent triangles.

And, the area of the hexagon $= 6 \times \frac{1}{2} \times 8 \times \sqrt{48}$
$$= 24 \times \sqrt{48} \qquad \text{Use a calculator.}$$
$$\doteq 166.28$$
The area of the hexagonal base is approximately 166 cm².

b) The volume of the prism is: $V =$ base area $\times$ length
Use the exact value of the base area: $24 \times \sqrt{48}$
The length of the prism is 12 cm.

So, $V = 24 \times \sqrt{48} \times 12$ Use a calculator.
$\doteq 1995.32$
The volume of the prism is approximately 1995 cm³.

c) The surface area A of the prism is the sum of the areas of
the 6 rectangular faces and the two bases.
The rectangular faces are congruent.
So, $A = 6 \times (8 \times 12) + 2 \times (24 \times \sqrt{48})$
$\doteq 576 + 332.55$
$= 908.55$
The surface area of the prism is approximately 909 cm².

Practice

1. Find each length indicated.
Sketch and label the triangle first.

a)

13 cm h 5 cm

b)

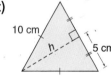

3 cm c

c)

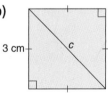

10 cm h 5 cm

2. Find each length indicated.
Sketch and label the triangle first.

a)

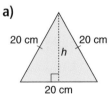

20 cm   20 cm h 20 cm

b)

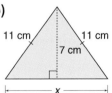

11 cm   11 cm   7 cm x

c)

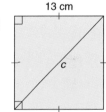

13 cm c

3. Find the area of each triangle.

a)

5 cm

b)

7 cm   4 cm   7 cm

c)

8 cm   5 cm

4. A prism has a base that is a
regular hexagon with side length 6 cm.
The prism is 14 cm long.

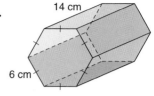

14 cm

6 cm

a) Find the area of the base
of the prism.
b) Find the volume of the prism.
c) Find the surface area of the prism.

5. (Assessment Focus)

Here is a tangram.
Its side length is 10 cm.

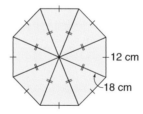

a) What is the area of Figure F?
How long is each side of the square?
b) What is the perimeter of Figure B?
c) What is the perimeter of Figure D?
d) How can you use Figure D to find
the perimeter of Figure E?
Show your work.

Number Strategies

A rectangular pool has
length 12 m and width 7 m.
A circular pool has the
same area as the
rectangular pool.
What is the circumference
of the circular pool?

6. Here is one base of an octagonal prism.
The prism is 30 cm long.

12 cm

18 cm

a) Find the volume of the prism.
b) Find the surface area of the prism.

Take It Further

7. Find the area and perimeter
of this right isosceles triangle.

12 cm

Reflect

How can the Pythagorean Theorem be used in isosceles and
equilateral triangles?
Include examples in your explanation.

Unit Review

Review any lesson with

online tutorial

What Do I Need to Know?

✓ **Side Length and Area of a Square**
The side length of a square is equal
to the square root of its area.
Length $= \sqrt{\text{Area}}$
Area $= (\text{Length})^2$

$A = 16\ cm^2$

4 cm

✓ **The Pythagorean Theorem**
In a right triangle, the area of the square
on the hypotenuse is equal to the sum of
the areas of the squares on the two legs.
$c^2 = a^2 + b^2$
Use the Pythagorean Theorem to find
the length of a side in a right triangle,
when two other sides are known.

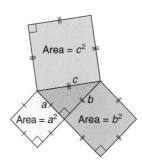

Area $= c^2$

c

a b

Area $= a^2$ Area $= b^2$

What Should I Be Able to Do?

For extra practice, go to page 495.

LESSON

8.1
8.2

1. Estimate each square root to the
nearest whole number.
a) $\sqrt{6}$ b) $\sqrt{11}$
c) $\sqrt{26}$ d) $\sqrt{35}$
e) $\sqrt{66}$ f) $\sqrt{86}$

2. Estimate each square root to
1 decimal place.
Show your work.
a) $\sqrt{55}$ b) $\sqrt{75}$
c) $\sqrt{95}$ d) $\sqrt{105}$
e) $\sqrt{46}$ f) $\sqrt{114}$

3. Use a calculator to write each
square root to 1 decimal place.
a) $\sqrt{46}$ b) $\sqrt{84}$
c) $\sqrt{120}$ d) $\sqrt{1200}$

4. A square blanket has an area of
16 900 cm². How long is each
side of the blanket?

8.3 **5.** Find the length of the unmarked side in each right triangle.

a)

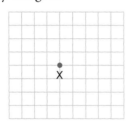
30 cm   16 cm

b)

35 cm
21 cm

c)

19 cm   25 cm

6. There is buried treasure at one of the points of intersection of the grid lines shown below.
Copy the grid.

X

The treasure is $\sqrt{13}$ units from the point marked X.
a) Where might the treasure be? Explain how you located it.
b) Could there be more than one position? Explain.

8.4 **7.** A boat travels due east at an average speed of 10 km/h.
At the same time, another boat travels due north at an average speed of 12 km/h.
After 2 h, how far apart are the boats? Explain your thinking.

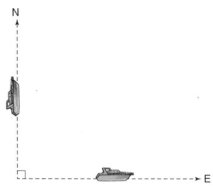
N

E

8.5 **8.** Find the perimeter of △ABC.

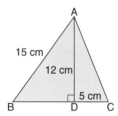
A
15 cm   12 cm
5 cm
B   D   C

9. Find the area of an equilateral triangle with side length 15 cm.

10. Here is one base of a pentagonal prism. It comprises five isosceles triangles, with the measures given. The prism is 7 cm long.

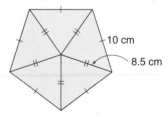
10 cm
8.5 cm

a) Sketch the prism.
b) Find the surface area of the prism.
c) Find the volume of the prism.

Practice Test

1. a) What is the area of square ABCD?
    b) What is the length of line segment AB?
       Explain your reasoning.

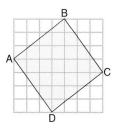

2. Find the side length of a square that has the same area
    as this rectangle.

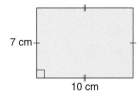

7 cm

10 cm

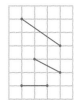

3. Draw these 3 line segments on 1-cm grid paper.
    a) Find the length of each line segment to the nearest millimetre.
    b) Could these segments be arranged to form a triangle?
       If your answer is no, explain why not.
       If your answer is yes, could they form a right triangle?
       Explain.

4. A parking garage has ramps from one level to the next.
    a) How long is each ramp?
    b) What is the total length of the ramps?

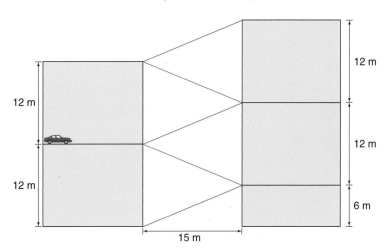

12 m

12 m

12 m

12 m

12 m

6 m

15 m

Throughout the ancient world, mathematicians were fascinated by right triangles. You will explore some of their discoveries.

Ancient Greece, 400 B.C.E.
Theodorus was born about 100 years after Pythagoras.
Theodorus used right triangles to create a spiral.
Today it is known as the Wheel of Theodorus.

➤ Follow these steps to draw the Wheel of Theodorus. You will need a ruler and protractor. Your teacher will give you a copy of a 10-cm ruler.

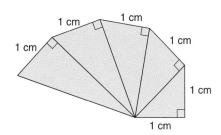

Step 1 Draw a right triangle with legs 1 cm.
Step 2 The hypotenuse of this triangle is one leg of the next triangle.
Draw the other leg of the next triangle 1 cm long.
Draw the hypotenuse.
Step 3 Repeat *Step 2* until you have at least ten triangles.

➤ Use the Pythagorean Theorem to find the length of each hypotenuse.
Label each hypotenuse with its length as a square root.
What patterns do you see?

➤ Use a ruler to measure the length of each hypotenuse to the nearest millimetre.
Use the copy of the 10-cm ruler. Mark the point on the ruler that represents the value of each square root.
Compare the two ways to measure the hypotenuse.
What do you notice?

➤ Without using a calculator or extending the Wheel of Theodorus, estimate $\sqrt{24}$ as a decimal.
Label $\sqrt{24}$ cm on your ruler.
Explain your reasoning and any patterns you see.

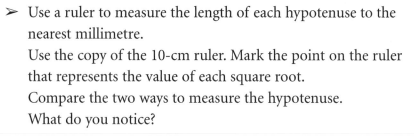

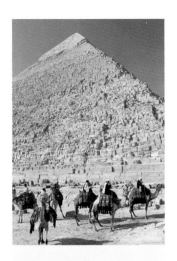

Ancient Egypt, 2000 B.C.E.

In ancient Egypt, the Nile River overflowed every year and
destroyed property boundaries.
Because the land plots were rectangular,
the Egyptians needed a way to mark a right angle.

The Egyptians tied 12 evenly spaced
knots along a piece of rope and
made a triangle from it.
Explain how you think the
Egyptians used the knotted rope
to mark a right angle.

Ancient Babylon, 1700 B.C.E.

Archaeologists have discovered evidence that the ancient
Babylonians knew about the Pythagorean Theorem over
1000 years before Pythagoras!

The archaeologists found
this tablet.

When the tablet is translated,
it looks like this.

What do you think the diagram on the tablet means?
Explain your reasoning.

Reflect on the Unit

What is the Pythagorean Theorem? How is it used?
Include examples in your explanation.

UNIT

1 **1.** Evaluate.

a) $3.8 + 5.7 \div 1.9$

b) $2.4^2 - (4.2 - 3.7)^2$

c) $(1.5 + 4.2) + 2.8 \times 7.2$

d) $1.5 + (4.2 + 2.8) \times 7.2$

2 **2.** Which is the better buy?

a) 8 cheese slices for $1.49 or 24 cheese slices for $3.29

b) 1.89 L of cranberry-raspberry drink for $3.27 or 3.78 L for $5.98

c) 100 g of iced tea mix for $0.29 or 500 g for $1.69

d) 1 can of chicken soup for $0.57 or a 12-pack for $5.99

3 **3.** Use linking cubes.

a) Build the object for the set of views below.

b) Sketch the object on isometric dot paper.

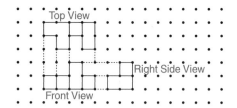

4 **4.** Write as many division questions as you can that have each fraction below as their quotient.

a) $\frac{1}{2}$ b) $\frac{2}{3}$ c) $\frac{4}{5}$ d) $\frac{5}{6}$

Write about the strategies you used to find the division questions.

5 **5.** In each case, identify the collected data as from a sample or a census. Justify your answer.

a) To find the mean number of chocolate chips in a cookie, every 100th cookie was tested.

b) To find what type of movie the family wanted to rent, all 6 family members were asked.

c) To find the number of Ontario families that enjoy skiing, one in 10 households was surveyed by phone.

d) To find out if the city should increase funding for snow removal, a questionnaire was enclosed in every tax bill.

6. Adam recorded the heights of a bean plant and sunflower plant.

Time (days)	Height (cm)	
	Bean	Sunflower
0	2	6
3	6	8
6	9	11
9	13	14
12	20	17
15	28	21
18	35	26
21	41	33
24	47	38
27	56	42
30	63	48
33	68	53

a) Display the data using the most suitable method. Justify your choice.

b) Describe any trends in the graph.

c) Predict the height of each plant after 39 days. Explain the method you used to make your prediction.

7. Draw a circle. Label its centre C. Choose two points G and H on the circle that are not the endpoints of a diameter. Join CG, GH, and CH. What type of triangle is △CGH? How do you know?

8. Which has the greater area: a circle with circumference 1 m or a circle with radius 30 cm? Justify your answer.

9. A cardboard tube is used to send a poster by mail. The tube is 0.8 m long with diameter 7 cm. The ends of the tube are closed with tape. What is the area of cardboard in the tube?

10. a) Name the complement of ∠ABE.

b) Name the supplement of ∠ABE.

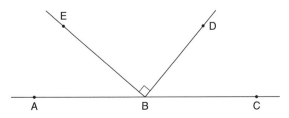

11. Look at this diagram.

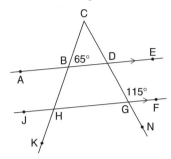

a) Name two pairs of:
 i) interior angles
 ii) alternate angles
 iii) corresponding angles

b) Find the measures of ∠CHG, ∠JHK, ∠CGH, and ∠FGN.

c) Find the measure of ∠BCD. What kind of triangle is △BCD?

12. Estimate each square root to 1 decimal place. Show your work. Then check with a calculator.
 a) $\sqrt{52}$ **b)** $\sqrt{63}$
 c) $\sqrt{90}$ **d)** $\sqrt{76}$

13. The area of the square on each side of a triangle is given. Is the triangle a right triangle? How do you know?
 a) 16 cm², 8 cm², 30 cm²
 b) 16 cm², 8 cm², 24 cm²

14. The dimensions of a rectangle are 3 cm by 4 cm. What is the length of a diagonal? Explain your reasoning.

UNIT 9

Integers

In golf, the 2003 US Masters Championship was held in Georgia, USA. Mike Weir, a Canadian golfer, tied with Len Mattiace after 72 holes. Here are seven players, in alphabetical order, and their leader board entries.

- What was Weir's leader board entry?
- Order the entries from greatest to least.
- What other uses of integers do you know?

Player	Over/Under Par
Jim Furyk	−4
Retief Goosen	+1
Jeff Maggert	−2
Phil Mickelson	−5
Vijay Singh	−1
Mike Weir	−7
Tiger Woods	+2

Par for the tournament is 288.
Jim Furyk shot 284.
His score in relation to par is 4 under, or −4.

What You'll Learn

- Compare and order integers.
- Add, subtract, multiply, and divide integers.
- Use the order of operations with integers.
- Graph integers on a grid.
- Graph transformations on a grid.
- Solve problems involving integers.

Why It's Important

We use integers in everyday life, when we deal with weather, finances, sports, geography, and science.

Key Words

- positive integer
- negative integer
- opposite integers
- zero pair
- rational number
- coordinate grid
- quadrants
- ordered pair
- *x*-axis
- *y*-axis
- origin

Skills You'll Need

Comparing and Ordering Integers

Positive integers, such as +5, +9, +1, are greater than 0.
Negative integers, such as −5, −9, −1, are less than 0.
Positive and negative integers can be shown on a number line:

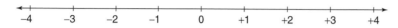

Any positive integer is greater than any negative integer.
For example, +1 is greater than −1000.

We use the symbols > and < to show order.

−3 is to the left of +1; so, −3 is less than +1, and we write: −3 < +1

+3 is to the right of −4; so, +3 is greater than −4, and we write: +3 > −4

Example 1

Order these integers from least to greatest: +5, −6, +3, −8, 0, −1, +8

Solution

+5, −6, +3, −8, 0, −1, +8
Sketch a number line from −8 to +8. Mark a point on the line for each integer.

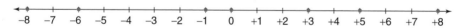

For least to greatest, read the integers from left to right: −8, −6, −1, 0, +3, +5, +8

✓ Check

1. Sketch a number line. Use it to order these integers:
0, +2, −1, +4, −5, −7, +10

2. Copy each statement. Use < or > to show which number in each pair is greater. Use a number line if it helps.
a) +2 ☐ +8   **b)** 0 ☐ −5   **c)** −7 ☐ 0
d) +250 ☐ −251 **e)** −100 ☐ −70 **f)** −361 ☐ −360

Using Models to Add Integers

On a number line, **opposite integers** are the same distance from 0, but are on opposite sides of 0.

+3 and −3 are opposite integers.

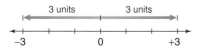

We can use coloured tiles to model integers.

One red tile models −1.   One yellow tile models +1.

+1 and −1 are opposite integers.
They combine to form a **zero pair**.
$(+1) + (−1) = 0$
The sum of any two opposite integers is 0 because they form a zero pair:
$(+3) + (−3) = 0$ because +3 and −3 form 3 zero pairs.

We can use zero pairs to add integers with opposite signs.
Recall that when we add 0 to a number, the sum is equal to the number.

Example 2

Add: $(+5) + (−3)$

Solution

Model +5 with 5 yellow tiles:
Model −3 with 3 red tiles:
Circle zero pairs.
There are 2 yellow tiles left. They represent +2.
So, $(+5) + (−3) = +2$

To add integers with the same sign, we combine the tiles
that represent the integers.
Then, we count the tiles.

Example 3

Add: $(-6) + (-4)$

Solution

Model -6 with 6 red tiles:

Model -4 with 4 red tiles:

There are 10 red tiles altogether. They represent -10.
So, $(-6) + (-4) = -10$

✓ Check

3. Add.

 a) $(+3) + (+5)$ **b)** $(-8) + (-11)$ **c)** $(+6) + (+3)$
 d) $(-5) + (-6)$ **e)** $(+5) + (+1)$ **f)** $(-3) + (-6)$
 g) $(+4) + (-2)$ **h)** $(-8) + (+5)$ **i)** $(-5) + (+8)$
 j) $(-4) + (+2)$ **k)** $(-9) + (+9)$ **l)** $(-7) + (+2)$

4. Write each scenario as a sum of two integers.
Then, find the sum to answer each question.

 a) The temperature was $-5°C$. It then rose $8°C$.
   What was the final temperature?

 b) Keera earned $8 and spent $6. How much money did Keera have left?

Using Models to Subtract Integers

To subtract integers, we model the first integer with coloured tiles,
then remove the tiles that represent the second integer.
When we do not have enough tiles to remove, we can add zero pairs.

Example 4

Subtract: $(-3) - (-8)$

Solution

$(-3) - (-8)$
Model -3 with 3 red tiles:

To take away −8, we need 5 more red tiles.

Add 5 zero pairs of tiles.

That is, add 5 red tiles and 5 yellow tiles. They represent 0.

Take away 8 red tiles.

5 yellow tiles remain. They represent +5.

So, $(-3) - (-8) = +5$

Example 5

Subtract: $(+2) - (-9)$

Solution

$(+2) - (-9)$

Model +2 with 2 yellow tiles:

To take away −9, we need 9 red tiles.

Add 9 zero pairs of tiles.

That is, add 9 red tiles and 9 yellow tiles. They represent 0.

Take away 9 red tiles.

11 yellow tiles remain. They represent +11.

So, $(+2) - (-9) = +11$

✓ Check

5. Subtract.

a) $(+1) - (+5)$ b) $(+4) - (+1)$ c) $(-5) - (-9)$

d) $(-8) - (-1)$ e) $(+10) - (+5)$ f) $(-10) - (-3)$

6. Subtract.

a) $(+3) - (-8)$ b) $(+7) - (-2)$ c) $(-9) - (+3)$

d) $(-5) - (-11)$ e) $(+8) - (-8)$ f) $(-5) - (+5)$

Focus Use a number line and calculator to add integers.

You have used coloured tiles to add integers.
We will now investigate other ways to add.

Explore

Work with a partner.
Use these integers: $+5, -9, -16, +28, -34, +41$
Choose two integers. Add them.
Choose two different integers. Add them.
Repeat this activity.
Add as many different pairs of integers as you can.
Sketch a number line to show each sum.
Write each addition equation.

Reflect & Share

Compare your addition equations with those of another
pair of classmates.
How can you add two integers when their signs are the same?
When their signs are different?

Connect

Recall how to use a number line to add integers.
The integer $+11$ is represented by an arrow, 11 units long,
pointing right.
The integer -5 is represented by an arrow, 5 units long, pointing left.

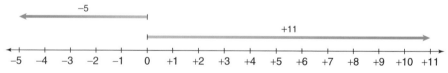

To add a positive integer, move right on a number line.

To add: $(-5) + (+11)$
Start at -5 on a number line. Move 11 units right.

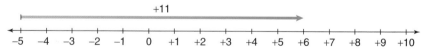

The arrow head is at $+6$.
So, $(-5) + (+11) = +6$

To add a negative integer, move left on a number line.

And, to add: $(-12) + (-3)$

Start at -12 on a number line and move 3 units left.

$(-12) + (-3) = -15$

When you use a calculator to add,
the keystrokes depend on the type of calculator.
When your calculator has this key $\boxed{(-)}$,
you use it to input the negative sign.
When your calculator has this key $\boxed{+/-}$,
you use it to change an input number to a negative number.

When you key in a positive number, you do not need to key in the positive sign.

Example 1

Use a calculator to add. $(-325) + (-428)$

Solution

$(-325) + (-428)$

For a calculator with $\boxed{(-)}$:

Input:

$\boxed{(-)}$ 325 $\boxed{+}$ $\boxed{(-)}$ 428 $\boxed{=}$

to display -753

So, $(-325) + (-428) = -753$

For a calculator with $\boxed{+/-}$:

Input:

325 $\boxed{+/-}$ $\boxed{+}$ 428 $\boxed{+/-}$ $\boxed{=}$

to display -753

Example 2

The temperature in Calgary, Alberta, was $-2°C$. A chinook came through and the temperature rose $15°C$. At nightfall, it fell $7°C$.

a) Write an addition expression to show the temperature changes.

b) Use a number line to find the final temperature.

Solution

a) The addition expression is: $(-2) + (+15) + (-7)$

b) $(-2) + (+15) + (-7) = +6$

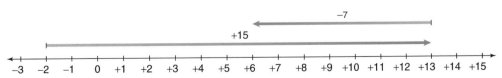

The final temperature is $+6°C$.

1. Use a number line to add.
   a) $(+5) + (-6)$ **b)** $(-8) + (+6)$
   c) $(-2) + (-4)$ **d)** $0 + (-5)$
   e) $(-2) + (-4)$ **f)** $(-5) + (+5)$

2. Use a number line to add.
   a) $(-3) + (+4) + (-6)$ **b)** $(+3) + (-5) + (+7)$
   c) $(+6) + (-8) + (-1)$ **d)** $(-10) + (+6) + (-2)$

3. a) Write the opposite integer.
   i) $+8$ **ii)** -5 **iii)** $+2$ **iv)** -8
   b) Add each integer to its opposite in part a.
   c) What do you notice about the sum of two opposite integers?

4. a) Add.
   i) $(+8) + (+6)$ **ii)** $(+3) + (+7)$
   iii) $(+5) + (+9)$ **iv)** $(+1) + (+12)$
   b) Look at the integer expressions and sums in part a.
   How are they related?
   How can you use this relationship to add two positive integers
   without a number line or a calculator?
   c) Check the relationship in part b by adding two positive integers
   of your choice.

5. a) Add.
   i) $(-8) + (-6)$ **ii)** $(-3) + (-7)$
   iii) $(-5) + (-9)$ **iv)** $(-1) + (-12)$
   b) Look at the integer expressions and sums in part a.
   How are they related?
   How can you use this relationship to add two negative integers
   without a number line or a calculator?
   c) Check the relationship in part b by adding two negative integers
   of your choice.

6. a) Find 4 pairs of integers that have the sum -5.
   b) Find 4 pairs of integers that have the sum $+4$.

Number Strategies

A jacket costs $125.

Its price is reduced by 15%. Then, 15% sales tax is applied.

Is the final price $125? Explain.

7. Assessment Focus When you add two integers with opposite signs, the sum may be 0, a positive integer, or a negative integer. When you look at an addition expression, how can you tell which of these sums it will be, without adding? Include examples in your explanation.

8. Add.

a) $(+513) + (-182)$ b) $(+560) + (-266)$

c) $(+793) + (-1089)$ d) $(-563) + (+182) + (+363)$

e) $(-412) + (+382) + (-79)$ f) $(-114) + (+483) + (-293)$

9. The value of a stock on the Toronto Stock Exchange has changed each week for six weeks as shown below:

Week 1	Week 2	Week 3	Week 4	Week 5	Week 6
Up $5	Down $6	Down $2	Up $4	Up $6	Down $2

a) Write an integer addition expression to represent the change in the value of the stock at the end of Week 6.

b) At the end of Week 6, how does the value of the stock compare with its value at the beginning of Week 1?

c) At the beginning of Week 1, the stock was worth $40. How much was the stock worth at the end of Week 3? At the end of Week 6?

Take It Further

10. Find each missing integer.

a) $-5 = (-2) + \square$ b) $\square + (-8) = +2$

11. Use only single-digit integers. How many ways can you complete each equation? How do you know you have found all possible ways?

a) $\square + \triangle = -2$ b) $\square + \triangle = -4$

Reflect

Write an addition expression that has each sum: a positive sum, a negative sum, and a sum of 0. Show how to calculate each sum.

Focus Use a calculator and number line to subtract integers.

Explore

Work with a partner.
You will need a calculator.

- Choose two positive integers between $+150$ and $+300$.
  Use a calculator to subtract them.
  Show the subtraction on a number line.
- Repeat the activity for a positive integer between $+150$ and $+250$, and a negative integer between -150 and -300.
- Repeat the activity for two negative integers between -150 and -300.

Reflect & Share

Compare expressions and number lines with those of another pair of classmates.
Suppose two integers are subtracted in reverse order.
What happens to their difference? Explain.
List the keystrokes to subtract two negative integers.

Connect

To subtract two integers, first think about how we subtract two whole numbers.
For example, to subtract: $13 - 6$,
we think, "What do we add to 6 to get 13?"
The answer is 7, so: $13 - 6 = 7$
To subtract two integers: $(-6) - (+13)$,
we think, "What do we add to $+13$ to get -6?"

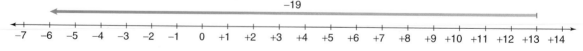

We add -19; that is, $(+13) + (-19) = -6$
So, $(-6) - (+13) = -19$
We also know that: $(-6) + (-13) = -19$

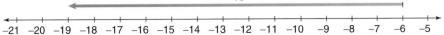

This example shows that subtracting an integer is the same as adding the opposite integer.

$$(-6) - (+13) = -19 \qquad\qquad (-6) + (-13) = -19$$

Subtract $+13$.   Add -13.

Example 1

Use a number line to subtract.

a) $(+14) - (+30)$ **b)** $(-18) - (-12)$

Solution

You can use coloured tiles to check.

To subtract, add the opposite.

a) Write $(+14) - (+30)$ as $(+14) + (-30)$.
Use a number line.
Start at $+14$. Move 30 units left.

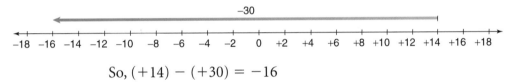

So, $(+14) - (+30) = -16$

b) Write $(-18) - (-12)$ as $(-18) + (+12)$.
Use a number line.
Start at -18. Move 12 units right.

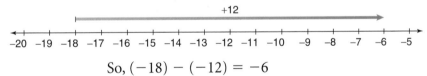

So, $(-18) - (-12) = -6$

Example 2

The mean temperature in January for Victoria, BC, is $+4°C$.
In Thunder Bay, the mean January temperature is $-11°C$.
Which temperature is lower and by how much?

Solution

Since $-11 < +4$, then $-11°C$ is the lower temperature.
Subtract the temperatures to find their difference.

$$(-11) - (+4) = (-11) + (-4)$$
$$= -15$$

The temperature in Thunder Bay is $15°C$ lower than the temperature in Victoria.

Some integer expressions require addition and subtraction.

Example 3

Evaluate. $(+5) + (-3) - (+7)$

Solution

$(+5) + (-3) - (+7)$ Write the subtraction as an addition.

$= (+5) + (-3) + (-7)$ Add the first two integers.

$= (+2) + (-7)$ Then add.

$= -5$

So, $(+5) + (-3) - (+7) = -5$

We can use a calculator to subtract directly, without adding the opposite.

To use a calculator to subtract: $(-137) - (+542)$

For a calculator with $\boxed{(-)}$:   For a calculator with $\boxed{+/-}$:

Input: $\boxed{(-)}137\boxed{-}542\boxed{=}$ Input: $137\boxed{+/-}\boxed{-}542\boxed{=}$

to display -679 to display -679

Practice

1. Rewrite as an addition statement. Then evaluate.
   a) $(+8) - (+4)$ **b)** $(-13) - (-8)$ **c)** $(-5) - (-5)$
   d) $(+20) - (-16)$ **e)** $(+30) - (-13)$ **f)** $(+21) - (-18)$

2. Use a number line to subtract.
   a) $(+7) - (+5)$ **b)** $(+9) - (-3)$
   c) $(-11) - (-4)$ **d)** $(-14) - (+8)$

3. Subtract.
   a) $(+4) - (+8)$ **b)** $(-9) - (-5)$ **c)** $(-7) - (+1)$
   d) $(+10) - (-3)$ **e)** $(+5) - (-5)$ **f)** $(-18) - (-3)$

 4. Subtract.
   a) $(-256) - (+125)$ **b)** $(-103) - (-214)$
   c) $(+213) - (+133)$ **d)** $(+148) - (-222)$

5. For each scenario:

  i) Write each number as an integer.

  ii) Subtract the second integer from the first integer.
  Explain each answer.

  a) A temperature 7°C above zero and
    a temperature 5°C below zero

  b) A temperature 15°C below zero and
    a temperature 8°C below zero

  c) A height 51 m above sea level and
    a depth 17 m below sea level

  d) A golf score of 2 over par and a golf score of 6 under par

  e) A rise of $21 in the value of a stock, then a fall of $14

Number Strategies

Write the next 3 terms
in each pattern.

Write each pattern rule.

- 1, 4, 9, 16, …

- 1, 2, 4, 7, 11, …

- 1, 8, 27, 64, …

6. The table shows the mean temperatures in January and July
for several cities in a certain year.

  a) Find the difference between the temperatures in July and
  January for each city. Show your work.

	City	January Temperature (°C)	July Temperature (°C)
i)	Victoria	+6	+21
ii)	Miami, US	+22	+31
iii)	Winnipeg	−18	+20
iv)	Perth, Australia	+25	+9
v)	Calgary	−4	+25

  b) Which city has the greatest difference in temperatures?
  How do you know?

  c) In Perth, why is the July temperature less than the
  January temperature?

7. **Assessment Focus** For each integer below, write a subtraction
expression that has this integer as its answer.

  a) −3   **b)** +2   **c)** 0

Where possible, do this 4 different ways:

- positive integer − positive integer
- positive integer − negative integer
- negative integer − positive integer
- negative integer − negative integer

Show your work.

8. Use the table below.

City	Record High Temperature (°C)	Record Low Temperature (°C)
Halifax, NS	+37	−29
Regina, SK	+43	−50
Thunder Bay, ON	+40	−41
Victoria, BC	+36	−16

a) Which city has the greatest record high temperature?
   The least record low temperature?
b) Find the difference between the temperatures for each city.
c) Which city has the greatest difference in temperatures?
d) What is the median record high temperature?
e) What is the range of record low temperatures?
f) Make up your own problem about these temperatures.
   Solve your problem.
Use a calculator to check your answers.

9. Evaluate.
a) $(-2) - (-8) - (+4)$ b) $(+5) - (-1) - (-3)$
c) $(+10) - (+3) - (-7)$ d) $(-5) - (+8) - (+6)$
e) $0 + (-5) + (+8) + (-3)$ f) $(-42) + (-65) - (+28)$
g) $(-1) - (+2) - (+3) - (+4)$ h) $(-241) - (+356) + (-5)$

10. Write the next 3 terms in each pattern; then, write
the pattern rule.
a) $+5, +12, +19, \ldots$ b) $-4, -2, 0, \ldots$
c) $-21, -17, -13, \ldots$ d) $+1, 0, -1, \ldots$

Take It Further

11. a) Find two integers with a sum of -12 and a difference of $+2$.
   b) Create and solve a similar integer problem.

Reflect

When you subtract two integers, the answers can be
positive, negative, or zero.
How can you predict the type of answer before you subtract?
Use examples in your explanation.

Focus   Add and subtract integers.

The whole numbers are 0, 1, 2, 3, 4, …, and so on.
The integers are …, −4, −3, −2, −1, 0, 1, 2, 3, 4, …, and so on.
We do not need to include the + sign to indicate a positive integer.
So, all whole numbers are integers.

Up until now, when we added and subtracted integers,
the integers were written in brackets.
For example, $(-3) + (+7)$ and $(-5) - (+8)$

We will now interpret sums and differences of numbers
such as $-3 + 7$ and $-5 - 8$.

Explore

Work with a partner.
Use a number line to find each answer.

$(+3) + (-7)$ and $3 - 7$
$(-3) + (-7)$ and $-3 - 7$
$(+3) + (+7)$ and $3 + 7$
$(-3) + (+7)$ and $-3 + 7$

What patterns do you see?
Write your own two sets of expressions like these, then evaluate.

Reflect & Share

Compare solutions with those of another pair of classmates.
How could you find each answer without a number line?

Connect

➤ To add: $-6 + 4$, use a number line.

$-6 + 4 = -2$
We can also use mental math.
One number is negative: -6
One number is positive: $+4$

The answer to $-6 + 4$ is the difference of 6 and 4, and the sign of the difference matches the sign of -6, the numerically larger number:
$-6 + 4 = -2$

➤ To subtract: $4 - 10$, use a number line.

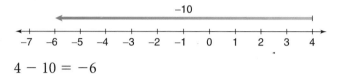

$4 - 10 = -6$

Use mental math.
One number is positive: 4
One number is negative: -10
The answer to $4 - 10$ is the difference of 10 and 4, and the sign of the difference matches the sign of -10, the numerically larger number:
$4 - 10 = -6$

➤ To evaluate: $-4 - 10$, use a number line.

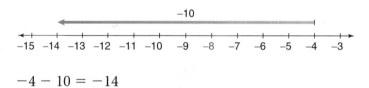

$-4 - 10 = -14$

Use mental math.
Both numbers are negative.
The answer is the sum of 4 and 10, and the sign of the answer matches the sign of both numbers.
So, $-4 - 10 = -14$

Example

Evaluate. $8 - 6 - 9$

Solution

$8 - 6 - 9$

$8 - 6 = 2$
So, $8 - 6 - 9 = 2 - 9$
$= -7$

1. Evaluate.
   a) $-5 + 7$ b) $3 + 4$ c) $-3 + 2$
   d) $-6 + 8$ e) $-11 + 13$ f) $-21 + 36$
   g) $5 - 12$ h) $-12 - 4$ i) $-6 - 9$
   j) $11 - 13$ k) $-5 - 18$ l) $15 - 3$

2. Evaluate.
   a) $6 - 1 + 3$ b) $-36 + 6 - 3$ c) $18 - 15 - 2$

3. **Assessment Focus**
   a) Evaluate each pair of expressions.
      i) $-6 + 4$; $4 - 6$ ii) $-7 + 3$; $3 - 7$
      iii) $-8 + 2$; $2 - 8$ iv) $-9 + 1$; $1 - 9$
   b) What patterns do you see in the expressions and the answers?
      Explain why these patterns occur.
   c) Write two more expressions that are related in the same way
      as those in part a.
   Show your work.

4. Evaluate each term. Each pattern continues.
   Write the next 3 terms in each pattern.
   Explain how you found each answer.
   a) $(-6 + 5), (-7 + 4), (-8 + 3), \ldots$
   b) $(-3 - 1), (-4 - 2), (-5 - 3), \ldots$

5. Last Monday, Suneel had $283 in her bank account.
   On Tuesday, Suneel withdrew $120 in cash, and wrote
   a cheque for $200. On Thursday, Suneel deposited $53.
   How much money did Suneel have in her account then?
   Show your work.

Number Strategies

Order from least to greatest.

$3.2, \frac{11}{3}, 3.6, \frac{13}{4}, 3.02$

Reflect

Do you prefer to use a number line or mental math
to evaluate an expression involving integers?
Use an example in your explanation.

Focus Use patterns to develop the rules for multiplying integers.

We can write the number of tiles in this array in two ways.

As a sum:
5 + 5 + 5 = 15

As a product:
(+3) × (+5) = +15

How can you use integers to write the number of tiles in this array in two ways?

In *Explore*, you will use patterns to find products such as (−3) × (+5) and (−3) × (−5), which cannot be represented as arrays.

Explore

Work on your own.
Your teacher will give you a large copy of this multiplication table.

Start at the bottom right of the table.
Multiply the positive integers.
Then complete the bottom left of the table.
Multiply a positive integer by a negative integer.
Use patterns or any method you wish to complete the top right, then the top left of the table.

Second Number

×	−5	−4	−3	−2	−1	0	+1	+2	+3	+4	+5
−5											
−4											
−3											
−2											
−1											
0											
+1											
+2											
+3											
+4											
+5											

First Number

Reflect & Share

Compare your completed table with that of a classmate.
Use the patterns in your table.
How could you multiply any negative integer by a positive integer?
How could you multiply any two negative integers?

These properties of whole numbers are also properties of integers.

Multiplying by 0

$3 \times 0 = 0$ and $0 \times 3 = 0$

So, $(-3) \times 0 = 0$ and $0 \times (-3) = 0$

Multiplying by 1

$3 \times 1 = 3$ and $1 \times 3 = 3$

So, $(-3) \times (+1) = -3$ and $(+1) \times (-3) = -3$

Order Property

$3 \times 4 = 12$ and $4 \times 3 = 12$

So, $(-3) \times (+4) = -12$ and $(+4) \times (-3) = -12$

Distributive Property

$$3 \times (4 + 5) = 3 \times 4 + 3 \times 5$$
$$= 12 + 15$$
$$= 27$$

So, $(+3) \times [(-4) + (-5)] = [(+3) \times (-4)] + [(+3) \times (-5)]$
$$= (-12) + (-15)$$
$$= -27$$

From the introduction on page 380, you know that

$$(+3) \times (+5) = +15$$

and $\quad (+3) \times (-5) = -15$

We can use the order property to show that:

Since $(+3) \times (-5) = -15$, then $(-5) \times (+3) = -15$

We can use the distributive property to investigate the product of two negative integers.

Here are two ways to calculate: $(-5) \times [(+3) + (-3)]$

Method 1

$(-5) \times [(+3) + (-3)]$
$= (-5) \times (+3) + (-5) \times (-3)$

$= \quad (-15) + (-5) \times (-3)$

Method 2

$(-5) \times [(+3) + (-3)]$

$= (-5) \quad \times \quad (0)$
$= 0$

These answers must be equal.

So, $(-15) + (-5) \times (-3) = 0$
But: $(-15) + (+15) = 0$
So, $(-5) \times (-3) = +15$
We could do this with any pair of opposite integers
in the square brackets.
So, the product of two negative integers is positive.

The results above can be used to write these rules for
multiplying integers:
- The product of two integers with the same sign is positive.
  That is, $(+7) \times (+6) = +42$ and $(-7) \times (-6) = +42$
- The product of two integers with opposite signs is negative.
  That is, $(+8) \times (-9) = -72$ and $(-8) \times (+9) = -72$

To multiply more than two integers, we use the order of operations.
That is, we multiply the integers in pairs, in the order they appear.

Example

Find each product.
a) $(+3) \times (-6) \times (-2)$
b) $(-2) \times (-4) \times (-5)$

Solution

a) $(+3) \times (-6) \times (-2)$ Multiply the first two integers.

$= (-18) \times (-2)$ Then, multiply.
$= +36$

b) $(-2) \times (-4) \times (-5)$ Multiply the first two integers.

$= (+8) \times (-5)$ Then, multiply.
$= -40$

When we write the product of integers, we do not need to write
the multiplication sign.
That is, we may write $(-8) \times (-9)$ as $(-8)(-9)$.
And, $(+3) \times (-6) \times (-2)$ may be written as $(+3)(-6)(-2)$.

1. Will each product be positive or negative? How do you know?
   a) $(-6) \times (+2)$ b) $(+6) \times (+4)$
   c) $(+4) \times (-2)$ d) $(-7) \times (-3)$

2. Find each product.
   a) $(+8)(-3)$ b) $(-5)(-4)$ c) $(-3)(+9)$
   d) $(+7)(-6)$ e) $(+10)(-3)$ f) $(-7)(-6)$
   g) $(0)(-8)$ h) $(+10)(-20)$ i) $(-14)(-30)$

3. Find each product.
   a) $(-1)(-8)(-2)$
   b) $(-11)(-12)(-1)$
   c) $(-1)(-1)(-1)(-1)(-1)$
   d) $(-2)(-3)(-4)(-5)$

4. Copy each equation.
   Replace $\square$ with an integer to make the equation true.
   a) $(+5) \times \square = +20$ b) $\square \times (-9) = +27$
   c) $(-9) \times \square = -54$ d) $\square \times (-3) = +18$
   e) $\square \times (+5) = -20$ f) $\square \times (-12) = +144$
   g) $\square \times (-6) = +180$ h) $(+3) \times \square \times (-4) = +24$

5. Write the next 3 terms in each pattern.
   Then write the pattern rule.
   a) $+1, +2, +4, +8, \ldots$ b) $+1, -6, +36, \ldots$
   c) $-1, +3, -9, \ldots$ d) $-4, -8, -12, \ldots$

6. a) Find the product of each pair of integers.
   i) $(+3)(-7)$ and $(-7)(+3)$ ii) $(+4)(+8)$ and $(+8)(+4)$
   iii) $(-5)(-9)$ and $(-9)(-5)$ iv) $(-6)(+10)$ and $(+10)(-6)$
   b) Use the results of part a. Does the order in which
   integers are multiplied affect the product? Explain.

7. Use these integers: $-5, +9, -8, +4, -2$
   a) Which two integers have the greatest product?
   b) Which two integers have the least product?
   c) How do you know there is not a greater product
   or a lesser product?

8. Assessment Focus

a) Find each product. Then use a calculator to extend the pattern 4 more rows.
  i) $(-2)(-3)$ **ii)** $(-2)(-3)(-4)$
  iii) $(-2)(-3)(-4)(-5)$ **iv)** $(-2)(-3)(-4)(-5)(-6)$

b) Use the results of part a.
  i) What is the sign of a product when it has an even number of negative factors? Explain.
  ii) What is the sign of a product when it has an odd number of negative factors? Explain.

c) Investigate what happens when a product has positive and negative factors. Do the rules in part b still apply? Explain.

9. Explain why the product of an integer multiplied by itself can never be negative.

Take It Further

10. How many different ways can you write -36 as the product of two or more integer factors?

11. When you multiply two natural numbers, the product is never less than either of the two numbers. Is the same statement true for the product of any two integers? Investigate, then write what you find out.

12. The product of two integers is -144. The sum of the integers is -7. What are the two integers?

13. The product of two integers is between $+160$ and $+200$. One integer is between -20 and -40.
  a) What is the greatest possible value for the other integer?
  b) What is the least possible value for the other integer?

Number Strategies

Estimate each square root to 1 decimal place.
$\sqrt{58}$, $\sqrt{47}$, $\sqrt{83}$, $\sqrt{31}$
Do not use the $\boxed{\sqrt{}}$ key on a calculator.

The natural numbers are 1, 2, 3, 4, …, and so on.

Reflect

Suppose your friend missed this lesson.
How would you explain to her how to multiply two integers?
Use examples in your explanation.

Dividing Integers

Recall that, for any multiplication fact with two different factors, you can write two related division facts. For example:

$9 \times 7 = 63$

So, $63 \div 9 = 7$ and, $63 \div 7 = 9$

We can apply the same rules to the product of two integers.

Explore

Work with a partner.

Write each product below as many different ways as you can.

For each product, write two related division facts.

➤ Write 75 as the product of two positive integers.

➤ Write 126 as the product of two negative integers.

➤ Write -72 as the product of a negative integer and a positive integer.

➤ Write -80 as the product of a positive integer and a negative integer.

Reflect & Share

Compare your division facts with those of another pair of classmates. Work together to develop rules for:

• dividing two positive integers
• dividing two negative integers
• dividing a negative integer by a positive integer
• dividing a positive integer by a negative integer

Connect

To divide integers, we use the fact that division is the inverse of multiplication.

➤ We know that: $(+5) \times (+3) = +15$

So, $(+15) \div (+5) = +3$ and $(+15) \div (+3) = +5$

      ↓   ↓   ↓

    dividend   **divisor** **quotient**

When the dividend and divisor are positive, the quotient is positive.

➤ We know that: $(-5) \times (+3) = -15$

So, $(-15) \div (+3) = -5$ and $(-15) \div (-5) = +3$

When the dividend is negative and the divisor is positive, the quotient is negative.

When both the dividend and divisor are negative, the quotient is positive.

➤ We know that: $(-5) \times (-3) = +15$

So, $(+15) \div (-5) = -3$ and $(+15) \div (-3) = -5$

When the dividend is positive and the divisor is negative, the quotient is negative.

The results above are true for all integers related in the different ways illustrated.

We can use these results to write rules for dividing integers:
- The quotient of two integers with the same sign is positive. That is, $(+56) \div (+8) = +7$ and $(-56) \div (-8) = +7$
- The quotient of two integers with opposite signs is negative. That is, $(+63) \div (-9) = -7$ and $(-63) \div (+9) = -7$

A division expression can be written with a division sign: $(-48) \div (-6)$; or as a fraction: $\frac{-48}{-6}$

When the expression is written as a fraction, we do not need to use brackets.

Example

Divide.

a) $(-100) \div (-20)$ b) $\frac{-30}{+5}$

Solution

a) Since the signs are the same, the quotient is positive.
$(-100) \div (-20) = +5$

b) Since the signs are different, the quotient is negative.
$\frac{-30}{+5} = -6$

In *Example*, part b, the integer -6 can be written as $-\frac{6}{1}$. When the integer is written this way, it is written as a **rational number**.

Practice

1. Copy and continue each pattern until you have 8 rows.
   Which rules for division of integers does each pattern illustrate?

   a) $(-12) \div (+3) = -4$
   $(-9) \div (+3) = -3$
   $(-6) \div (+3) = -2$
   $(-3) \div (+3) = -1$

   b) $(+25) \div (-5) = -5$
   $(+15) \div (-3) = -5$
   $(+5) \div (-1) = -5$
   $(-5) \div (+1) = -5$

   c) $(+8) \div (+2) = +4$
   $(+6) \div (+2) = +3$
   $(+4) \div (+2) = +2$
   $(+2) \div (+2) = +1$

   d) $(+14) \div (+7) = +2$
   $(+10) \div (+5) = +2$
   $(+6) \div (+3) = +2$
   $(+2) \div (+1) = +2$

   e) $(-14) \div (+7) = -2$
   $(-10) \div (+5) = -2$
   $(-6) \div (+3) = -2$
   $(-2) \div (+1) = -2$

   f) $(-10) \div (-5) = +2$
   $(-5) \div (-5) = +1$
   $(0) \div (-5) = 0$
   $(+5) \div (-5) = -1$

2. **a)** Use each multiplication fact to find a related quotient.

   i) Given $(+8) \times (+3) = +24$,
   find $(+24) \div (+3) = \square$

   ii) Given $(-5) \times (-9) = +45$,
   find $(+45) \div (-9) = \square$

   iii) Given $(-7) \times (+4) = -28$,
   find $(-28) \div (+4) = \square$

   iv) Given $(+11) \times (-6) = -66$,
   find $(-66) \div (+11) = \square$

   b) For each division fact in part a, write a related division fact.

3. Divide.

   a) $(+12) \div (-6)$ **b)** $(-9) \div (-3)$ **c)** $\frac{-20}{-5}$

   d) $\frac{+21}{-7}$ **e)** $(-32) \div (-8)$ **f)** $(-144) \div (+12)$

   g) $(-250) \div (+10)$ **h)** $0 \div (-8)$ **i)** $(+125) \div (+5)$

4. Nirmala borrowed $7 every day.
   She now owes $56.
   For how many days did Nirmala borrow money?
   a) Write this problem as a division expression using integers.
   b) Solve the problem.

5. Write the next three terms in each pattern.
What is each pattern rule?
a) $-3, +9, -27, \ldots$
b) $+6, -12, +18, -24, \ldots$
c) $+5, +20, -10, -40, +20, +80, \ldots$
d) $-64, +32, -16, \ldots$
e) $+100\,000, -10\,000, +1000, \ldots$

6. **Assessment Focus** Suppose you divide two integers.
When is the quotient:
a) less than both integers?
b) greater than both integers?
c) between the two integers?
d) equal to $+1$?
e) equal to -1?
f) equal to 0?
Use examples to illustrate your answers.
Show your work.

 7. Divide.
a) $(+624) \div (-52)$
b) $(-2231) \div (-23)$
c) $(-1344) \div (+16)$
d) $(-2068) \div (-47)$

Take It Further

8. Evaluate.
a) $(-32) \div (+4) \div (-2)$
b) $(-81) \div (-9) \div (-9)$
c) $(+56) \div (-4) \div (-2)$

9. Find as many examples as you can of three different 1-digit numbers that are all divisible by $+2$ and have a sum of $+4$.

Reflect

How do you divide two integers?
Include an example for each different possible division.

Mid-Unit Review

9.1 **1. a)** Write these integers in order from least to greatest:
$+20, -4, -6, +13, 0, +2, -1$

b) Show these integers on a number line.

2. Use a number line to add.
a) $(-5) + (-7)$
b) $(-10) + (+7)$
c) $(-5) + (+12)$
d) $(-3) + (+5) + (-4)$

3. a) Add: $(-18) + (+5)$
b) Find three other pairs of integers that have the same sum as part a.

9.2 **4.** Evaluate.
a) $(-7) - (+2)$
b) $(-5) - (-2)$
c) $(+4) - (-3)$
d) $(-41) - (-17)$
e) $(-3) - (+4) + (-5)$

5. Evaluate.
a) $(-146) - (-571)$
b) $(-365) + (-198) - (+118)$

6. The price of a new car is $27 599. The value of the car decreases by $2600 each year for five years. What will the value of the car be after five years?
a) Write an integer expression to represent this problem.
b) Solve the problem.

9.3 **7.** Evaluate.
a) $10 - 8 - 11$ **b)** $-3 + 5 + 9$
c) $-11 - 10 - 9$ **d)** $12 + 15 - 3$

9.4 **8.** Multiply.
a) $(+8) \times (-4)$
b) $(-120) \times (-10)$
c) $(-4) \times (+7)$
d) $(+6)(-12)$
e) $(5)(0)(-1)$
f) $(-4)(-8)(-1)$

9.3
9.4 **9.** Use integers to answer each question. Show your work.
a) The temperature drops 5°C, then drops another 3°C. What was the total drop in temperature?
b) A swimming pool drains at a rate of 35 L/min for 30 min. How much water drained out of the pool?
c) The price of a house rose $25 000, dropped $28 999, then rose $14 500. What was the total change in the price of the house?

9.5 **10.** Divide.
a) $(-81) \div (+9)$
b) $(-12) \div (-6)$
c) $0 \div (-9)$
d) $(+650) \div (-25)$
e) $(-1288) \div (-28)$
f) $(-100) \div (-100)$

Recall the order of operations with whole numbers.

- Do the operations in brackets first.
- Do any work with exponents.
- Multiply and divide, in order, from left to right.
- Add and subtract, in order, from left to right.

It may help you to remember the order of operations when you think BEDMAS.

The same order of operations applies to all integers.

Explore

Work with a partner.
Use these integers: $-2, 4, -24, 7$
Use any operations or brackets.
Write the expression that has the greatest value.

Reflect & Share

Share your expression with that of another pair of classmates.
If the expressions are different, check that the expression with the greater answer is correct.
Work together to write an expression that has the least value.

Connect

Since we use curved brackets to show an integer; for example, (-2), we will use square brackets to group terms.

Example 1

Evaluate. $100 - 3[20 \div (-2)]$

Solution

$100 - 3[20 \div (-2)]$

$= 100 - 3 \times [20 \div (-2)]$ Do the operation in brackets first.

For clarity, we write positive integers as whole numbers.

$= 100 - 3(-10)$

$= 100 + (-3)(-10)$ Multiply.

$= 100 + (+30)$ Add.

$= 130$

When an expression is written as a fraction,
the fraction bar has two meanings.

- The fraction bar indicates division.
- The fraction bar acts like brackets. That is, the operations in the numerator and denominator must be done before dividing the numerator by the denominator.

Example 2

Evaluate.

a) $\dfrac{4 \times (-8) + 2}{-6}$ b) $2 - (-3)^2$

Solution

a) $\dfrac{4 \times (-8) + 2}{-6}$ Multiply first.

$= \dfrac{(-32) + 2}{-6}$ Add the integers in the numerator.

$= \dfrac{-30}{-6}$ Divide.

$= 5$

b) $2 - (-3)^2$ Do the exponent first:
                                               $(-3)^2$ means $(-3)(-3)$

$= 2 - (-3)(-3)$ Multiply.

$= 2 - (+9)$ Subtract.

$= -7$

Practice

1. a) Evaluate.

     i) $12 \div (2 \times 3) - 2$ **ii)** $12 \div 2 \times (3 - 2)$

  b) Why are the answers different? Explain.

2. Evaluate. State which operation you do first.

  a) $(+7)(+4) - (+5)$ **b)** $(+6)[(+2) + (-5)]$

  c) $(-3) + (+4)(+7)$ **d)** $(-6) + (+4) \times (-2)$

  e) $(+15) \div [(+10) \div (-2)]$ **f)** $(+18) \div (-6) \times (+2)$

3. Evaluate.

  a) $(-1)^3$ **b)** $(-3)^2 + 9$ **c)** $(5)^3 \times (-4)$

  d) $\dfrac{(-2)^3}{-4}$ **e)** $\dfrac{(-6)(-8)}{4}$ **f)** $\dfrac{(-12)(-3)}{-6}$

4. Evaluate. Show all the steps.

a) $(-3)(-2) + 4$

b) $(-8)(-2) + (-1)$

c) $3(-4) - 2$

d) $-2(5 + 3)$

e) $10 \div 2 + 4 \times (-3)$

f) $\dfrac{(-7)(4) + 8}{(-2)^2}$

5. **Assessment Focus** Robert, Brenna, and Christian got different answers for this problem: $-40 - 2[(-8) \div 2]$
Robert's answer was -32, Christian's answer was -48, and Brenna's answer was 168.

a) Which student had the correct answer?

b) Show and explain how the other two students got their answers. Where did they go wrong?

6. Which expression has a value that is closest to -500? Explain.
$(-2)^2 \times (-100) \div 4 \times 5$
$376 \div 4 \times (-5)$
$(-1360) \div 8 \times (-3)$

7. Keisha had $405 in her bank account.
Over the summer, she made 4 withdrawals of $45 each.
What is the balance in Keisha's account?
Write an integer expression to represent this problem.
Solve the problem.

8. The daily highest temperatures for one week in February were:
$-2°C, +5°C, -8°C, -4°C, -11°C, -10°C, -5°C$
Find the mean temperature.

Number Strategies

Write each number in expanded form.
- 654
- 7258
- 83 507
- 901 472

Take It Further

9. Write an expression for each statement.
Evaluate each expression.

a) Divide the sum of -24 and 4 by -5.

b) Multiply the sum of -4 and 10 by -2.

c) Subtract 4 from -10, then divide by -2.

Reflect

Make up an integer expression that has three operations.
Evaluate the expression. Show your work.

Focus Locate and graph points in four quadrants on a coordinate grid.

You have plotted points with whole-number coordinates on a grid.
Point A has coordinates (3, 2).
What are the coordinates of point B? Point C? Point D?

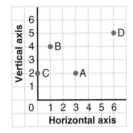

A vertical and a horizontal number line intersect at right angles at 0.
This produces a grid on which you can plot points with integer coordinates.

Explore

Work with a partner.
You will need grid paper and a ruler.
Copy this grid.

Draw a figure on the grid.
Make sure there is at least one vertex in each of the 4 parts on the grid.
Each vertex should be at a point where grid lines meet.
Label each vertex with a letter and the coordinates of the point.
List the vertices, with their coordinates, in order.
Trade lists with your partner.
Use the list to draw your partner's figure.

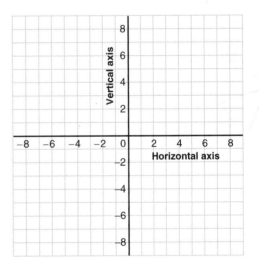

Reflect & Share

Compare the figures you and your partner drew.
If they do not match, try to find which figure is incorrect, and why.

A vertical number line and a horizontal number line that intersect at right angles at 0 form a **coordinate grid**.

The horizontal axis is the **x-axis**.

The vertical axis is the **y-axis**.

The axes meet at the **origin**.

The axes divide the plane into four **quadrants**.

They are numbered counterclockwise.

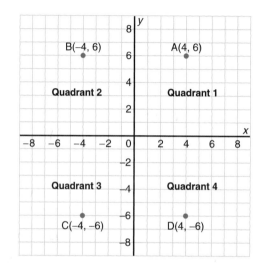

We do not need arrows on the axes.

A pair of coordinates is called an ordered pair.

We do *not* include a + sign for a positive coordinate.

In Quadrant 1, point A has coordinates $(4, 6)$.

In Quadrant 2, point B has coordinates $(-4, 6)$.

In Quadrant 3, point C has coordinates $(-4, -6)$.

In Quadrant 4, point D has coordinates $(4, -6)$.

Math Link

History

René Descartes lived in the 17th century.
He developed the coordinate grid.
It is named the Cartesian grid in his honour.
There is a story that René was lying in bed and watching a fly on the ceiling.
He invented coordinates as a way to describe the fly's position.

Example

a) Write the coordinates of each point.

   i) P   **ii)** Q   **iii)** R   **iv)** S

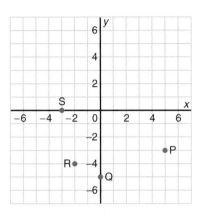

b) Plot each point on a grid.

   i) D$(-1, 3)$ **ii)** E$(-3, -5)$ **iii)** F$(0, -2)$ **iv)** G$(-4, 0)$

Solution

Remember, first move left or right, then up or down.

a) Start at the origin each time.

   i) To get to P, move 5 units right and 3 units down.
   So, the coordinates of P are $(5, -3)$.

   ii) To get to Q, move 0 units right and 5 units down.
   So, the coordinates of Q are $(0, -5)$

   iii) To get to R, move 2 units left and 4 units down.
   So, the coordinates of R are $(-2, -4)$.

   iv) To get to S, move 3 units left and 0 units down.
   So, the coordinates of S are $(-3, 0)$.

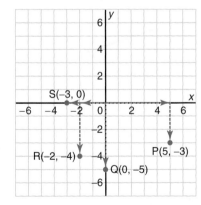

b) **i)** D(−1, 3)

Start at −1 on the *x*-axis.

Move 3 units up. Mark point D.

ii) E(−3, −5)

Start at −3 on the *x*-axis.

Move 5 units down. Mark point E.

iii) F(0, −2)

Start at the origin.

Move 2 units down the *y*-axis. Mark point F.

iv) G(−4, 0)

Start at −4 on the *x*-axis.

Since there is no movement up or down,

point G lies on the *x*-axis. Mark point G.

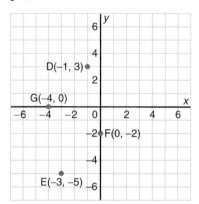

Practice

1. Write the coordinates of each point from A to K.

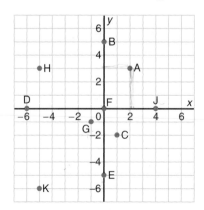

2. Use the coordinate grid in question 1. Which points have:
   a) x-coordinate 0?   **b)** y-coordinate 0?
   c) the same x-coordinate?   **d)** the same y-coordinate?
   e) equal x- and y-coordinates?   **f)** y-coordinate $+2$?

3. Draw a coordinate grid. Label the axes. Plot each point.
   a) $A(6, -6)$ **b)** $B(5, 0)$ **c)** $C(-2, 7)$
   d) $D(-3, 8)$ **e)** $E(3, 1)$ **f)** $F(0, -4)$
   g) $O(0, 0)$ **h)** $H(-4, -1)$ **i)** $J(-8, 0)$

4. Suppose you are given the coordinates of a point.
   You do not plot the point.
   How can you tell which quadrant the point will be in?

5. Draw a scalene triangle on a coordinate grid.
   Each vertex should be in a different quadrant.
   a) Label each vertex with its coordinates.
   b) What is the area of the triangle?

6. **Assessment Focus**

   Use a coordinate grid.
   How many different rectangles can you draw that have area 12 units²?
   For each rectangle you draw, label its vertices.

7. a) Plot these points: $K(-3, 4), L(1, 4), M(1, -2)$
   b) Find the coordinates of point N that forms rectangle KLMN.

Take It Further

8. a) Plot these points on a grid: $A(5, -7)$, $B(-3, 3)$, and $C(8, 8)$.
   Join the points.
   b) Find the area of $\triangle ABC$.

9. Plot the points $C(-5, 0)$ and $D(-2, -3)$.
   E is a point such that $\triangle CDE$ is a right triangle.
   Find at least three possible positions for E.
   Write the coordinates of each point.

Reflect

Choose four points, one in each quadrant. Write instructions to plot each point. Draw a grid to show your work.

Focus   Graph translation and reflection images on a coordinate grid.

Recall that a translation moves a figure in a straight line.
When the figure is on a square grid, the translation is described by movements right or left, and up or down.

A translation and a reflection are transformations.

Which translation moved this figure to its image?

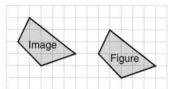

A figure can also be reflected in a mirror line. Where is the mirror line that relates this figure and its image?

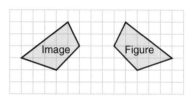

Explore

Work on your own.
You will need 0.5-cm grid paper and a ruler.
Draw axes on the grid paper to get 4 quadrants. Use the whole page.
Label the axes.
Draw and label a quadrilateral.
Each vertex should be where the grid lines meet.

➤ Translate the quadrilateral. Draw and label the translation image.
  What do you notice about the figure and its image?

➤ Choose an axis.
  Reflect the quadrilateral in this axis.
  Draw and label the reflection image.
  What do you notice about the figure and its image?

➤ Trade your work with that of a classmate.
  Identify your classmate's translation.
  In which axis did your classmate reflect?

Reflect & Share

Did you correctly identify each transformation? Explain.
If not, work with your classmate to find the correct transformations.

➤ To translate △ABC 5 units right and 6 units down:
Begin at vertex A(−2, 5).

We read A′ as "A prime."

Move 5 units right and 6 units down to point A′(3, −1).
From vertex B(2, 3), move 5 units right and
6 units down to point B′(7, −3).
From vertex C(−5, 1), move 5 units right and
6 units down to point C′(0, −5).

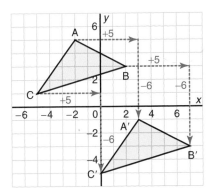

Then, △A′B′C′ is the image of △ABC after a translation
5 units right and 6 units down.
△ABC and △A′B′C′ are congruent.

➤ To reflect △ABC in the y-axis:
Reflect each vertex in turn.
The reflection image of A(−2, 5) is A′(2, 5).
The reflection image of B(2, 3) is B′(−2, 3).
The reflection image of C(−5, 1) is C′(5, 1).

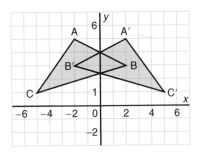

Then, △A′B′C′ is the image of △ABC
after a reflection in the y-axis.
△ABC and △A′B′C′ are congruent.
The triangles have different orientations:
we read △ABC clockwise; we read
△A′B′C′ counterclockwise.

Example

a) Plot these points: A(4, −4), B(6, 8), C(−3, 5), D(−6, −2)
   Join the points to draw quadrilateral ABCD.
   Reflect the quadrilateral in the *x*-axis.
   Draw and label the reflection image A′B′C′D′.

b) What do you notice about the line segment joining each point to its reflection image?

Solution

a) After a reflection in the *x*-axis:
   A(4, −4) → A′(4, 4)
   B(6, 8) → B′(6, −8)
   C(−3, 5) → C′(−3, −5)
   D(−6, −2) → D′(−6, 2)

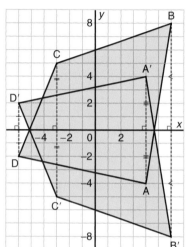

b) The line segments AA′, BB′, CC′, DD′ are vertical.
   The *x*-axis is the perpendicular bisector of each line segment.
   That is, the *x*-axis divides each line segment into 2 equal parts, and the *x*-axis intersects each line segment at right angles.

In *Practice* question 6, you will investigate a similar reflection in the *y*-axis.

Practice

1. Identify each transformation. Explain your reasoning.

   a)

   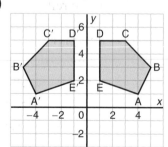

   b)

   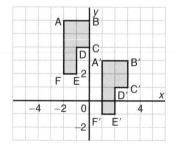

2. The diagram shows 4 parallelograms.

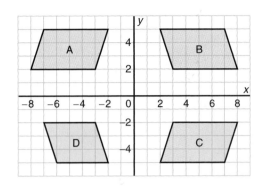

a) Are any 2 parallelograms related by a translation? Explain.

b) Are any 2 parallelograms related by a reflection? Explain.

3. Copy this pentagon on grid paper.

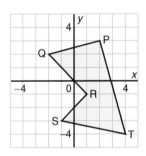

a) Draw the image after a translation 3 units left and 2 units up.

b) Draw the image after a reflection in the *x*-axis.

c) Draw the image after a reflection in the *y*-axis.

4. Plot these points on a coordinate grid:
$A(1, 3)$, $B(3, -2)$, $C(-2, 5)$, $D(-1, -4)$, $E(0, -3)$, $F(-2, 0)$

a) Reflect each point in the *x*-axis.
Write the coordinates of each point and its reflection image.
What patterns do you see in the coordinates?

b) Reflect each point in the *y*-axis.
Write the coordinates of each point and its reflection image.
What patterns do you see in the coordinates?

c) How could you use the patterns in parts a and b to check that you have drawn the reflection image of a figure correctly?

5. a) Plot the points in question 4.
Translate each point 4 units left and 2 units down.

b) Write the coordinates of each point and its translation image.
What patterns do you see in the coordinates?

c) How could you use these patterns to write the coordinates of an image point after a translation, without plotting the points?

6. a) Plot these points on a coordinate grid:
P(1, 4), Q(−3, 4), R(−2, −3), S(5, −1)
Join the points to draw quadrilateral PQRS.
Reflect the quadrilateral in the y-axis.

b) What do you notice about the line segment joining each point to its image?

7. On a coordinate grid, draw a line through A(10, 10), O(0, 0), and B(−10, −10).
Use this line as the reflection line.
Draw a quadrilateral on one side of the line.
Draw its reflection image.
What patterns do you see in the coordinates of each point and its image?

8. a) Draw a figure and its image that could represent a translation and a reflection.

b) What attributes does the figure have?

9. Assessment Focus Draw a figure on a coordinate grid.

a) Choose a translation and/or a reflection that you could repeatedly apply to the figure and its images to make a design or pattern.

b) Does the figure tessellate? If your answer is no, could you change the transformation to make it tessellate? Explain.

10. You have transformed figures on a coordinate grid.
Think about transformations in the real world.

a) Where do you see examples of translations?

b) Where do you see examples of reflections?

Number Strategies

What is the capacity of a cylindrical mug with height 9 cm and diameter 7.5 cm?

Reflect

How is a translation different from a reflection?
How are these transformations alike?
How can coordinate grids be used to illustrate these differences and similarities?

Focus | Graph rotation images on a coordinate grid.

Recall that a rotation turns a figure about the turn centre.
The rotation may be clockwise or counterclockwise.
The turn centre may be:

On the figure 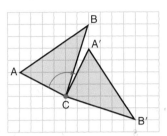   Off the figure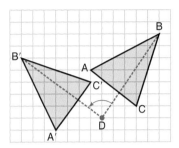

How would you describe each rotation?

Explore

Work with a partner.
You will need 0.5-cm grid paper, tracing paper, a protractor, and a ruler.
Draw axes on grid paper to get 4 quadrants.
Place the origin at the centre of the paper.
Label the axes.
Draw and label a figure in the 1st quadrant.
Use the origin as the turn centre.

➤ Rotate the figure 90° counterclockwise.
   Draw its image.
➤ Rotate the original figure 180° counterclockwise.
   Draw its image.
➤ Rotate the original figure 270° counterclockwise.
   Draw its image.

What do you notice about the figure and its 3 images?

Reflect & Share

Compare your work with that of another pair of classmates.
What strategies did you use to measure the rotation angle?
Would the images have been different if you had rotated clockwise
instead of counterclockwise? Explain.

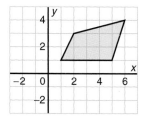

To rotate the figure at the left clockwise:

➤ Trace the figure and the axes.
Label the positive *y*-axis on the tracing paper.
Rotate the tracing paper clockwise about the origin
until the positive *y*-axis coincides with the positive *x*-axis.
With a sharp pencil, mark the vertices of the image.
Join the vertices to draw the image after a 90° clockwise
rotation about the origin, below left.

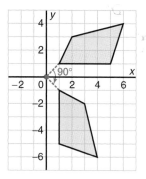   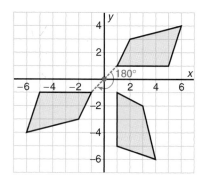

➤ Place the tracing paper so the
figure coincides with its image.
Rotate the tracing paper clockwise about the origin until
the positive *y*-axis coincides with the negative *y*-axis.
Mark the vertices of the image.
Join the vertices to draw the image of the original figure
after a 180° clockwise rotation about the origin, above right.

➤ Place the tracing paper so the figure
coincides with its second image.
Rotate the tracing paper
clockwise about the origin
until the positive *y*-axis coincides
with the negative *x*-axis.
Mark, then join, the vertices of
the image.
This is the image after a 270°
clockwise rotation about the origin.

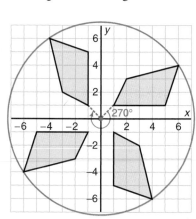

All 4 quadrilaterals are congruent.
A point and all its images lie on a circle, centre the origin.

Example

A counterclockwise rotation is shown by a positive angle such as +90°, or 90°. A clockwise rotation is shown by a negative angle such as −90°.

a) Plot these points: B(−5, 6), C(−3, 4), D(−8, 2)
Join the points to draw △BCD.
Rotate △BCD 90° about the origin, O.
Draw and label the rotation image △B′C′D′.

b) Join C, D, C′, D′ to O.
What do you notice about these line segments?

Solution

A rotation of 90° is a counterclockwise rotation.

a) Use tracing paper to draw
the image △B′C′D′.
Rotate the paper counterclockwise
until the positive y-axis coincides
with the negative x-axis.

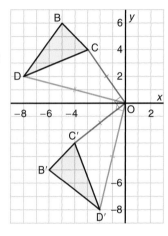

We could use the Pythagorean Theorem to verify that these line segments are equal.

b) From the diagram,
OC = OC′ and OD = OD′
∠COC′ = ∠DOD′ = 90°

The *Example* illustrates these properties of a rotation.
- A point and its image are the same distance from the rotation centre.
- The angle between the segments joining a point and its image to the rotation centre is equal to the rotation angle.

In the *Practice* questions, you will verify these properties for other angles of rotation.

Practice

1. Each grid shows a figure and its rotation image.
Identify the angle of rotation and the rotation centre.

a)

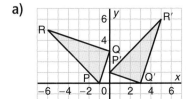

b)

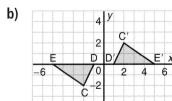

2. Identify each transformation. Explain how you know.

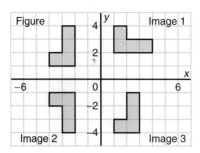

3. a) Copy △DEF on grid paper.

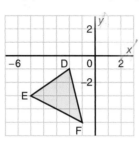

When there are 2 rotation images, we use a "double" prime notation for the vertices of the second image.

b) Rotate △DEF −90° about the origin to its image △D′E′F′.

c) Rotate △DEF +270° about the origin to its image △D″E″F″.

d) What do you notice about the images in parts b and c?
Do you think you would get a similar result with any figure that you rotate −90° and +270°? Explain.

4. Plot each point on a coordinate grid:
A(2, 5), B(−3, 4), C(4, −1)

a) Rotate each point 180° about the origin O to get image points A′, B′, C′.

b) Draw and measure:
   i) OA and OA′   **ii)** OB and OB′   **iii)** OC and OC′
   What do you notice?

c) Measure each angle.
   i) ∠AOA′   **ii)** ∠BOB′   **iii)** ∠COC′
   What do you notice?

d) What other rotation of A, B, and C would result in the image points A′, B′, C′? Explain.

5. Repeat question 4 for a rotation of −90° about the origin.

Number Strategies

Find each percent of $375.00.

- 1%
- 10%
- 0.1%
- 15%
- 150%
- 1.5%

6. (Assessment Focus) Draw and label 6 points on a grid, one in each quadrant and one on each axis.

a) Rotate each point $-90°$ about the origin.
Write the coordinates of each point and its image.
What patterns do you see in the coordinates?

b) Repeat part a for a rotation of $180°$ about the origin.

c) Repeat part a for a rotation of $-270°$ about the origin.

d) How could you use the patterns in parts a, b, and c to draw a rotation image without using tracing paper?

7. You have rotated figures on a grid.
Think about rotations in the real world.
Where do you see examples of rotations outside of math class?

8. Draw a quadrilateral in the 3rd quadrant.

a) Rotate the quadrilateral $180°$ about the origin.

b) Reflect the quadrilateral in the x-axis.
Then reflect the image in the y-axis.

c) What do you notice about the image in part a and the second image in part b?
Do you think you would get a similar result if you started:
i) with a different figure?   **ii)** in a different quadrant?
Investigate to find out.
Write about what you discover.

Take It Further

9. Plot these points: C(2, 6), D(3, −3), E(5, −7)

a) Reflect △CDE in the x-axis to its image △C′D′E′.
Rotate △C′D′E′ $-90°$ about the origin to its image △C″D″E″.

b) Rotate △CDE $-90°$ about the origin to its image △PQR.
Reflect △PQR in the x-axis to its image △P′Q′R′.

c) Do the final images in parts a and b coincide? Explain.

Reflect

When you see a figure and its transformation image on a grid, how can you identify the transformation?
Include examples in your explanation.

Creating a Study Sheet

A study sheet helps you to review important math ideas. Your study sheet may be different from a classmate's study sheet, but all study sheets should include the most important information from a unit. Add to and review your study sheet several times during a unit.

Here are some things to include on a study sheet.

- **Key Words**

Record *Key Words* from the unit.

Use a definition, a picture or example, and a problem.

If any word is difficult to remember, create a word card. This card contains something to help you remember the meaning of the word. Use *Connect* and the *Illustrated Glossary* to help.

- **Formulas**

Note any formulas or procedures that you may need to remember. Look in *Connect* for these.

- **Main Ideas**

Select important main ideas for each lesson.

Use lesson titles and *Focus* to organize the topics for the study sheet.

Use *Reflect* from your notebook to help remember the main learning from the lessons. Highlight and summarize key points on your study sheet.

- **Sample Questions**

Select sample questions for each lesson, such as the *Assessment Focus*, to review. Try doing the questions again.

- **Journal Notes**

If you keep a journal, review your notes, then highlight and summarize things that you should remember on your study sheet.

- **Review Questions**

Select sample questions from the *Unit Review* to practise.

Here is a study sheet for
Unit 8 Square Roots and Pythagoras.

Key Words

Square Numbers	Square Roots
1, 4, 9, 16, 25, 36, 49, 64, 81, 100	$\sqrt{1} = 1$ $\sqrt{16} = 4$ $\sqrt{36} = 6$ $\sqrt{81} = 9$

Formulas

$c^2 = a^2 + b^2$

Pythagorean Theorem: In a right triangle, the area of the square on the hypotenuse is equal to the sum of the areas of the squares on the legs.

Main Idea

The Pythagorean Theorem can be used to find the length of one side in a right triangle when two other sides are known.

Sample Question

The size of a TV set is described by the length of a diagonal of the screen. One TV is labelled as size 70 cm.
The screen is 40 cm high.
What is the width of the screen?
Draw a diagram to illustrate your answer.

Journal Notes

To use the Pythagorean Theorem in an isosceles triangle, I draw a perpendicular from the vertex of the angle that is different from the equal angles, to the opposite side. Then I have two right triangles.

Review Questions

I need to review question 10, page 356, on finding the surface area of a pentagonal prism.

Unit Review

What Do I Need to Know?

✓ Adding Integers

Use a number line.

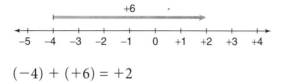

$(-4) + (+6) = +2$

✓ Subtracting Integers

Add the opposite.
Write $(-5) - (+4)$ as $(-5) + (-4)$, then add.
$(-5) + (-4) = -9$

✓ Multiplying Integers

The product of two integers with the same sign is a positive integer.
$(+6) \times (+4) = +24; \ (-18) \times (-3) = +54$

The product of two integers with different signs is a negative integer.
$(-8) \times (+5) = -40; \ (+9) \times (-6) = -54$

✓ Dividing Integers

The quotient of two integers with the same sign is a positive integer.
$(+56) \div (+8) = \frac{+56}{+8} = +7; \ (-24) \div (-6) = \frac{-24}{-6} = +4$

The quotient of two integers with different signs is a negative integer.
$(-30) \div (+6) = \frac{-30}{+6} = -5; \ (+56) \div (-7) = \frac{+56}{-7} = -8$

✓ Graphing on a Coordinate Grid

The coordinates of each point are:
$A(3, 2), B(-3, 2), C(-3, -2)$, and $D(3, -2)$

✓ Transformations on a Coordinate Grid

A point or figure can be:
- translated
- reflected in the x-axis or the y-axis
- rotated about the origin

LESSON

9.1 **1. a)** List these integers from least to greatest: $+8, -10, -3, +1, -7$

   b) Mark each integer in part a on a number line.

9.2 **2.** Use a number line to add or subtract.

   a) $(-8) + (+5)$

   b) $(+14) + (-8)$

   c) $(-5) - (+3)$

   d) $(-7) - (-2)$

   e) $(+4) - (-3) + (-5)$

   f) $(-3) + (-8) - (-7)$

   g) $(+6) - (+10) + (-2)$

   h) $(-9) - (-11) - (-6)$

3. Here are the results of a golf game.

   Al: -3;   Lana: $+2$;   Kirima: 0;
   Earl: $+1$;   Reg: -5;   Jody: -4

   a) Who won the game?

   b) How many fewer strokes did the winner take than the person who came last?

4. At midnight in Winnipeg, the temperature was $-23°C$. During the next 24 h, the temperature rose $12°C$, then dropped $8°C$. What was the final temperature? Explain.

 5. Evaluate.

   a) $(+512) + (-173)$

   b) $(-879) - (-1092)$

   c) $(-243) + (+987)$

   d) $(+1591) - (-847)$

9.3 **6.** Evaluate.

   a) $3 - 5$ **b)** $-1 + 10$

   c) $-5 - 6$ **d)** $3 - 5 + 7$

   e) $-4 + 3$ **f)** $-3 + 5 - 7$

7. The temperature change in a chemistry experiment was $-2°C$ every 30 min. The initial temperature was $6°C$. What was the temperature after 4 h?

9.4 **8.** Multiply.

   a) $(-7)(-5)$

   b) $(+10)(-6)$

   c) $(-3)(-9)(-1)$

   d) $(-2)(-2)(-2)$

   e) $(-7)(-8)(0)$

   f) $(-11)(+13)(-2)$

9. Evaluate.

   a) $21 - 5 - 5 + 6 - 2$

   b) $0 - 6 + 4 + 8 - 1 + 2 + 7$

9.1
9.2 **10.** Answer true or false to each statement. Explain your answer.
9.4

   a) The sum of an integer and its opposite is always 0.

   b) When you subtract two positive integers, their difference is always a positive integer.

   c) The product of a positive integer and a negative integer is always positive.

   d) The product of an integer and its opposite is always 0.

9.5 **11.** Divide.

a) $(-56) \div (-7)$

b) $(+40) \div (-5)$

c) $(-121) \div (+11)$

d) $\dfrac{-36}{-4}$

e) $\dfrac{+72}{-4}$

f) $\dfrac{-28}{+2}$

9.6 **12.** Evaluate.

a) $(-8) \div (-4) + 6 \times (-3)$

b) $(-5) + (-12) \div (-3)$

c) $18 + 3[10 \div (-5)]$

d) $[(-16) \div 8]^2 - 12$

e) $\dfrac{4 \times (-5) - 4}{-6}$

f) $\dfrac{(-3)^2 + 5}{(-2)^2 - (-3)}$

13. Evaluate.

a) $4[(-3) + 16]$

b) $3 - 2(10 \div 2)$

c) $5 \times (-2) - 2[4 \div (-2)]$

d) $(-3)(-2)(4) + 3(-5)$

e) $\dfrac{3 \times (-6) - 3}{-7}$

f) $9 - 3[(-2)^3 + 4]$

14. In a darts game, Suzanne and Corey each threw the darts 10 times.

Corey had: three $(+2)$ scores; three (-3) scores; and four $(+1)$ scores. Suzanne had four $(+2)$ scores; four (-3) scores; and two $(+1)$ scores.

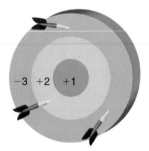

a) What was each person's final score?

b) Who won the game? Explain.

9.1 **15.** For each number below, find two
9.2 integers for which that number is:
9.4   i) the sum
9.5   ii) the difference
   iii) the quotient
   iv) the product

a) -8 b) -2

c) -12 d) -3

9.7 **16. a)** On a coordinate grid, plot each point. Join the points in order. Then join D to A.

A$(-2, -2)$ B$(6, -2)$
C$(3, 7)$ D$(-5, 7)$

b) Name the quadrant in which each point is located.

c) Identify the figure. Find its area.

9.8 **17. a)** Plot these points on a
9.9 coordinate grid:

A$(-2, 3)$, B$(-4, 0)$,
C$(-2, -3)$, D$(2, -3)$

Join the points to draw quadrilateral ABCD.

b) Draw the image of quadrilateral ABCD after each transformation:

   i) a translation 7 units left and 8 units up

  ii) a reflection in the x-axis

iii) a rotation of 90° counterclockwise about the origin

c) How are the images alike? Different?

Practice Test

1. Use a number line to order these integers from least to greatest.
   $+5, -3, 0, -11, -8, +7$

2. Evaluate.
   a) $(-4) + (-8)$ b) $9 + (-17)$
   c) $(-8) \times 6$ d) $(-56) + (-61)$
   e) $(-10) - (-3)$ f) $(4)(-2)$
   g) $(-2)^4$ h) $(-36) \div 9$
   i) $(-3) \times (-5) \times (-11)$

3. Continue each pattern. Write the next 4 terms.
   Write the pattern rule.
   a) $-4, 8, -16, 32, \ldots$ b) $-9, -2, -5, 2, -1, 6, \ldots$

4. Evaluate.
   a) $(-20) \times (-5) + 16 \div (-8)$
   b) $\dfrac{14 - 10 \div 2}{-3}$
   c) $(-3)^2 + 2 \times (-4)$

5. A number is multiplied by -4.
   Then 3 is subtracted from the product.
   The answer is 13.
   What is the number?

6. The temperature on Sunday was 4°C. The temperature dropped 8°C on Monday and dropped twice as much on Tuesday.
   What was the temperature on Tuesday?

7. a) On a coordinate grid, draw a triangle with area 12 square units. Place each vertex in a different quadrant.
   b) Write the coordinates of each vertex.
   c) Explain how you know the area is 12 square units.
   d) Translate the triangle 6 units right and 3 units down.
   e) Reflect the triangle in the y-axis.
   f) Rotate the triangle 90° clockwise about the origin.

A Grade 8 class and a local bank are partners to sponsor
a golf tournament to raise money for local charities.
The bank provides these prizes:

> **1st place—$5000 to a charity of the player's choice**
> **2nd and 3rd places—$1000 to a charity of the player's choice**

Golf Terms
Recall that "par" is the number of strokes it should take for a player
to reach the hole.
If par is 3 and you take 5 strokes, then your score in relation
to par is +2, or 2 over.
If par is 3 and you take 2 strokes, then your score in relation to
par is −1, or 1 under.

A bogey is 1 stroke more than par, or 1 over par.
A double bogey is 2 strokes more than par, or 2 over par.
A birdie is 1 stroke less than par, or 1 under par.
An eagle is 2 strokes less than par, or 2 under par.

Here are the top 6 golfers:
Chai Kim, Delaney, Hamid, Hanna, Kyle, and Weng Kwong

1. The golf course has 9 holes. Here is one person's results:
par on 3 holes, a bogey on 2 holes, a birdie on 1 hole,
an eagle on 2 holes, and a double bogey on 1 hole
a) Write an integer expression to represent these results.
b) Evaluate the expression in part a to calculate the score in
relation to par.

2. Chai Kim wrote his results in a table like this.

Hole	1	2	3	4	5	6	7	8	9
Par	3	4	3	3	5	4	4	3	3
Under/Over Par	0	−1	+2	0	−1	0	0	−1	0
Score	3	3	5	3					

a) Copy and complete the table. Use the following information:

Hole 1, 4, 6, 7, 9	Par
Hole 2, 5, 8	Birdie
Hole 3	Double bogey

b) What was Chai Kim's final score?

c) What was his final score in relation to par?

3. For each person below, make a table similar to the table in question 2.

Use the information below. What is each golfer's final score?

a) Kyle:
- Bogey holes 1, 3, 5, 9
- Birdie hole 6
- Par holes 2, 4, 7, 8

b) Delaney:
- Bogey holes 3, 4, 6
- Birdie holes 1, 2, 7, 8, 9
- Eagle hole 5

c) Hamid:
- Birdie every hole except hole 8
- Double bogey hole 8

4. a) Hannah had a score of −5 in relation to par.
Weng Kwong had a score of +3 in relation to par.
Use the information in questions 2 and 3.
Rank the players in order from least to greatest score.

b) Who won the tournament and the $5000 prize?
What was the score in relation to par?

c) Who won the $1000 prizes?
What were the scores in relation to par?

5. Use a table similar to that in question 2.
Complete the table with scores of your choice.
Calculate the final score, and the final score in relation to par.

Reflect on the Unit

What did you find easy about working with integers? What was difficult for you?
Give examples to illustrate your answers.

Telecommunication companies offer telephone services.

These tables show the plans for cell phones for two companies. Each plan includes 200 free minutes.

What patterns do you see in the tables?

Write a pattern rule for each pattern. Describe each plan.

Assume the patterns continue. How could you find the total cost for 60 additional minutes for each plan?

Company A

Number of Additional Minutes	Total Cost ($)
0	35
4	36
8	37
12	38
16	39
20	40

Company B

Number of Additional Minutes	Total Cost ($)
0	40
5	41
10	42
15	43
20	44
25	45

What You'll Learn

- Investigate number properties.
- Write an expression for the nth term of a pattern.
- Evaluate algebraic expressions by substituting fractions and integers.
- Read, write, and solve equations.
- Represent algebraic relationships using tables, graphs, and equations.

Why It's Important

- Algebra is used to communicate with symbols. It can be used to describe patterns.
- Patterns and equations are used to investigate changes in our world. For example, urban planners use equations to investigate population growth.

417

Skills You'll Need

Writing Expressions and Equations

We use a letter, such as x or n, to represent a number.

We can write an algebraic expression to represent a word statement.

For example, "a number plus five," or "five more than a number" can be written as $n + 5$.

When we write an algebraic expression as equal to a number or another expression, we have an equation.

For example, $n + 5 = 8$ is an equation.

Example 1

a) Write an algebraic expression for this statement:
   Three more than four times a number

b) Write an equation for this sentence:
   A number divided by four is 5.

Solution

a) Three more than four times a number
   Let x represent the number.
   Then, four times a number is $4x$.
   Three more than $4x$ is:
   $4x + 3$ or $3 + 4x$

b) A number divided by four is 5.
   Let z represent the number.
   z divided by four is: $\frac{z}{4}$
   The equation is: $\frac{z}{4} = 5$

✓ Check

1. Write an algebraic expression for each statement.
   a) a number multiplied by seven
   b) six less than a number
   c) five more than three times a number
   d) three less than five times a number

2. Write an equation for each sentence.
   a) A number divided by seven is 6.
   b) The sum of eight and a number is 17.
   c) Five more than two times a number is 11.

Evaluating Expressions

To evaluate an algebraic expression for a particular value of the variable, replace the variable with a number. Then, find the value of the expression. The number we substitute can be a fraction or an integer.

Example 2

Evaluate the expression $2x + 3y + 4z$ for $x = -1$, $y = \frac{1}{3}$, and $z = \frac{1}{2}$.

Solution

$2x + 3y + 4z$

Substitute: $x = -1$, $y = \frac{1}{3}$, and $z = \frac{1}{2}$

$2x + 3y + 4z = 2(-1) + 3\left(\frac{1}{3}\right) + 4\left(\frac{1}{2}\right)$

$\qquad = 2 \times (-1) + 3 \times \frac{1}{3} + 4 \times \frac{1}{2}$ Multiply first.

$\qquad = -2 + 1 + 2$ Then add.

$\qquad = -2 + 3$

$\qquad = 1$

✓ Check

3. Evaluate each expression.
   a) $3 + x$ for $x = \frac{1}{2}$ **b)** $3 - x$ for $x = -2$ **c)** $3x$ for $x = \frac{1}{4}$

4. Evaluate each expression for $p = \frac{2}{3}$ and $q = \frac{1}{4}$.
   a) $p + q$ **b)** $p - q$ **c)** pq

5. Evaluate each expression for $m = \frac{2}{5}$ and $n = \frac{1}{2}$.
   a) $2m + n$ **b)** $2n + m$ **c)** $2m + 2n$
   d) $2m - n$ **e)** $2n - m$ **f)** $2n - 2m$
   g) mn **h)** $2mn$ **i)** $\frac{1}{2}mn$

6. Evaluate each expression in question 5 for $m = -3$ and $n = -6$.

7. Evaluate each expression.
   a) $3x - 2y + 4z$, when $x = \frac{3}{4}$, $y = \frac{1}{5}$, $z = \frac{5}{4}$
   b) $3x + 5y - 3z$, when $x = \frac{5}{4}$, $y = \frac{1}{6}$, $z = \frac{2}{3}$
   c) $3x + 3y - 2z$, when $x = \frac{1}{5}$, $y = \frac{4}{3}$, $z = \frac{1}{15}$

8. Evaluate each expression in question 7 for $x = 2$, $y = -4$, and $z = -1$.

10.1 Number Properties

Focus Relate the distributive property and other properties to algebra.

Recall how you used a diagram to multiply: 4×37

This diagram shows:

$$\begin{aligned} 4 \times 37 &= 4 \times (30 + 7) \\ &= 4 \times 30 + 4 \times 7 \\ &= 120 + 28 \\ &= 148 \end{aligned}$$

Explore

Work with a partner. Use 0.5-cm grid paper if it helps.

$5(n + 8)$ means $5 \times (n + 8)$.

➢ Draw a diagram to illustrate 5×28.
What is the product?

➢ Draw a diagram to illustrate $5(n + 8)$.
What is the product?

➢ Draw a diagram to illustrate $5(n + m)$.
What is the product?

➢ Draw a diagram to illustrate $d(n + m)$.
What is the product?

Reflect & Share

Compare your diagrams and products with another pair of classmates.
What patterns do you see in the products?
How can you use the patterns to write $d(n + m)$ without brackets?

Connect

When we use symbols to represent numbers, the following properties are still true.

Adding 0

Adding 0 does not change the number.

$4 + 0 = 4$ and $n + 0 = n$

$0 + 135 = 135$ and $0 + n = n$

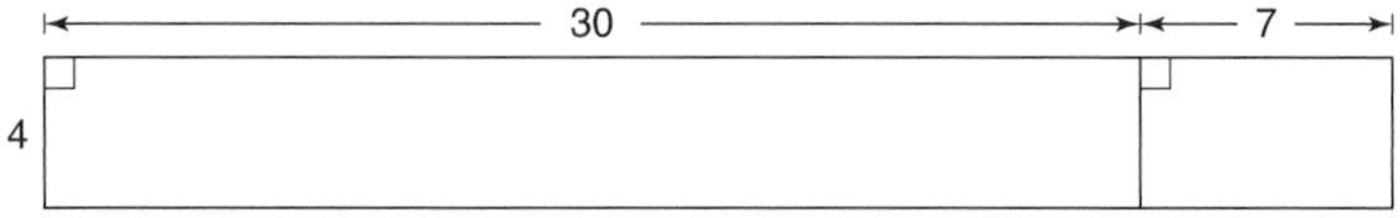

Multiplying by 1

When 1 is a factor, the product is always the other factor.

$1 \times 11 = 11$ and $1 \times n = n$

$256 \times 1 = 256$ and $n \times 1 = n$

Multiplying by 0

When 0 is a factor, the product is always 0.

$15 \times 0 = 0$ and $n \times 0 = 0$

$0 \times 137 = 0$ and $0 \times n = 0$

Order of addition and multiplication

When you add, the order does not matter.

$9 + 4 = 13$ and $4 + 9 = 13$ $\qquad a + b = b + a$

When you multiply, the order does not matter.

$6 \times 8 = 48$ and $8 \times 6 = 48$ $\qquad ab = ba$

Distributive Property

We will investigate $a(b + c)$ and $ab + ac$ for different values of a, b, and c.

Recall that $a(b + c)$ means $a \times (b + c)$, ab means $a \times b$, and ac means $a \times c$.

a	b	c	(b + c)	a(b + c)	ab	ac	ab + ac
2	4	7	11	22	8	14	22
3	6	2	8	24	18	6	24
7	1	1	2	14	7	7	14
12	8	3	11	132	96	36	132
0	7	5	12	0	0	0	0

We can illustrate this property with a diagram.

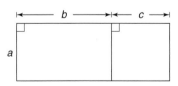

The numbers in these columns are the same.

This table illustrates the **distributive property** of multiplication:

$a(b + c) = ab + ac$

That is, the product of $a(b + c)$ is the same as the sum $ab + ac$.

Example

Use the distributive property to write each expression as a sum of terms.

a) $7(c + 2)$

b) $2(2a + 3b + 4)$

Solution

a) $7(c + 2) = 7(c) + 7(2)$
$= 7c + 14$

b) $2(2a + 3b + 4) = 2(2a) + 2(3b) + 2(4)$
$= 4a + 6b + 8$

In the *Example*, when we use the distributive property, we **expand**.

Practice

1. Draw a rectangle to show that $5(x + 2)$ and $5x + 10$ are equivalent.

2. Expand.
   a) $2(x + 10)$ b) $5(x + 1)$ c) $10(x + 2)$
   d) $6(12 + 6y)$ e) $8(8 + 9y)$ f) $5(7y + 6)$

3. Write two formulas for the perimeter, P, of a rectangle. Explain how the formulas illustrate the distributive property.

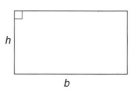

4. Explain how you know $hb = bh$. Use an example to justify your answer.

5. Expand.
   a) $5(2x + 2y + 2)$ b) $4(3x + 5y + 1)$ c) $8(7x + 3y + 2)$

6. **Assessment Focus** Which expressions in each pair are equivalent? Explain your reasoning.
   a) $2x + 20$ and $2(x + 20)$ b) $3x + 7$ and $10x$
   c) $6 + 2t$ and $2(t + 3)$ d) $9 + x$ and $x + 9$

Reflect

What is the distributive property?
Include a diagram with your explanation.

Describing Number Patterns

Focus Write an expression for the *n*th term of a number pattern.

Explore

Work with a partner.

You will need grid paper.

Charla has juvenile diabetes.

She needs five injections of insulin per day.

Each needle can be used only once.

Charla wants to go to camp.

She must take all her needles with her.

She must always have at least
6 extra needles available.

Number of Days	Number of Needles
1	
2	
3	
4	
5	
6	

➤ Copy and complete this table.
Find the number of needles Charla
needs to take with her for up to 6 days.

➤ Graph the data.

➤ Write an algebraic expression for the number of needles required
for any number of days.
Use the expression to find the number of needles required for
7 days, 14 days, and 30 days.

Reflect & Share

Compare your results with those of another pair of classmates.
Work together to explain how the table, the graph, and the
expression are related.

Connect

We can use a table, a graph, and algebra to describe and
extend a number pattern.
Look at the pattern: 1, 3, 5, 7, …
To find the 20th term, use one of these three methods.

➤ Make a table, then extend the table to find the 20th term.
The term value increases by 2 each time.
The pattern rule is: Start at 1. Add 2 each time.
From the table on the next page, the 20th term is 39.

Term Number	Term Value
1	1
2	3
3	5
4	7
5	9
20	39

The 20th term is 39.

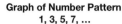

Graph of Number Pattern 1, 3, 5, 7, ...

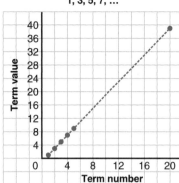

➤ Graph the pattern, then extend the graph to find the 20th term.
The points lie on a straight line.
To get from one point to another,
move 1 unit right
and 2 units up.

Use a ruler to draw a broken line through the points to show the trend.
Extend the broken line to the right to find that the 20th term is 39.

➤ The term values are consecutive odd numbers: 1, 3, 5, 7, 9, ...
The algebraic expression $2n$ produces even numbers,
when we substitute $n = 1, 2, 3, 4, ...$
That is, $2(1) = 2$
$\qquad 2(2) = 4$
$\qquad 2(3) = 6$
$\qquad 2(4) = 8$, and so on
Each odd number is 1 less than the following even number.
So, the expression $2n - 1$ produces odd numbers,
when we substitute $n = 1, 2, 3, 4, ...$
That is, $2(1) - 1 = 2 - 1 = 1$
$\qquad 2(2) - 1 = 4 - 1 = 3$
$\qquad 2(3) - 1 = 6 - 1 = 5$
$\qquad 2(4) - 1 = 8 - 1 = 7$
This table shows how the term value relates to the term number.

In each case, the term value is equal to:
The term number multiplied by 2, then subtract 1

Term Number	Term Value	Pattern Rule for Term Value
1	1	$1 = 2(1) - 1$
2	3	$3 = 2(2) - 1$
3	5	$5 = 2(3) - 1$
4	7	$7 = 2(4) - 1$
5	9	$9 = 2(5) - 1$

Let t represent the term number.

Then an expression for the term value is $2t - 1$,
where t is any natural number.

To check that the expression for the term value is correct,
substitute a number for t.

Substitute $t = 2$.

$$2t - 1 = 2 \times 2 - 1$$
$$= 4 - 1$$
$$= 3$$

So, the 2nd term is 3, which matches the 2nd term
in the pattern given.

This method allows us to find the value of any term in the pattern.
For example, the 20th term has value: $2(20) - 1 = 39$

Example

Here is a number pattern.

8, 12, 16, 20, …

a) Complete a table for the first 5 terms of this pattern.
Extend the table to find the 10th term.
Describe the pattern.
Write a pattern rule.

b) Graph the pattern.

c) Write an expression for the nth term.

d) Use the expression in part c to verify the 10th term.

Solution

a) 8, 12, 16, 20, …
The pattern begins with 8.
To get the next term, add 4 each time.
The pattern rule is: Start at 8. Add 4 each time.
Extend the table to find the 10th term is 44.

Term Number	1	2	3	4	5	6	7	8	9	10
Term Value	8	12	16	20	24	28	32	36	40	44

b) Graph the pattern.
The points lie on a straight line.
Use a ruler to draw a broken line through the
points to show the trend.

Graph of Number Pattern

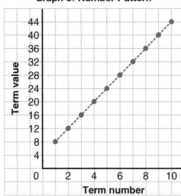

c) Find a pattern rule that relates the term value
to the term number.
Each term is 4 more than the previous term.
Look for patterns that involve multiples of 4.

In each case, the term
value is equal to:
**Four more than
four times the
term number**

Term Number	Term Value	Pattern Rule for Term Value
1	$8 = 4 + 4$	$8 = 4(1) + 4$
2	$12 = (4 + 4) + 4$	$12 = 4(2) + 4$
3	$16 = (4 + 4 + 4) + 4$	$16 = 4(3) + 4$
4	$20 = (4 + 4 + 4 + 4) + 4$	$20 = 4(4) + 4$

To write an expression for the nth term, let n represent any term
number. Then, the nth term is: $4n + 4$

d) To find the 10th term, substitute $n = 10$ into $4n + 4$.
$$4n + 4 = 4(10) + 4$$
$$= 44$$
The 10th term is 44. This verifies the value in the table in part a.

Practice

1. Substitute $n = 1, 2, 3, 4, 5,$ and 6 to generate a number pattern.
Describe each pattern, then write a pattern rule.
a) $2n + 1$ **b)** $3n - 1$ **c)** $2n + 2$ **d)** $4n - 2$

2. For each number pattern, write an expression for the nth term.
a) $\dfrac{1}{1}, \dfrac{1}{2}, \dfrac{1}{3}, \dfrac{1}{4}, \dfrac{1}{5}, \dfrac{1}{6}, \cdots$ **b)** $\dfrac{1}{2}, \dfrac{2}{3}, \dfrac{3}{4}, \dfrac{4}{5}, \dfrac{5}{6}, \dfrac{6}{7}, \cdots$

3. For each number pattern below:

   i) Describe the pattern. Write the pattern rule.

   ii) Use a table to find the 12th term.

   iii) Write an expression for the nth term.

   iv) Use the expression to find the 100th term.

   a) 1, 2, 3, 4, 5, …   **b)** 2, 3, 4, 5, 6, …

   c) 3, 4, 5, 6, 7, …   **d)** 4, 5, 6, 7, 8, …

4. For each number pattern below:

   i) Write a pattern rule. Justify your rule.

   ii) Graph the pattern. Use the graph to find the 9th term.

   iii) Write an expression for the nth term.

   iv) Use the expression to find the 60th term.

   a) 2, 4, 6, 8, 10, …   **b)** 6, 9, 12, 15, 18, …

   c) 3, 7, 11, 15, 19, …   **d)** 10, 15, 20, 25, 30, …

5. Here are two number patterns.

   • 1, 4, 9, 16, 25, …   • 4, 8, 16, 32, 64, …

Does the number 512 appear in either pattern? Both patterns? Justify your answer.

6. **Assessment Focus** Here is the beginning of a number pattern.

10, 20, …

   a) Extend the pattern in two different ways.

   b) Describe each pattern. Write a pattern rule for each.

   c) Write an expression for the nth term for one pattern.

   d) Can you write an expression for the nth term of the other pattern? Explain.

Take It Further

7. For each number pattern below:

   i) Write a pattern rule. Justify your rule.

   ii) Find the 15th term.

   iii) Write an expression for the nth term.

   iv) Use the expression to find the 30th term.

   a) $\frac{2}{2}, \frac{3}{5}, \frac{4}{8}, \frac{5}{11}, \ldots$ **b)** 1, 3, 6, 10, 15, …

Number Strategies

The product of two fractions is $\frac{1}{2}$.

Find four different pairs of fractions that have a product of $\frac{1}{2}$.

Reflect

Name three ways to describe and extend a number pattern.
Which way is the most efficient? Explain.

10.3 Describing Geometric Patterns

Focus Write an expression for the *n*th term of a geometric pattern.

Explore

Work in a group.
Here is a pattern with squares.

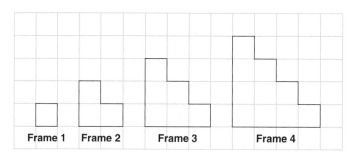

The pattern continues.
Find the perimeter of each frame.
What pattern do you see in the perimeters?
Use a table to show the pattern.
Graph the pattern.
Write a rule for the pattern.
Use a variable.
Write an algebraic expression you could use to find
the perimeter of any frame.
Use the expression to find the perimeters of Frame 5, Frame 10,
and Frame 100.

Reflect & Share

Share your algebraic expression with that of another group.
Are the expressions the same?
If not, how can you check if either expression is correct?
Could both expressions be correct? Explain.

Connect

We can use algebra to describe and extend a geometric pattern.
Here is a pattern of equilateral triangles drawn on isometric paper.

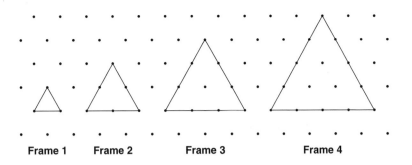

Frame 1   Frame 2   Frame 3   Frame 4

Frame	Perimeter (units)
1	3
2	6
3	9
4	12

This table shows the perimeter of each frame.

The pattern rule for the perimeters is:
Start at 3. Add 3 each time.

If we use this pattern rule to find the perimeter of Frame 40, we would need to know the perimeter of all the frames from Frame 1 to Frame 39.

Instead, we look for a pattern rule for the perimeter in terms of the frame number.

The perimeters are multiples of 3, so write each perimeter as a product, with one factor of 3.

Frame	Perimeter (units)	Perimeter as a Product
1	3	$3 = 3 \times 1$
2	6	$6 = 3 \times 2$
3	9	$9 = 3 \times 3$
4	12	$12 = 3 \times 4$

In each case, the perimeter is equal to 3 times the frame number.
We can use this pattern to find the perimeter of Frame 40:
$3 \times 40 = 120$
The perimeter of Frame 40 is 120 units.
We write the pattern using algebra.
Let f represent the frame number.
Then, an algebraic expression for the perimeter of Frame f is: $3f$
f is any natural number.

To check that the expression is correct, substitute a number for f.
Substitute $f = 4$.
$$3f = 3(4)$$
$$= 12$$
So, Frame 4 has perimeter 12 units.
This is verified by the table on page 429.

Example

Picture frames are decorated with square tiles in the pattern shown.
Each tile has side length 1 cm.
The pattern continues.

   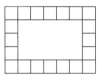

| Frame 1 | Frame 2 | Frame 3 | Frame 4 |

a) Find the area of the picture in each frame.
   What pattern do you see in the areas?

b) Graph the pattern in part a.
   How does the graph illustrate the pattern?

c) Use a variable.
   Write an algebraic expression for the area of the
   picture in any frame.

d) Use the expression in part c.
   Find the area of the picture in Frame 99.

Solution

a) Each picture is a rectangle.
   Its area is length × width.
   Write the areas in a table.

Frame	Area of Picture (cm²)
1	3 × 2 = 6
2	3 × 3 = 9
3	3 × 4 = 12
4	3 × 5 = 15

The areas are multiples of 3.
The pattern rule is:
Start at 6. Add 3 each time.

b) The graph starts at (1, 6).
   To get the next point each time,
   move 1 right and 3 up.
   Moving 1 right is the increase
   in the frame number.
   Moving 3 up is the increase in the area.

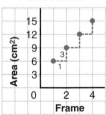

Area of Picture against
Frame Number

c) For an algebraic expression, look at each area in terms of the frame number.
Adding 3 each time indicates a pattern where the term number is multiplied by 3.
So, multiply each term number by 3 and find out what needs to be added each time to get the area.

Frame	Area of Picture (cm²)	Area in Terms of Frame Number
1	6	$3 \times 1 + 3$
2	9	$3 \times 2 + 3$
3	12	$3 \times 3 + 3$
4	15	$3 \times 4 + 3$

Each area is: 3 times the frame number, then add 3
Use the variable n.
An algebraic expression for the area of the picture in Frame n is:
3 times n, then add 3
This is written: $3n + 3$

d) For the area of the picture in Frame 99,
substitute $n = 99$ in $3n + 3$.

$$3n + 3 = 3(99) + 3$$
$$= 297 + 3$$
$$= 300$$

The picture in Frame 99 has area 300 cm².

Practice

1. Use the pattern of frames in the *Example*.
Each frame has the same height of 5 cm.
   a) Find the length of each frame.
   Make a table.
   What patterns do you see in the lengths?
   b) Graph the pattern.
   How does the graph illustrate the pattern?
   c) Write an algebraic expression for the length of the *n*th frame.
   d) Use the expression in part c.
   Find the length of Frame 50.

2. Here is a pattern of triangles made with congruent toothpicks.

Frame 1   Frame 2   Frame 3   Frame 4

The pattern continues.

a) Find the number of toothpicks in each frame.
   What patterns do you see?

b) Graph the data in part a.

c) Write an algebraic expression for the number of toothpicks
   in the nth frame.

d) Find the number of toothpicks in Frame 45.

3. Here is a pattern of squares.

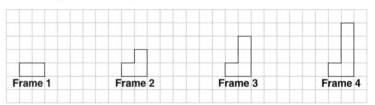

Frame 1   Frame 2   Frame 3   Frame 4

Each square has side length 1 cm.

The pattern continues.

a) Find the perimeter of each frame. Make a table.
   What pattern do you see in the perimeters?

b) Graph the pattern. Explain how the graph illustrates the pattern.

c) Write an algebraic expression for the perimeter
   of the nth frame.

d) Find the perimeter of Frame 75.

4. Here is a pattern made from congruent square tiles.
   Each tile has side length 1 cm.
   The pattern continues.

Frame 1   Frame 2   Frame 3   Frame 4

a) Find the area of each frame.
   What patterns do you see in the areas?

b) Use a pattern to find the area of Frame 8.

c) Write an algebraic expression for the area of the nth frame.

d) Which frame has an area of 625 cm^2? Justify your answer.

5. Hexagonal tables are arranged as shown below.
One person sits at each side of a table.
The pattern continues.

Frame 1   Frame 2   Frame 3   Frame 4

a) How many people can sit at the tables in each frame?
What pattern do you see in the number of people?

b) How many people can sit at the tables in Frame 9?

c) Explain how you could find the number of people who could
be seated at any table arrangement in this pattern.

6. **Assessment Focus** Use grid paper.

a) Draw the first four frames of a growing pattern.

b) Describe the patterns in the frames.

c) Describe or draw Frame 5, Frame 10, and Frame 100.

d) Choose one aspect of your pattern;
for example, area, perimeter, and so on.
Write an algebraic expression for the *n*th frame
of your pattern.

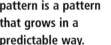

Recall that a growing
pattern is a pattern
that grows in a
predictable way.

Take It Further

7. Bryn has a sheet of paper. He cuts the paper in half to produce
two pieces. Bryn places one piece on top of the other. He then cuts
these pieces in half. The pattern continues. The table shows some
of the results.

Number of Cuts	1	2	3	4	5	6	7	8	9	10
Number of Pieces	2	4	8							

a) Copy and complete this table.

b) What patterns do you see in the number of pieces?

c) Use a pattern to find the number of pieces after 15 cuts.

d) Write an algebraic expression for the number of pieces
after *n* cuts.

Reflect

Explain the meaning of the term "*n*th frame."

Mid-Unit Review

10.1 **1.** Write two expressions for the area of the shaded rectangle.

2. Draw a rectangle to show that:
$6(3 + a) = 18 + 6a$

3. Expand.
a) $3(x + 11)$
b) $5(12 + y)$
c) $4(x + 5y + 9)$
d) $8(5x + 2y + 3)$

10.2 **4.** For each number pattern below:
a) Use a table to find the 8th term.
   Describe the pattern.
   Write a pattern rule.
b) Graph the pattern. Use the graph to find the 12th term.
c) Write an expression for the nth term.
d) Use the expression to find the 40th term.
   i) 1, 7, 13, 19, 25, …
   ii) 2, 7, 12, 17, 22, …
   iii) 4, 7, 10, 13, 16, …

5. Laurel buys a box of mechanical pencils, and a tube of 8 refill leads. Each pencil contains 3 leads. Laurel puts the tube of refill leads into her pencil case, then adds one pencil at a time.

a) Make a table to show the number of leads in the pencil case for up to 7 pencils.
   Describe the pattern.
   Write a pattern rule.
b) Graph the data in the table.
c) Write an algebraic expression for the number of leads in the pencil case for any number of pencils.
d) Use the expression in part c to find the number of leads in the pencil case for 21 pencils.

10.3 **6.** Here is a pattern made with congruent square tiles.

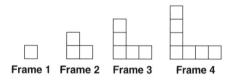

Frame 1   Frame 2   Frame 3   Frame 4

a) Count the number of tiles in each frame.
   What pattern do you see?
b) Make a table to show the pattern.
c) Graph the pattern.
d) Write an algebraic expression for the number of tiles in the nth frame.
e) Use the expression in part d to find the number of tiles in Frame 30.
f) Will any frame have each number of tiles?
   i) 31   ii) 32   iii) 33
   How do you know?

10.4 Solving Equations with Algebra Tiles

Focus Use algebra tiles to solve equations involving whole numbers.

Recall that one red unit tile and one yellow unit tile combine to model 0.
These two unit tiles form a zero pair.

−1   +1

The yellow variable tile represents x.
The opposite of x is $-x$.
So, the red variable tile represents $-x$.
One red variable tile and one yellow variable tile also combine to model 0.
These two variable tiles form a zero pair.

x

$-x$

Flip the yellow tile to get a red tile.

Explore

Work with a partner.
You will need algebra tiles.

➤ For the equation: $2x = 9 - x$
- Interpret the equation in words.
- Use algebra tiles to solve the equation.
- Sketch the tiles you used.

➤ Repeat the activity for this equation: $2 - 3x = 2x - 8$

Reflect & Share

Compare the solutions for the equations with those of another pair of classmates.
What strategies did you use to solve the equations?
How did you use zero pairs?

Connect

Recall how we used algebra tiles to solve equations in *Unit 1*.
Remember that to keep the balance of an equation, what you do to one side you must also do to the other side.

To solve the equation $3x - 8 = -x$,
isolate the variable tiles on one side of the equation.

On the left side, put algebra tiles to represent $3x - 8$.

On the right side, put algebra tiles to represent $-x$.

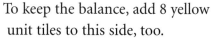

To isolate the x-tiles on the left side, add 8 yellow unit tiles to make zero pairs.

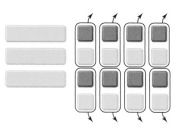

To keep the balance, add 8 yellow unit tiles to this side, too.

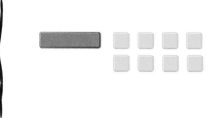

To isolate the unit tiles on the right side, add 1 yellow x-tile to each side.

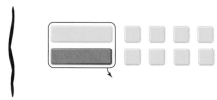

There are 4 x-tiles. So, arrange the unit tiles into 4 equal groups.

The tiles above show the solution $x = 2$.

When you solve an equation, you should always verify the solution. To do this, substitute the solution into the equation to check that it satisfies the equation. Substitute $x = 2$ into $3x - 8 = -x$.

Left side $= 3x - 8$ 　　　　　　　　Right side $= -x$
$= 3(2) - 8$ 　　　　　　　　　　　　$= -2$
$= 6 - 8$
$= -2$

Since the left side equals the right side, $x = 2$ is correct.

Example

a) Use algebra tiles to solve the equation $2x + 3 = 4x - 3$.

b) Verify the solution.

c) Interpret the equation in words.

Solution

a) $2x + 3 = 4x - 3$

Isolate the x-tiles on the left side.
Add 3 red unit tiles to each side.

Isolate the unit tiles on the right side.
Add 4 red x-tiles to each side.

There are 2 x-tiles. So, arrange the unit tiles into 2 equal groups.

The tiles show that one red x-tile equals 3 red unit tiles.
Flip the tiles on each side.
One yellow x-tile equals 3 yellow unit tiles.
So, $x = 3$

b) To verify the solution, substitute $x = 3$ into $2x + 3 = 4x - 3$.

Left side $= 2x + 3$	Right side $= 4x - 3$
$= 2(3) + 3$	$= 4(3) - 3$
$= 6 + 3$	$= 12 - 3$
$= 9$	$= 9$

Since the left side equals the right side, $x = 3$ is correct.

c) $2x + 3 = 4x - 3$

This means: two times a number plus three is equal to
four times the number minus three.

The *Example* shows what you do if you end up with
red variable tiles.
Flip the tiles on both sides of the equation.

Practice

1. Interpret each equation in words.
   Then use algebra tiles to solve the equation.
   a) $2x = x + 5$ **b)** $3x - 2 = x$
   c) $7x - 9 = 4x$ **d)** $6 - x = 2x$

2. Use algebra tiles to solve each equation.
   a) $7 - 3x = -4x + 13$ **b)** $4x + 3 = 2x + 7$
   c) $3x - 4 = x + 2$ **d)** $5 - x = 7 - 2x$

3. **a)** Interpret each equation in words.
   b) Use algebra tiles to solve each equation.
   c) Verify each solution.
      i) $2x + 2 = 3x - 5$
      ii) $5x - 6 = 8 - 2x$
      iii) $3x - 13 = x - 7$

4. One less than two times a number is equal to
   three more than the number.
   Let x represent the number.
   Then, an equation is: $2x - 1 = x + 3$
   Use algebra tiles to solve the equation. What is the number?

5. Five times a number is equal to
   two more than three times the number.
   Let n represent the number.
   Then, an equation is: $5n = 2 + 3n$
   a) Use algebra tiles to solve the equation. What is the number?
   b) Verify your solution.

6. The sum of a number and three more than the number is 23.
Let t represent the number.
Then, an equation is: $t + t + 3 = 23$
 a) Use algebra tiles to solve the equation. What is the number?
 b) Verify your solution.

7. **Assessment Focus** Two times the edge length
of a cube is 6 cm longer than the edge length.
Let l centimetres represent the edge length of the cube.
An equation for the edge length is: $2l = 6 + l$
 a) Use algebra tiles to solve the equation.
 What is the edge length of the cube?
 b) Verify the solution.
 c) What are the surface area and the volume of the cube?

Take It Further

8. The sum of three consecutive numbers is 63.
 a) Write an equation you could use to solve this problem.
 b) Solve the equation. What are the numbers?
 c) Verify your solution.

9. Solve these equations. Verify your solutions.
 a) $7x + 4 = 3x - 8$ **b)** $3 - 2x = 13 + 3x$

Math Link

Science
Pressure is force per unit area.
Pressure is measured in pascals (Pa).
A formula for pressure is:
Pressure = $\frac{\text{Force}}{\text{Area}}$
When we know the pressure in pascals and the area
in square metres, we can solve this formula to find
the force in newtons (N).

Reflect

Explain how you can use algebra tiles to solve an equation
with variables on both sides of the equal sign.
Include an example in your explanation.

10.5 Solving Equations Algebraically

Focus Solve a problem by solving a related equation.

Explore

Work with a partner. Solve this problem.
My mother's age is 4 more than 2 times my brother's age.
My mother is 46 years old.
How old is my brother?

Reflect & Share

Discuss the strategies you used for finding the brother's age with those of another pair of classmates.
Did you use an equation?
If not, how could you represent this problem with an equation?

Connect

In *Unit 1*, you learned how to solve equations algebraically.
All the equations in *Unit 1* had solutions that were whole numbers.
We use the same method to solve an equation where the solution is a fraction or a decimal.

Example 1

Three more than two times a number is 4. What is the number?
a) Write an equation to represent this problem.
b) Solve the equation.
c) Verify the solution.

Solution

a) Let the number be n.
    Then, two times the number is: $2n$
    And, three more than two times the number is: $3 + 2n$
    The equation is $3 + 2n = 4$

b)

$$3 + 2n = 4$$
$$3 + 2n - 3 = 4 - 3 \qquad \text{To isolate } 2n, \text{ subtract 3 from each side.}$$
$$2n = 1$$
$$\frac{2n}{2} = \frac{1}{2} \qquad \text{Divide each side by 2.}$$
$$n = \frac{1}{2}$$

Using the inverse operation here is the same as using zero pairs.

c) To verify the solution, substitute $n = \frac{1}{2}$ into $3 + 2n = 4$.

Left side $= 3 + 2n$ Right side $= 4$
$$= 3 + 2\left(\tfrac{1}{2}\right)$$
$$= 3 + 1$$
$$= 4$$

Since the left side equals the right side, $n = \frac{1}{2}$ is correct.
The number is $\frac{1}{2}$.

In *Example 1*, we could write the solution $n = \frac{1}{2}$ as a decimal,
$n = 0.5$.
However, some fractions, such as $\frac{1}{3}$, are repeating decimals.
Do not convert a fraction of this type to a decimal.

We can use an equation to solve problems related to number
patterns.
When we know the nth term and the term value,
we can solve an equation to find the term number.

Example 2

The nth term of a number pattern is $5n - 2$.
What is the term number when the term value is 348?

Solution

The nth term is $5n - 2$.
The term value of an unknown term number is 348.
Write the equation: $5n - 2 = 348$
Solve this equation for n.

$$5n - 2 = 348$$
$$5n - 2 + 2 = 348 + 2 \qquad \text{To isolate } 5n, \text{ add 2 to each side.}$$
$$5n = 350$$
$$\frac{5n}{5} = \frac{350}{5} \qquad \text{Divide each side by 5.}$$
$$n = 70$$

The 70th term has value 348.

In *Example 2*, the equation could have been solved by inspection:
$5n - 2 = 348$
Think: what do you subtract 2 from to get 348?
Answer: you subtract 2 from 350.
Think: what do you multiply 5 by to get 350?
Answer: you multiply 5 by 70.
So, $n = 70$

The equation could also have been solved by systematic trial:

$5n - 2 = 348$

Use a calculator to substitute different numbers for n until the left side of the equation equals 348.

In *Example 2*, there is only one value of n that makes the equation true. If $n = 69$, or if $n = 71$, or if n equals any number other than 70, the equation is not true.

Practice

Use algebra, systematic trial, or inspection to solve an equation.

1. Solve each equation.
   a) $2x = 3$ **b)** $3x = 2$ **c)** $4x = 6$ **d)** $5x = 12$

2. Solve each equation. Verify the solution.
   a) $2x - 1 = 5$ **b)** $7 = 1 + 3n$
   c) $10 = 4a - 1$ **d)** $5 + 2m = 6$

3. Write, then solve, an equation to answer each question. Verify the solution.
   a) Ten more than three times a number is 25.
      What is the number?
   b) Ten less than three times a number is 25.
      What is the number?
   c) Twenty-five subtracted from one-half a number is 10.
      What is the number?
   d) One-half of a number is subtracted from 25.
      The answer is 10.
      What is the number?

4. Navid has $72 in her savings account.
   Each week she saves $24.
   When will Navid have a total savings of $288?
   a) Write an equation you can use to solve the problem.
   b) Solve the equation.
      When will Navid have $288 in her savings account?
   c) How can you check the answer?

5. Assessment Focus The Grade 8 students had an end-of-the-year dance. The disc jockey they hired charged a flat rate of $85, plus $2 for each student who attended the dance. The disc jockey was paid $197. How many students attended the dance?
   a) Write an equation you can use to solve the problem.
   b) Solve your equation. Verify the solution.

6. The nth term of a number pattern is $4n - 3$.
   a) What is the term value for each term?
      i) the 10th term   **ii)** the 20th term
   b) What is the term number for each term value?
      i) 53   **ii)** 97

7. The nth term of a number pattern is $9n + 1$.
   What is the term number for each term value?
   a) 154   **b)** 118   **c)** 244

8. Use this information:
   Water flows into a bathtub at a rate of 15 L/min.
   a) Write a problem that can be solved using an equation.
   b) Write, then solve, the equation.

9. Use this information:
   Boat rental: $300   Fishing rod rental: $20
   a) Write a problem that can be solved using an equation.
   b) Write the equation, then solve the problem.
   c) How could you have solved the problem without writing an equation? Explain.

Calculator Skills

The area of a square is 225 m².

Calculate the length of a diagonal of the square to the nearest centimetre.

Take It Further

10. Two more than the square of a number is 123.
   What is the number?
   a) Write an equation you could use to find the number.
   b) Solve the equation. What is the number?
   c) Verify the solution.

Reflect

Choose one of the word problems in this section.
Explain the steps you used to write the equation,
then to solve the equation.

Writing a Journal

A journal is a place to record ideas, observations, illustrations, and responses.
The responses to *Reflect* in each lesson are often recorded in a journal.
Here are some ideas for other items to include.

➤ Comment on thoughts and feelings, successes and challenges:
- I worked well in the group today because…
- I could improve my skills with integer operations by …

➤ Explain key math ideas, formulas, and words:
- Write the word followed by a definition, picture, and example. Here is an example for *Unit 3*.

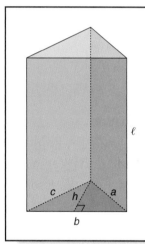

Triangular Prism

A triangular prism is a polyhedron with two congruent triangular bases, and its other faces are rectangles.

- The volume is:
  $V = base\ area \times height$
  $V = \frac{1}{2}bh\ell$

- The surface area is:
  $SA = sum\ of\ the\ areas\ of\ the\ faces$
  $SA = a\ell + b\ell + c\ell + bh$

➤ Write the steps you would use to do a math task:
- The steps I would follow to draw a circle graph…

➤ Create a math problem that uses the ideas from the lesson or unit:
- Create a problem you could use an equation to solve.

➤ Make a list of examples of a math topic. Use headings to organize the list:
- List the different types of problems that involve percents.
- Draw different kinds of polygons with the same attributes.

➤ Explain or justify a solution, pattern, or choice of strategy:
  – This solution makes sense because…
  – I chose to make a model because …

➤ Explain how you could apply the math:
  – Who would need to calculate the area of a circle? Why?
  – How does the media use charts and graphs when they want to persuade?
  – Math was in the news today…

➤ Summarize what you learned:
  – The main ideas I learned today (this week) are…
  – Draw a concept map to show the key ideas today (this week). Here is a concept map for *Unit 10, Lesson 10.1.*

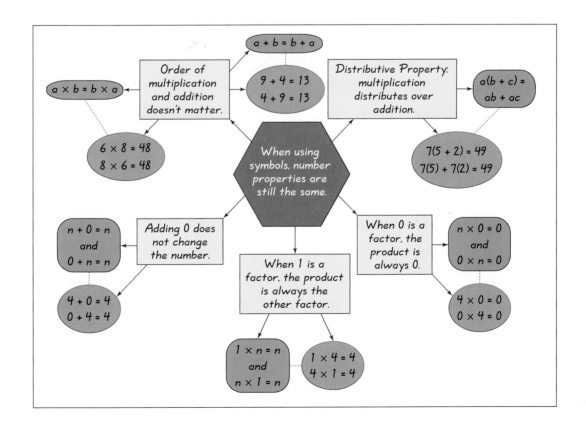

Health Care Professionals

In the 16th century, a person who needed medical attention often went to the local hair-cutting shop for treatment. The patient would be seen by a *barber-surgeon*, someone who was not only skilled at cutting hair, but also trained to cut into the human body. In fact, a surgeon was often nicknamed "sawbones." The tool most often used by a barber-surgeon was a leech!

Today, hospitals and emergency rooms are staffed with medical experts. From the paramedic who may treat the patient on the way to the hospital, to the nurse in the recovery ward, everyone has extensive training. Mathematics is an important part of this training.

Suppose a doctor prescribes a patient 30 mg of a certain drug. The medicine is in a bottle, with 150 mg of the drug diluted in 20 mL of liquid. How many millilitres of the medicine must the nurse give the patient? The nurse must be precise because too much or too little of the drug could harm or even kill the patient. The nurse uses this equation to determine the dosage, in millilitres:

$$\text{Dosage} = \frac{\text{amount of drug needed}}{\text{amount of drug diluted in bottle}} \times \text{the amount of liquid in bottle}$$

$$= \frac{30}{150} \times 20$$

$$= \frac{1}{5} \times 20$$

$$= 4$$

The correct dose is 4 mL.

Calculating dosages for children is often based on their body mass, and the dosage will be a fraction of a typical adult dose.

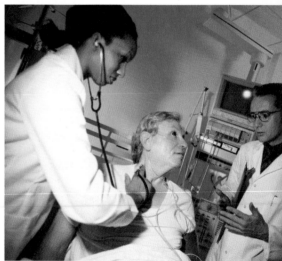

Unit Review

What Do I Need to Know?

☑ **Distributive Property**

The product of a number and the sum of two
numbers can be written as a sum of two products:
$a(b + c) = ab + ac$

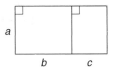

☑ **The *n*th Term of a Number Pattern**

- The *n*th term can be used to find the term value
  of any term in a pattern.
  For example, for the pattern with *n*th term $3n + 2$,
  the 9th term is: $3(9) + 2 = 29$
- The *n*th term can also be used to find the term number
  when the term value is known.
  For example, for the pattern with *n*th term $3n + 2$,
  to find the term number with the term value 23,
  solve the equation $3n + 2 = 23$, to get $n = 7$.
  The 7th term has value 23.

What Should I Be Able to Do?

For extra practice, go to page 497.

LESSON

10.1
1. Expand.
    a) $6(x + 9)$
    b) $3(11 + 4x)$
    c) $5(7x + 6y + 5)$
    d) $4(3a + 5b + 7c)$

10.2
2. For each algebraic expression,
substitute $n = 1, 2, 3, 4$, and 5
to generate a number pattern.
Describe each pattern,
then write a pattern rule.
    a) $3n + 5$ **b)** $5n + 15$

3. For each number pattern below:
    a) Write a pattern rule.
        Justify your rule.
    b) Graph the pattern. Use the
        graph to find the 7th term.
    c) Write an expression for
        the *n*th term.
    d) Find the 70th term.
        i) 8, 12, 16, 20, 24, …
        ii) 5, 7, 9, 11, 13, …

10.3 **4.** Here is a pattern drawn on isometric dot paper.

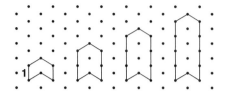

Frame 1   Frame 2   Frame 3   Frame 4

The distance between two adjacent dots is 1 unit.
The pattern continues.

a) Find the perimeter of each frame. What pattern do you see in the perimeters?

b) Use a pattern to find the perimeter of Frame 9.

c) Write an expression for the perimeter of the nth frame.

d) Find the perimeter of Frame 50.

10.4 **5.** Interpret each equation in words. Then use algebra tiles to solve the equation.
Verify each solution.

a) $12 - x = 3x$

b) $4x - 7 = 2x + 3$

c) $3x - 8 = x$

d) $3 - 7x = 7 - 9x$

6. Five less than two times a number is equal to one less than the number.
Let n represent the number.
Then, an equation is:
$2n - 5 = n - 1$

a) Use algebra tiles to solve the equation.
What is the number?

b) Verify your solution.

10.5 **7.** Solve each equation.
Verify the solution.

a) $3x + 2 = 4$

b) $4x = 10$

c) $11 = 3x + 1$

d) $4x - 7 = x + 1$

8. The school's sports teams hold a banquet. The teams are charged $125 for the rental of the hall, plus $12 for each meal served. The total bill was $545. How many people attended the banquet?

a) Write an equation you could use to solve the problem.

b) Solve your equation.

c) Verify the solution.

9. The nth term of a number pattern is $4n - 1$.

a) Write the first 5 terms of the pattern.

b) Which term number has each term value?
   i) 79   ii) 139   iii) 395

10. a) Write an expression for the nth term of this number pattern:
7, 13, 19, 25, …

b) Use the expression in part a. Which term number has each term value?
   i) 151   ii) 307   iii) 433

Practice Test

1. Interpret each equation in words.
   Solve the equation.
   Verify the solution.
   a) $x + 5 = 3x - 9$
   b) $2x - 5 = 10$

2. Whoopi saves pennies. She has 10¢ in her jar at the start.
   Whoopi starts on January 1st. She saves 3¢ every day.
   a) How many pennies does Whoopi have in the jar on
      each of January 1st, 2nd, 3rd, 4th, 5th, and 6th?
      Record the results in a table.
      What pattern do you see? Write a pattern rule.
   b) Write an expression for the nth term.
   c) Use the expression to find the 25th term.
   d) How could you find how much money Whoopi
      saved in January?

3. Anoki is holding a skating party.
   The rental of the ice is $75, plus $3 per skater.
   a) Write an expression for the cost in dollars for n skaters.
   b) Use the expression in part a to find the total cost for
      25 skaters.
   c) What if Anoki has a budget of $204. Write an equation
      you can solve to find how many people can skate.
      Solve the equation.

4. Two number patterns have these nth terms.
   Pattern A: $6n + 4$
   Pattern B: $5n - 3$
   a) Find the 48th term of Pattern A.
   b) Use the term value from part a.
      Which term number in Pattern B has this term value?
      How do you know?

Suppose your older sister has bought a cell phone.
She asks for your help to find the best cell phone plan.

Part 1

1. Here are three cell phone plans.
Each plan includes 200 free minutes.

CanTalk: $30.00 per month, plus $0.30 per additional minute
Connected: $35.00 per month, plus $0.25 per additional minute
In-Touch: $40.00 per month, plus $0.20 per additional minute

Copy and complete this table.

Number of Additional Minutes	40	80	120	160	200
CanTalk					
Connected					
In-Touch					

2. Which plan would you choose if your sister uses
40 additional minutes per month? 120 additional minutes?
200 additional minutes? Explain.

3. Graph the data in the table.
Use a different colour for each plan.
Join each set of points with a broken line.
Label each line with the name of the plan.
What patterns do you see?
What happens to the lines when the number of additional
minutes is 100?
What does this represent?
Which plan would you choose if your sister uses 100 additional
minutes per month? Explain.

Part 2

For each plan, write an expression for the monthly cost of n additional minutes.

Use each expression to find the total monthly cost for 85 additional minutes for each plan.

Suppose your sister can spend $80 a month on her cell phone. Write an equation you can solve to find how many additional minutes she can afford with each plan.

Solve each equation. Explain what each solution means.

Convert each money amount to cents before you write the equations.

Part 3

Write a paragraph to explain what decisions you have made about choosing the best cell phone plan.

Reflect on the Unit

Explain how patterns, expressions, and equations are used to solve problems. Include an example in each case.

Probability

- A quiz has two questions. Each question provides two answers. You guess each answer. What is the probability that you guess both answers correctly?

- Jason's golden retriever is about to have two puppies. Jason wants to sell the puppies. He gets more money for female puppies. What is the probability that both puppies are female?

  How are these two problems alike?

What You'll Learn

- Identify 0 to 1 as a range of probabilities.
- List possible outcomes of experiments.
- Identify possible and favourable outcomes.
- Calculate probabilities from tree diagrams and lists.
- Compare theoretical and experimental probability.
- Apply probability to sports, weather predictions, and political polling.

Why It's Important

You need to be able to make sense of comments you read or hear in the media relating to odds, chance, and probability.

Key Words

- experimental probability
- relative frequency
- theoretical probability
- simulation
- odds

Skills You'll Need

Experimental Probability

The **experimental probability** of an event occurring is:

$$\frac{\text{Number of times the event occurs}}{\text{Total number of trials}}$$

The experimental probability may be written as a fraction, decimal, or percent. Experimental probability is also called **relative frequency**.

Example 1

In baseball, a batting average is a relative frequency of the number of hits. The number of times a player goes to bat is the player's "at bats."

This table shows data for 3 girls from a baseball team.

Name	At Bats	Hits
Abba	29	11
Gina	35	17
Stacy	42	23

Calculate the batting average for each player.

Solution

To calculate each player's batting average, divide the number of hits by the number of at bats.

Abba $= \frac{11}{29}$ Gina $= \frac{17}{35}$ Stacy $= \frac{23}{42}$

$\doteq 0.379$ $\doteq 0.486$ $\doteq 0.548$

A batting average is always written with 3 decimal places.

✓ Check

1. a) A quality control inspector tested 235 CDs and found 8 defective. What is the relative frequency of finding a defective CD?

 b) A rug cleaning service uses telemarketing to get business. For every 175 telephone calls made, on average, there will be about 28 new customers. What is the experimental probability of getting a new customer over the phone?

Theoretical Probability

When the outcomes of an experiment are equally likely,

the **theoretical probability** of an event is: $\dfrac{\text{Number of outcomes favourable to that event}}{\text{Number of possible outcomes}}$

Usually we simply refer to the *probability*.
We can use theoretical probability to predict how many times a particular event will occur when an experiment is repeated many times.

Example 2

A number cube is labelled from 1 to 6.

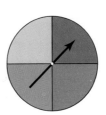

a) The number cube is rolled.
   What is the probability of getting a number less than 4?
b) The cube is rolled 40 times.
   Predict how many times a number less than 4 will show.

Solution

When a number cube is rolled, there are 6 possible outcomes: 1, 2, 3, 4, 5, 6
The outcomes are equally likely.

a) A number less than 4 is 1, 2, or 3.
   So, 3 outcomes are favourable to the event
   "a number less than 4."
   The probability of rolling a 1 or 2 or 3 is: $\frac{3}{6} = \frac{1}{2}$
b) The predicted number of times a number less than 4 will show is:
   $\frac{1}{2} \times 40 = 20$

✓ Check

2. The pointer on this spinner is spun 80 times.
   Is each statement true or false? Justify your answer.
   The pointer will land:

   a) on blue exactly 20 times
   b) on red or green about 40 times
   c) on each colour an equal number of times
   d) on each colour approximately 20 times

Focus Investigate the range of 0 to 1 for probability.

When your favourite hockey team is playing, you may ask: Will the team win?
You may try to predict the outcome of the game.
By making a prediction, you are estimating the probability that the event will occur.

Explore

Work on your own.
Use one of these words to describe the probability of each event below: impossible, unlikely, likely, certain

A   Get a tail when a coin is tossed.
B   Roll a 3 or 5 on a number cube labelled 1 to 6.
C   Not roll a 3 or 5 on a number cube labelled 1 to 6.
D   The same team will win the Grey Cup three years in a row.
E   The sun will set tomorrow.
F   A card drawn from a standard deck is a diamond.
G   A card drawn from a standard deck is not a diamond.
H   Roll a 4 on a number cube labelled 1 to 6.
I   January immediately follows June.
J   You will listen to a CD today.
Calculate the probability of each event where you can.
If you cannot calculate a probability, estimate it. Explain your estimate.

Reflect & Share

Compare your results with those of a classmate.
Arrange the events in order from impossible to certain.
What is the probability of an impossible event?
What is the probability of a certain event?

Connect

On page 455, you reviewed the method to calculate the theoretical probability of an event. When *all* the outcomes are favourable to that event, then the fraction:

$$\frac{\text{Number of outcomes favourable to the event}}{\text{Number of possible outcomes}}$$

has numerator equal to denominator, and the probability is 1.

When *no* outcomes are favourable to that event, then the fraction:

$$\frac{\text{Number of outcomes favourable to the event}}{\text{Number of possible outcomes}}$$

has numerator equal to 0, and the probability is 0.

So, the probability of an event occurring can be marked on a scale from 0 to 1.
When an event is impossible, the probability that it will occur is 0, or 0%.
When an event is certain, the probability that it will occur is 1, or 100%.
All other probabilities lie between 0 and 1.

Impossible   Certain

| 0.0 | 0.1 | 0.2 | 0.3 | 0.4 | 0.5 | 0.6 | 0.7 | 0.8 | 0.9 | 1.0 |
0%   100%

Example

Twenty cans of soup were immersed in water.
Their labels came off. The cans are identical.
There are: 2 cans of chicken soup; 3 cans of celery soup;
4 cans of vegetable soup; 5 cans of mushroom soup;
and 6 cans of tomato soup
One can is opened.
a) What is the probability of each event?
   i) The can contains celery soup.
   ii) The can contains fish.
   iii) The can contains celery soup or chicken soup.
   iv) The can contains soup.
b) State which event in part a is:
   i) certain   ii) impossible

Solution

a) There are 20 cans, so there are 20 possible outcomes.
   i) Three cans contain celery soup.
      The probability of opening a can of celery soup is:
      $\frac{3}{20} = \frac{15}{100} = 0.15$, or 15%
   ii) None of the cans contains fish.
       The probability of opening a can of fish is: 0, or 0%

iii) Three cans contain celery soup and two contain chicken soup. This is 5 cans in all.

The probability of opening a can of celery soup or chicken soup is:

$\frac{5}{20} = \frac{25}{100} = 0.25$, or 25%

iv) Since all the cans contain soup, the probability of opening a can of soup is:

$\frac{20}{20} = 1$, or 100%

b) i) The event that is certain to occur is opening a can that contains soup.

This event has the greatest probability, 1.

ii) The event that is impossible is opening a can that contains fish.

This event has zero probability, 0.

In the *Example*, there are 17 cans that do not contain celery soup. So, the probability of not opening a can of celery soup is $\frac{17}{20}$. Then,

the probability + the probability $= \frac{3}{20} + \frac{17}{20} = 1$, or 100%
of opening   of not opening
a can of   a can of
celery soup   celery soup

The events "opening a can of celery soup" and "not opening a can of soup" are complementary events. The sum of their probabilities is 1.

This is true in general:

the probability + the probability of $= 1$, or 100%
of an event   that event
occurring   not occurring

Practice

1. Copy this number line. Place the letter of each event on page 459 on the number line at the number that best matches the probability of the event.

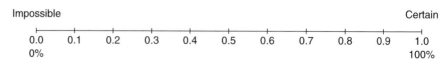

Impossible   Certain

0.0   0.1   0.2   0.3   0.4   0.5   0.6   0.7   0.8   0.9   1.0
0%   100%

a) Without looking, Jodi picks a pink block from a bucket with 5 pink blocks, 7 blue blocks, and 8 red blocks.

b) December immediately follows November.

c) Roll a 1 or 2 on a number cube labelled 1 to 6.

d) It will be warm in January in Ontario.

e) The colour of an apple is blue.

2. A TV weather channel reports a 35% chance of rain today. What is the probability that it will not rain today?

3. An election poll predicts that, in one riding, 4 out of 7 voters will vote for a certain political party. There are 25 156 voters. If the prediction is true, how many voters will not vote for that political party?

4. A magazine subscription service uses telemarketing to get business. For every 400 telephone calls, on average, there will be 88 orders for which 80 are actually paid.

a) What is the probability that, on any given call, a telemarketer will receive an order?

b) What is the probability that a telemarketer will receive an order that will be paid for?

5. Give an example of an event with each probability. Explain your choice.

a) 1 b) $\frac{1}{6}$ c) $\frac{1}{2}$ d) $\frac{3}{4}$ e) 0

6. Light bulbs from a production line are tested. Of the 150 bulbs tested, 7 were defective.

a) What is the probability of a bulb not being defective? Describe two different ways to calculate this.

b) What if 60 356 bulbs are produced in a week. How many may be defective?

7. To play a board game, Marty and Shane roll a number cube labelled 1 to 6. The person with the greater number goes first. Marty rolls a 4.

a) What is the probability Marty will go first?

b) What is the probability Marty will not go first?

c) What is the probability that both Marty and Shane have to roll the number cube again? Explain.

+	1	2	3	4
2				
3				
4				
5				

8. **Assessment Focus** A regular tetrahedron has 4 faces.
The faces of one tetrahedron are labelled 1 to 4.
The faces of the other tetrahedron are labelled 2 to 5.
The two tetrahedrons are rolled.
The numbers on the faces that land are added.
a) Copy and complete this table for the possible outcomes.
b) List the possible outcomes for the sum.
c) Calculate the probability of each outcome.
d) Add the probabilities in part c.
What do you notice?
Explain your result.
e) Which event has each probability?
   i) 25%   ii) 75%   iii) 1   iv) 0

Take It Further

9. In the card game, *In Between*, a deck of cards is shuffled.
Two cards are placed face up. A third card is dealt to the player.
If the third card falls in between the two cards that are face up,
the player wins. What is the probability that the player will
not win for each pair of cards that are face up?

a) a 2 of hearts and
   a 7 of spades

b) a 5 of diamonds and
   a king of clubs

c) an ace of clubs
   and a 4 of spades

d) a 10 of diamonds and
   a jack of clubs

Reflect

Can the probability of an event be less than 1? Greater than 1?
Explain.

Explore

Work with a partner.

You will need a number cube labelled 1 to 6 and a coin.

➤ One of you tosses the coin and one rolls the cube. Record the results.

➤ Calculate the experimental probability of the event "a head and a 2 show" after each number of trials.

- 10 trials
- 20 trials
- 50 trials
- 100 trials

➤ List the possible outcomes of this experiment.

➤ What is the theoretical probability of the event "a head and a 2 show"?

➤ How do the experimental and theoretical probabilities compare?

Reflect & Share

Compare your outcomes and probabilities with those of another pair of classmates.

Did you use a tree diagram to list the outcomes?

If not, work together to make a tree diagram.

Combine your results to get 200 trials.

What is the experimental probability of a head and a 2?

How do the experimental and theoretical probabilities compare?

Connect

Recall that a tree diagram can be used to find the outcomes of an experiment when the outcomes are equally likely.

Example

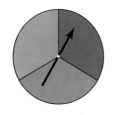

Carina and Paolo play the *Same/Different* game with the spinner at the left. A turn is two spins.

Player A scores 1 point if the colours are the same.

Player B scores 1 point if the colours are different.

a) Use a tree diagram to list the possible outcomes of this game.

b) Find the probability of getting the same colours.

c) Find the probability of getting different colours.

d) Is this a fair game? Explain.

e) If you think this game is fair, give your reasons.
   If the game is not fair, how could you change the rules to make it fair?

Solution

a) The first branch of the tree diagram lists the equally likely outcomes of the first spin: blue, green, pink
   The second branch lists the equally likely outcomes of the second spin: blue, green, pink
   For each outcome from the first spin, there are 3 possible outcomes for the second spin.
   Follow the paths from left to right.
   List all the possible outcomes.

First Spin	Second Spin	Possible Outcomes
blue	blue	blue/blue
	green	blue/green
	pink	blue/pink
green	blue	green/blue
	green	green/green
	pink	green/pink
pink	blue	pink/blue
	green	pink/green
	pink	pink/pink

Start

b) From the tree diagram, there are 9 possible outcomes.
   Three outcomes have: blue/blue, green/green, pink/pink
   The probability of the same colour is:
   $\frac{3}{9} = \frac{1}{3} \doteq 0.33$, or about 33%

c) Six outcomes have different colours: blue/green, blue/pink, green/blue, green/pink, pink/blue, pink/green
The probability of different colours is:
$\frac{6}{9} = \frac{2}{3} \doteq 0.67$, or about 67%

d) The game is not fair. The chances of scoring are not equal.
Player A can score only 3 out of the possible 9 points, while Player B can score 6 out of the possible 9 points.

e) Here is one way to make the game fair: Player A should score 2 points when the colours are the same. When each player has 9 turns, the players have equal chances of winning.

Carina and Paolo played the *Same/Different* game 100 times.
There were 41 same colours and 59 different colours.
The experimental probability of the same colours is:
$\frac{41}{100} = 0.41$, or 41%
The experimental probability of different colours is:
$\frac{59}{100} = 0.59$, or 59%
These probabilities are different from the theoretical probabilities in the solution of the *Example*.
The greater the number of times the game is played, the closer the theoretical and experimental probabilities may be.

Practice

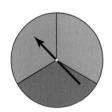

Spinner 1

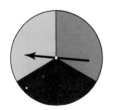

Spinner 2

1. Here is a spinner game called *Make Green*.
To play the game, a player spins the pointer on each spinner.
To win, a player must get blue on Spinner 1 and yellow on Spinner 2, because blue and yellow make green.
Your teacher will give you blank spinners.
Use an open paper clip as a pointer.

a) Make these spinners. Play the game 10 times.
Record your results.
How many times did you make green?

b) Combine your results with those of 9 classmates.
How many times did make green occur in 100 tries?
What is the experimental probability for make green?

c) Use a tree diagram to list the possible outcomes for make green.

d) What is the theoretical probability for make green?

e) How do the probabilities in parts b and d compare?

2. A number cube is labelled 1 to 6.
A coin is tossed and the cube is rolled.
Use the tree diagram you made in *Reflect & Share*.
Find the probability of each event.

a) a head and a 4

b) a number less than 3

c) a tail and a prime number

d) not a 5

3. **Assessment Focus** Tara designs the game *Mean Green Machine*. A regular tetrahedron has its 4 faces coloured red, pink, blue, and yellow.
A spinner has the colours shown.

When the tetrahedron is rolled,
the colour on its face down is recorded.
A player can choose to:

• roll the tetrahedron and spin the pointer, or

• roll the tetrahedron twice, or

• spin the pointer twice

To win, a player must make green by getting blue and yellow.
With which strategy is the player most likely to win?
Justify your answer.
Show your work.

4. Two tigers are born at the zoo each year for 2 years.
A tiger is male or female.

a) List the possible outcomes for the births after 2 years.

b) What is the probability of exactly 1 male tiger after 2 years?

c) What is the probability of at least 2 females after 2 years?

d) What is the probability of exactly 2 females after 2 years?

5. An experiment is: rolling a number cube labelled 1 to 6 and picking a card at random from a deck of cards.

The number on the cube and the suit of the card are recorded.

a) Use a deck of cards and a number cube.
   Carry out this experiment 10 times. Record your results.

b) Combine your results with those of 9 classmates.

c) What is the experimental probability of each event?
   i) a heart and a 4
   ii) a heart or a diamond and a 2
   iii) a red card and an odd number

d) Draw a tree diagram to list the possible outcomes.

e) What is the theoretical probability of each event in part c?

f) Compare the theoretical and experimental probabilities of the events in part c.
   What do you think might happen if you carried out this experiment 1000 times?

6. Neither Andrew nor David likes to set the table for dinner. They toss a coin to decide who will set the table. David always picks tails. What is the probability David will set the table 3 days in a row? Justify your answer.

Take It Further

7. Each letter of the word WIN is written on a piece of paper.
The pieces of paper are placed in a can.
You make three draws.
You place the letters from left to right in the order drawn.
The pieces of paper are not replaced after each draw.
If you draw the letters that spell WIN, you win a prize.

a) What is the probability of winning a prize?

b) What if the pieces of paper are replaced after each draw.
   Is the probability of winning increased or decreased? Explain.

c) Design a similar contest using the letters of CRUISE.
   Repeat parts a and b.

Reflect

How is a tree diagram useful when calculating probabilities?
Use an example in your explanation.

Mid-Unit Review

11.1 1. A TV news channel reports the results of a political poll. According to the poll, 53% of Canadians will reelect the current political party. What is the probability that the current political party will not be reelected?

2. Sal tossed a thumbtack 80 times. It landed point up 27 times. What is the experimental probability that the tack will land on its side?

3. Each number below is the probability of an event. Which of these words best describe the probability? certain, unlikely, likely, impossible
a) 0.8   **b)** 0   **c)** 0.2   **d)** 0.4

4. A dentist tested a new whitener for teeth. She found that, in 1050 tests, it worked 1015 times.
a) What is the experimental probability that the whitener works? That it does not work?
b) The whitener is used on 8500 people. About how many people can expect whiter teeth?

5. Fari was at bat 211 times and hit 97 times. Eric was at bat 234 times and hit 119 times. Which player has the better batting average? Explain.

11.2 6. Use a number cube labelled 1 to 6. Roll it twice. Record the results. Repeat this experiment until you have 10 results. Combine your results with those of 9 classmates. You now have 100 results.
a) Draw a tree diagram to list the possible outcomes.
b) What is the theoretical probability of each event?
  i) rolling a 4 and a 5
  ii) rolling two even numbers
  iii) rolling the same number twice
  iv) rolling a 3
  v) not rolling a 1 or a 6
c) What is the experimental probability of each event in part b?
d) Compare the two probabilities for each event. What do you notice?

7. Red and black combine to make brown.
a) Sketch two different spinners with pointers that could be spun to land on the two colours that combine to make brown.
b) Draw a tree diagram to show the possible outcomes of spinning the two pointers.
c) Calculate the probability of making brown.

Simulations

Focus Use simulation to estimate probabilities.

A real-life situation can be simulated by a probability experiment.
For certain events, a **simulation** is more practical than gathering data.

Explore

Work with a partner.
You will need a number cube labelled 1 to 6.
A store has a "Scratch and Save" day.
The store gives each customer a card with 6 circles on it.
Under the circles there are two matching percents and
four different percents.
You scratch two circles.

If you get the two matching percents, you have that discount
on all items you buy in the store that day.
The percents are arranged randomly on the card.
What is the probability you will scratch two matching percents?

➤ How can you use the number cube to estimate the probability
  you will scratch two matching percents?
➤ Conduct the experiment as many times as you can.
  Record each result.
➤ What is the experimental probability you will scratch two
  matching percents?

Reflect & Share

Compare your results with those of another pair of classmates.
Are the probabilities equal? Explain.
Work together to sketch a spinner you could use, instead of a
number cube, to carry out this experiment.

When we use a simulation to estimate probability, the model we use must have the same number of outcomes as the real situation. We use a coin when there are 2 equally likely outcomes.

We use a number cube, labelled 1 to 6, when there are 6 equally likely outcomes.

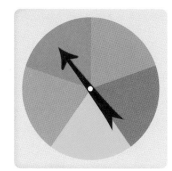

We use a spinner divided into congruent sectors, where the number of sectors matches the number of equally likely outcomes.

Example

In a Grade 8 class, for a group of 4 students, what is the probability that 2 or more students will have birthdays the same month? Design a simulation to find out.

Solution

There are 12 months in a year.

So, the probability of being born in a particular month is $\frac{1}{12}$.

We need a simulation that has 12 equally likely outcomes.

Use a number cube labelled 1 to 6, and a coin.

For each number on the cube, assign a head or a tail.

Then, let each month be represented by one of these pairs:

January H1, February T1,

March H2, April T2,

May H3, June T3,

July H4, August T4,

September H5, October T5,

November H6, December T6

Toss a coin and roll a number cube 4 times;

one for each student in the group.

Record if any month occurred two or more times.

Conduct the experiment 100 times.

An estimate of the probability is:

$$\frac{\text{The number of times a month occurred two or more times}}{100}$$

Practice

1. Work with a partner.
   You will need a coin and a number cube labelled 1 to 6.
   Conduct the experiment in the *Example* 25 times.
   Combine your results with those of 3 other pairs of students.
   Estimate the probability that, in a group of 4 people,
   2 or more people have birthdays in the same month.

2. What if you want to estimate the probability that, in a group of
   6 students, at least 3 students have birthdays in the same month.
   How could you change the experiment in the *Example* to do this?

3. a) When a child is born, the child is either female or male.
      What could you use to simulate this?
   b) You want to estimate the probability there are
      exactly 3 girls in a family of 4 children.
      Describe a simulation you could use.
   c) Conduct the simulation in part b.
      What is the estimated probability?
   d) Use a tree diagram to calculate the probability
      of exactly 3 girls.
   e) How do your answers to parts c and d compare? Explain.

4. What if you wanted to estimate or calculate the probability of
   exactly 1 boy in a family of 4 children.
   How could you use the results in question 3 to do this?

5. **Assessment Focus** A multiple-choice test has 5 questions.
   Each question has 4 answers.
   For each question, a student randomly chooses an answer.
   a) Design a spinner you could use to estimate the probability of
      getting 1 question correct.
   b) Make this spinner. Conduct a simulation to estimate the
      probability of getting 3 correct answers out of the 5 questions.
   c) How many times did you conduct the simulation?
      What is your estimate for the probability?
   Show your work.

6. Moira is on the school baseball team.
On average, Moira gets 1 hit every 3 times at bat.
Moira goes up to bat 4 times during a game.

a) How can this spinner
be used to simulate
Moira's batting average?
Justify your answer.

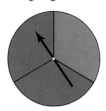

b) Conduct the simulation 30 times. Copy and complete this
table. Record the frequency of each number of hits per game.

Hits per Game	0	1	2	3	4
Frequency					

c) Estimate the probability that, in a game,
Moira will get each number of hits.
i) 0 hits   **ii)** 1 hit   **iii)** 2 hits   **iv)** 3 hits   **v)** 4 hits

7. The weather forecast for each of the next 6 days is
a 50% chance of rain.
Describe a simulation to estimate the probability that
it will rain on 3 of those 6 days.
Conduct the simulation. Explain the result.

Take It Further

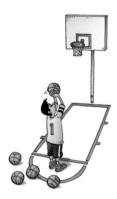

8. A basketball player has a 70% shooting average.
He is about to take 2 foul shots.
There is very little time remaining in the game,
and his team is behind by 1 point.

a) Describe a simulation to represent the player's
shooting ability.

b) Conduct 20 simulations of the player's 2 foul shots.

c) What is the experimental probability that the player's team
will win the game?

Reflect

When would you use a simulation to estimate a probability?
Include an example in your explanation.

Work with a partner.
You will need a number cube labelled 1 to 6, and a deck of cards.

➤ Suppose you roll the number cube.
What is the probability of getting a number greater than 2?
What is the probability of getting a number less than 2?
Which event is more likely?
How many times as likely is it?
➤ Suppose you remove a card, at random, from the deck.
What is the probability it is a heart?
What is the probability it is not a heart?
Which event is more likely?
How many times as likely is it?

Reflect & Share

How are the two events in each experiment alike?
How are they different?
Think of another experiment you could conduct that would have two events related the same way as those in *Explore*.

In the media, you may hear or read about the likelihood of the
Toronto Maple Leafs winning the Stanley Cup.
This likelihood may be referred to in terms of **odds**.
For example, if the Maple Leafs are likely to win, then the
odds in favour of their winning could be 5 to 1.
If the Maple Leafs are likely to lose, then the **odds against** their
winning could be 3 to 1.

In a bag, there are 12 counters.

A counter is picked at random.

The number of outcomes favourable to the counter being yellow is 4.

The number of outcomes not favourable to the counter being yellow is 8.

So, the odds in favour of the counter being yellow are 4 to 8.

This is a ratio, so it can be written in simplest form.

Divide each term by 4.

The odds in favour of the counter being yellow are 1 to 2.

The odds against the counter being yellow are 2 to 1.

In general,

odds in favour = number of favourable outcomes to number
of unfavourable outcomes

odds against = number of unfavourable outcomes to number
of favourable outcomes

Example

A card is drawn at random from a deck of cards.

a) What are the odds in favour of drawing a face card?

b) What are the odds against drawing a face card?

Solution

There are 3 face cards in each suit: Jack, Queen, King

There are 4 suits.

So, there are 3 × 4, or 12 favourable outcomes.

There are 52 cards in the deck.

So, there are 52 − 12, or 40 unfavourable outcomes.

a) The odds in favour of drawing a face card are 12 to 40.
Divide the terms by their common factor 4.
The odds in favour of drawing a face card are 3 to 10.

b) The odds against drawing a face card are 10 to 3.

Practice

1. What are the odds in favour of each event?

a) Getting a number greater than 1 when a number cube
labelled 1 to 6 is rolled

b) Getting a 2 when a card is randomly picked from a deck of playing cards

c) Getting the sum 7 when two number cubes labelled 1 to 6 are rolled and the numbers are added

2. What are the odds against each event in question 1?

3. What are the odds against each event?
   a) Getting a number less than 3 when a number cube labelled 1 to 6 is rolled
   b) Getting a black card when a card is randomly picked from a deck of playing cards
   c) Getting the sum 5 when two number cubes labelled 1 to 6 are rolled and the numbers are added

4. What are the odds in favour of each event in question 3?

5. Sera has 5 toonies, 8 loonies, 7 dimes, and 3 quarters in her piggy bank. One coin falls out at random.
   a) What are the odds in favour of the coin being a dime?
   b) What are the odds against the coin being a quarter?

6. The probability that David will be first in the 50-m backstroke is 25%. What are the odds in favour of his being first?

7. A weather report for Eastern Ontario was for a 40% probability of snow the next day.
   What are the odds against it snowing the next day?

8. **Assessment Focus** Use any materials or items you wish.
   a) Design an experiment where the odds in favour of one event are 3 to 7.
   b) Describe the experiment.
   List all possible outcomes and the odds in favour of each one.
   Show your work.

Number Strategies

There are six teams in a volleyball tournament. Each team plays every other team twice.

How many games will be played altogether?

Reflect

Explain how odds in favour and odds against are related.
Include an example in your explanation.

Extending a Problem

- Read each question 1 to 7 carefully.
- Solve the problem using numbers, materials, pictures, graphs, tables, equations, and/or words.
  Try to find a solution in two different ways.
- Communicate your different solutions.
- Explain the solutions to someone else.
  Try to use only words in your explanation.
  Point to the numbers and other math information you used while you explain.
- Make up a similar problem by changing something (context, math information, problem statement). Solve the new problem.

1. You earn $20 a week doing yard work for a neighbour.
   You can either be paid the $20, or you can pull two bills from a brown paper bag containing:
   - two $5 bills
   - two $10 bills
   - one $20 bill
   For example, you might pull out a $5 bill followed by a $10 bill, and earn only $15. Or you might pull out the $20 bill followed by a $10 bill, and earn $30.
   Which is the better way for you to be paid?
   Justify your answer.

2. Use squares to make the next similar figure in this pattern.
   a) How many squares would be needed to make the 10th figure? The 20th figure?
   b) Suppose the pattern had been made with congruent equilateral triangles or congruent parallelograms.
      Would the numbers of figures needed to make the pattern change? Explain.

3. Suppose you have an unlimited supply of three masses:
   3 kg, 11 kg, and 17 kg
   How many different ways can you combine at least one of each mass to make a total mass of exactly 100 kg?

4. Myles' car has an automatic transmission.
He drove 8 km. He spent 2 min at a stoplight.
Marlee's car has a manual shift.
She drove 9 km with no stops.
Use the table below. Who used more gas?

Action	Automatic	Manual
Idling	0.16 L/min	0.16 L/min
Starting	0.05 L	0.05 L
Driving	1 L/22 km	1 L/20 km

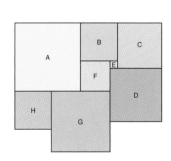

5. This staircase has 3 steps.
How many blocks would you need for a 20-step staircase?

6. The cost of renting a truck for a day is $45.
There is an additional charge of $0.20/km.
The cost, C dollars, for renting the truck and
driving it k kilometres is $C = 45 + 0.2k$.
a) The truck is driven 70 km. What was the rental cost?
b) Armand rented the truck. He paid a total of $65.
How far was the truck driven?
c) Jasmine rented the truck. She paid a total of $58.20.
How far did she drive the truck?

7. The area of Square F is 16 square units.
The area of Square B and of Square H is 25 square units.
The measure of each side is a whole number of units.
What is the area of each square?
a) A   **b)** C   **c)** D   **d)** E   **e)** G

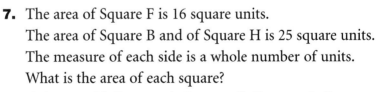

8. This table shows the temperatures on 5 consecutive days in four
Canadian cities during February. Create a math problem from
these data. Solve the problem.

City	Monday	Tuesday	Wednesday	Thursday	Friday
Kamloops, BC	4°C	6°C	8°C	9°C	5°C
Fredericton, NB	−10°C	−6°C	−3°C	−5°C	−6°C
Windsor, ON	3°C	2°C	−1°C	−4°C	0°C
Swift Current, SK	−5°C	4°C	6°C	8°C	6°C

Unit Review

Review any lesson with

What Do I Need to Know?

☑ **Experimental Probability**

The experimental probability of an event is:

$$\frac{\text{The number of times the event occurred}}{\text{The number of times the experiment is conducted}}$$

Experimental probability is also called relative frequency.

☑ **Theoretical Probability**

The theoretical probability of an event is:

$$\frac{\text{The number of outcomes favourable to that event}}{\text{Total number of outcomes}}$$

☑ **Probability Range**

All probabilities are greater than or equal to 0, and less than or equal to 1.

The probability of an event that is impossible is 0, or 0%.

The probability of an event that is certain is 1, or 100%.

$$\left(\begin{array}{c}\text{The probability that}\\\text{an event occurs}\end{array}\right) = 1 - \left(\begin{array}{c}\text{The probability that the}\\\text{event does not occur}\end{array}\right)$$

These events are complementary events.

The sum of the probabilities of all possible outcomes is 1.

☑ **Odds**

Odds in favour = number of favourable outcomes to number of unfavourable outcomes

Odds against = number of unfavourable outcomes to number of favourable outcomes

Your World

The Princess Margaret Hospital Foundation runs a lottery every year to raise money for the hospital. One year, the chances of winning were given as 1 in 15. So, the odds in favour of winning were 1 to 14.

LESSON

11.1 **1. a)** What is the probability of a certain event?

b) What is the probability of an impossible event?

c) Estimate or calculate the probability of each event.

   i) Chris randomly picks an orange out of a basket that contains 2 oranges, 6 apples, and 8 peaches.

   ii) March immediately follows April.

   iii) Roll a 1, 2, 3, or 4 on a number cube labelled 1 to 6.

   iv) It will be cold in January in the Arctic.

   v) You will have homework tonight.

2. A TV weather channel reports there is a 60% chance of snow tomorrow.
What is the probability that it will not snow tomorrow?

3. An election poll predicts that, in a certain riding, 5 out of 8 voters will vote for the Liberal Party in the next provincial election.
According to this prediction, how many voters out of 180 000 voters would not vote Liberal?

11.2 **4.** For the *Prime Numbers* game, roll two number cubes labelled 1 to 6. Then multiply the two numbers. When the product is a prime number, Player A gets 10 points. When the product is a composite number, Player B gets 1 point.

a) Draw a multiplication table to list the outcomes.

b) What if the number cubes are rolled 60 times.
Estimate the number of points each player might get.

c) Is this a fair game? Explain.

d) Could you draw a tree diagram to list the outcomes? Explain.

5. Merio's Deli offers sandwiches on rye or whole wheat bread with one choice of meat: turkey, ham, or pastrami; and one choice of cheese: mozzarella, Swiss, or cheddar.
Mya could not make up her mind. She requested a sandwich that would surprise her.

a) Draw a tree diagram to list the possible sandwiches.

b) What is the probability Mya will get each sandwich?

   i) Swiss cheese on rye bread

   ii) cheddar cheese but no turkey

Justify your answers.

11.2 6. In a board game, players take turns to spin pointers on these spinners. The numbers the spinners land on are multiplied. The player moves that number of squares on the board.

a) List the possible products.
b) What is the probability of each product in part a?
c) Which products are equally likely? Explain.
d) What is the probability of a product that is a perfect square? Explain.
e) What is the probability of a product that is less than 10? Explain.

7. Irina and Reanna share a dirt bike. They toss a coin to decide who rides first each day. Irina picks heads the first day, tails the second day, and tails the third day.
a) Draw a tree diagram to list the outcomes for 3 coin tosses.
b) What is the probability Irina will be the first to ride the dirt bike three days in a row?

11.3 8. Explain how to simulate the birth of a male or female puppy with each item. Justify your answer.
a) a coin
b) a number cube
c) a spinner

9. According to a news report, 1 out of every 4 new computers is defective.
a) Design an experiment to estimate the probability that, when 6 new computers are delivered to a store, 3 of them are defective.
b) Conduct the experiment. What is your estimate of the probability? Justify your answer.

11.4 10. Two number cubes are rolled. Each cube is labelled to 1 to 6. The numbers that show are subtracted.
a) What are the odds in favour of:
  i) the difference is 1?
  ii) the difference is greater than 3?
b) What are the odds against:
  i) the difference is an odd number?
  ii) the difference is 5?

11.

The probability of Anna hitting a home run is 25%. What are the odds against Anna hitting a home run?

Practice Test

1. A coffee shop states that if you buy an extra large coffee, you have a 15% chance of getting a free bagel. Out of 45 550 extra large coffee cups, 7280 cups have the word FREE BAGEL written under the rim.
   a) What is the probability of getting a free bagel if you buy an extra large coffee?
   b) Is the coffee shop's statement correct? Explain.

2. Weng-Wai and Sarojinee make up games that can be played with two tetrahedrons labelled 2 to 5.
   In Weng-Wai's game, players add the numbers rolled.
   If the sum is even, Player A gets 1 point.
   If the sum is odd, Player B gets 1 point.
   In Sarojinee's game, players multiply the numbers rolled.
   If the product is even, Player A gets 1 point.
   If the product is odd, Player B gets 1 point.
   a) What is the probability that Player A will win in Weng-Wai's game? Sarojinee's game?
   b) Is each game fair? Justify your answer.
   c) How could you use tree diagrams to solve this problem?

3. Keyna used this tangram as a dart board. Calculate the probability that a randomly thrown dart will land in each area. Justify your answers.
   a) orange
   b) purple
   c) blue or yellow
   d) green or orange

4. Describe a simulation you could conduct to estimate the probability that 3 people with birthdays in September were all born on an odd-numbered day.

A company that makes fortune cookies has 4 different messages in the cookies.

There are equal numbers of each message in a batch of cookies. Design a simulation you could use to solve these problems.

➢ What if you have 2 fortune cookies.
   What is the probability you will get 2 different messages?

➢ What if you have 3 fortune cookies.
   What is the probability you will get 3 different messages?

➢ What if you have 4 fortune cookies.
   What is the probability you will get 4 different messages?

➢ Estimate how many cookies you will need to get 4 different messages.

Conduct each simulation. Use any materials you think will help. Explain your choice of materials. Answer each problem. Show your work.

Reflect on the Unit

List the different methods you have learned to estimate and calculate probability. Include an example of how each method is used.

Integer Probability

Work with a partner.

Four integer cards, labelled −3, −2, +1, and +3, are placed in a bag.

James draws three cards from the bag, one card at a time. He adds the integers.

James predicts that because the sum of all four integers is negative, it is more likely that the sum of any three cards drawn from the bag will be negative.

In this *Investigation*, you will conduct James' experiment to find out if his prediction is correct.

Materials:
- four integer cards labelled −3, −2, +1, +3
- brown paper bag

Part 1

➤ Place the integer cards in the bag.
  Draw three cards and add the integers.
  Is the sum negative or positive?
  Record the results in a table.

Integer 1	Integer 2	Integer 3	Sum

➤ Return the cards to the bag. Repeat the experiment until you have 20 sets of results.

➤ Look at the results in your table. Do the data support James' prediction? Explain.

➤ Combine your results with
those of 4 other pairs of
classmates.
You now have 100 sets of
results.
Do the data support
James' prediction?
Explain.

➤ Use a diagram or other
model to find the
theoretical probability of
getting a negative sum.
Do the results match your
experiment?

➤ Do you think the values of the integers make a difference?
Find 4 integers (2 positive, 2 negative) for which
James' prediction is correct.

Part 2

Look at the results of your investigation in *Part 1*.
➤ If the first card James draws is negative, does it affect
the probability of getting a negative sum?
Use the results of *Part 1* to support your thinking.

➤ If the first card James draws is positive, does it affect
the probability of getting a negative sum?
Use the results of *Part 1* to support your thinking.

Take It Further

➤ Emma wonders if she will get a different result if she multiplies
the three integers instead of adding them.
She predicts that she is more likely to get a negative product
than a positive product.

➤ Design an experiment to find out if Emma's prediction
is correct.
Perform the experiment. Was Emma's prediction correct?
Explain.

UNIT

1 **1.** Write each number in expanded form and in scientific notation.
   a) 335
   b) 6272
   c) 24 242

2. Solve each equation.
   a) $x + 4 = 11$
   b) $x - 4 = 9$
   c) $4 + x = 7 + 9$
   d) $23 - 7 = x + 6$

2 **3.** a) The longest snake is the reticulated python.
   The record length for this python is 985 cm.
   Which scale would you use to draw this snake in your notebook? Justify your choice.
   i) 1:2   **ii)** 1:50
   iii) 1:100   **iv)** 1:1000

   b) Use the scale you chose in part a. Draw a line segment to represent a scale drawing of the snake.
   Is the scale you chose reasonable? Explain.

4. About 96% of the students at Westlake Middle School take at least one course from drama, music, or art.
   One thousand one hundred fifty-two students take at least one of these courses.
   How many students attend the school?

3 **5.** A rectangular sheet of cardboard is used to make a box.
   The cardboard is cut to make an open box with a base that measures 5 cm by 4 cm. The volume of the box is 60 cm³. Find the original area of the cardboard.

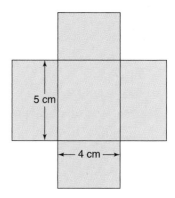

5 cm

← 4 cm →

4 **6.** Use the fact that $\frac{1}{4} = 0.25$ to write each number as a decimal.
   a) $\frac{1}{2}$ b) $\frac{3}{4}$ c) $1\frac{1}{2}$ d) $1\frac{3}{4}$

7. a) Find the number of hours in each fraction of a day.
   i) $\frac{2}{3}$ **ii)** $\frac{5}{6}$
   iii) $\frac{3}{4}$ **iv)** $\frac{3}{8}$

   b) Write each time in minutes as a fraction of an hour.
   i) 12 min   **ii)** 20 min
   iii) 6 min   **iv)** 136 min

5 **8.** The times, in minutes, that 14 students spent doing math homework over the weekend are:
   27, 36, 48, 35, 8, 40, 41,
   39, 74, 47, 44, 125, 37, 47

a) Organize the data in a stem-and-leaf plot.

b) Calculate the mean, median, and mode for the data.

c) What are the outliers? Calculate the mean without the outliers. What do you notice? Explain.

d) Which measure of central tendency best describes the data? Explain.

9. *Census at School*, Canada, surveyed elementary students in 2003 to find how much time they spent travelling to school. The results are shown in the table.

Number of Minutes	Number of Students
0–10	3531
11–20	2129
21–30	994
31–40	292
41–50	433
51–60	193
More than 60	111

a) Use a circle graph to display these data.

b) You will go to high school next year. Suppose the high school has an enrolment of 1200 students. How many students would you expect to take more than 30 min to get to school? What assumptions did you make?

6 **10.** **a)** Draw a circle with radius 12 cm. Calculate its circumference and area.

b) Draw a circle with diameter 6 cm. Calculate its circumference and area.

c) How are the circumferences in parts a and b related? Explain.

d) How are the areas in parts a and b related? Explain.

11. A can of tuna has height 3.5 cm and diameter 8.5 cm.

a) Calculate the volume of the can.

b) The label covers the curved surface of the can. Calculate the area of the label.

7 **12.** Find the measure of each angle labelled a, b, c. Explain your reasoning.

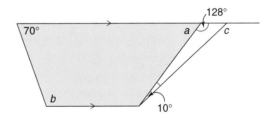

13. **a)** Draw a large triangle. Construct the perpendicular bisector of each side. Use this construction to draw a circle through the vertices of the triangle.

b) Explain why you can use this method to draw a circle through 3 points that are not on a line.

14. Use your knowledge of constructing a 60° angle and a 45° angle to construct a 105° angle.

8 **15.** Find each length indicated. Sketch and label the figure first.

a)

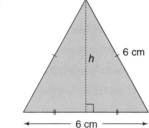

6 cm

h

6 cm

b)

20 cm

8 cm

x

c)

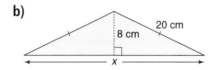

c

7 cm

d)

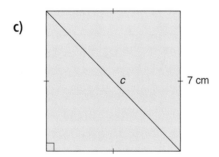

d

27.4 m

9 **16.** Evaluate.
a) $(+7) + (-12)$
b) $(-2) - (-12)$
c) $(-4) - (+13)$
d) $(-21) + (+16)$
e) $(+16) + (-9)$
f) $(-1) - (-9)$
g) $(-11) - (+11)$
h) $(+14) + (-18)$

17. Evaluate.
a) $(+3)(-8)$
b) $(+28) ÷ (-4)$
c) $(-5)(-6)(+7)$
d) $(-56) ÷ (-8)$

18. Evaluate.
a) $(-3)[(+5) - (-3)]$
b) $[(+8) ÷ (-4)] - (+10)(+3)$
c) $[(-6)(-8)] ÷ [(-10) ÷ (+5)]$

19. Draw and label a coordinate grid. Where on this grid are all points with:
a) first coordinate negative?
b) second coordinate positive?
c) first coordinate zero?
d) second coordinate −1?
e) equal coordinates?

20. Draw a triangle on a coordinate grid. Draw and label each image:
a) Translate the triangle 3 units left and 5 units down.
b) Reflect the triangle in the *x*-axis.
c) Rotate the triangle 90° clockwise above the origin.

10 **21.** For each number pattern below:
- **i)** Describe the pattern.
  Write the pattern rule.
- **ii)** Use a table to find the 11th term.
- **iii)** Write an expression for the nth term.
- **iv)** Use the expression to find the 70th term.
- **a)** 3, 5, 7, 9, 11, …
- **b)** 5, 8, 11, 14, 17, …
- **c)** 7, 11, 15, 19, 23, …
- **d)** 9, 14, 19, 24, 29, …

22. Four times the side length of an equilateral triangle is 9 cm longer than the side length.

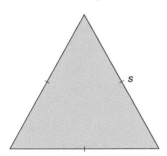

Let s centimetres represent the side length of the triangle.
An equation for the side length is:
$4s = s + 9$
- **a)** Use algebra tiles to solve the equation.
  What is the side length of the equilateral triangle?
- **b)** Verify the solution.
- **c)** What is the perimeter of the triangle?

11 **23.** A spinner has 3 congruent sectors labelled D, E, and F.
A bag contains 3 linking cubes: 2 green and 1 red
The pointer is spun and a cube is picked at random.
- **a)** Use a tree diagram to list the possible outcomes.
- **b)** What is the probability of:
  - **i)** spinning E?
  - **ii)** picking a green cube?
  - **iii)** spinning E and picking a green cube?
  - **iv)** spinning D and picking a red cube?

24. Conduct the experiment in question 23.
- **a)** Record the results for 10 trials.
  - **i)** State the experimental probability for each event in question 23, part b.
  - **ii)** How do the experimental and theoretical probabilities compare?
- **b)** Combine your results with those of 9 other students. You now have the results of 100 trials. Repeat part a, parts i and ii.
- **c)** What happens to the experimental and theoretical probabilities of an event when the experiment is repeated hundreds of times?

1. The table shows data about some of the top movies filmed in the Greater Toronto Area, as of November 2004.

Movie	Year	North American Gross (US$ millions)	International Gross (US$ millions)
Twister	1996	242.0	253.0
My Big Fat Greek Wedding	2002	241.4	115.0
Police Academy	1984	81.2	38.8
Good Will Hunting	1997	138.0	87.5
X-Men	2000	157.0	137.0

   a) What were the total gross earnings of each movie in the table?
   b) Order the movies from greatest to least total gross earnings.
   c) What were the total gross international earnings of these five movies?
   d) Write a problem that could be solved using these data.
      Solve your problem.

2. For each pair of numbers below:
   i) Find all the common factors.
   ii) Find the first 3 common multiples.
   a) 18, 126
   b) 40, 70
   c) 16, 40
   d) 3, 11

3. Find the missing prime factor in this equation: $1800 = 2^3 \times 3^2 \times \square$

4. Why can a prime number not be written as a product of prime factors?

5. Write in standard form.
   a) $5 \times 10^7 + 7 \times 10^2 + 2 \times 10^1 + 4$
   b) $6 \times 10^5 + 4 \times 10^4 + 8 \times 10^3 + 4 \times 10^2 + 5 \times 10^1 + 9$

6. These facts from *Guinness World Records 2005* involve large numbers. Write each number in scientific notation. Check with a calculator.
   a) The longest paperclip chain made by one person in 24 h was 162 760 cm long.
   b) The largest petition on record was signed by 21 202 192 people from 153 countries.
   c) The largest bouquet was made from 101 791 roses.

7. Write each number in expanded form.
   a) The typical person breathes 370 000 m³ of air in her or his lifetime.
   b) The average person makes about 1140 telephone calls each year.
   c) In 2002, Canada had 20 100 000 visitors.

8. Evaluate.
   a) $13.3 + 7.2 \times 3.5$
   b) $27.0 \div (1.5 \times 3) - 4.8$
   c) $41.3 - 2.5^2 \div 0.5$

9. Solve each equation. Verify the solution.
   a) $x + 6 = 11$
   b) $x - 6 = 11$
   c) $6 - x = 5$
   d) $6 + x = 5 + 9$

10. One DVD rents for $4. How many DVDs can be rented for $32? Let x represent the number of DVDs. Then, an equation is $4x = 32$. Solve the equation. Answer the question.

1. Brian received these marks on three tests: Geography: $\frac{18}{25}$ Math: $\frac{30}{40}$ Science: $\frac{40}{50}$ On which test did Brian receive the highest mark? How do you know?

2. Find each missing term.
   a) $\frac{h}{16} = \frac{5}{4}$
   b) $\frac{k}{3} = \frac{30}{45}$
   c) $b:6 = 35:42$
   d) $r:54 = 2:9$

3. The ice-cream parlour sells 5 chocolate cones for every 3 vanilla cones it sells. On Saturday, 30 vanilla cones were sold. How many chocolate cones were sold that day?

4. On a road map of Ontario, the scale is 1:1 000 000. The map distance between Toronto and Peterborough is 12 cm. What is the actual distance between these towns?

5. Find the average speed for each climb. Which average speed is greater?
   a) The world record for the fastest pole climb is 10.75 s to climb a 24-m pole.
   b) The world record for the fastest coconut tree climb is 4.88 s to climb an 888-cm tree.

6. Raj paid $35.00 for 4 m of fabric. How much would he pay for 6.5 m of the same fabric?

7. In a class of 33 Grade 8 students, 11 students were 14 years old when they graduated. The other students were under 14. What percent of students were 14 years old?

8. Forty-six members of the community volunteered to help clean up the ravine. About 26% of the volunteers were retired. How many of the volunteers were retired?

9. Hillary bought an outfit for her graduation. The outfit was on sale for 30% off. She saved $36.00.
   a) What was the original price of the outfit?
   b) How much did she pay for the outfit, including 15% sales tax?

10. In Toronto, the price of a new bungalow in 1968 was $25 000. In 2004, the price of this bungalow had increased to $400 000. What was the percent increase in price?

11. Only 24% of tree seedlings planted grew. Three hundred sixty trees grew. How many trees were planted?

12. To get advance sales, a school dance ticket sells for $4.00. This is 80% of the ticket price at the door. What will the ticket cost at the door?

13. Amal borrows $5000 from her mother to pay for a one-year college program. She plans to pay the money back in equal monthly payments over 3 years, including interest at a rate of 4.5% per year.
    a) What simple interest does Amal pay?
    b) What is each monthly payment?
    c) How much will Amal have paid at the end of 3 years?

1. An object is made with 6 linking cubes. Here are its views.

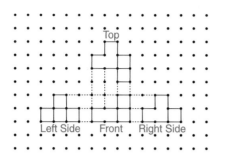

a) Build the object with linking cubes.
b) Make an isometric drawing of the object.

2. Calculate the area of this net.
Write the area in square metres and square centimetres.

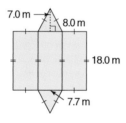

3. Calculate the surface area of each prism. Draw a net first if it helps. Write the surface area in square centimetres and square metres.

a)

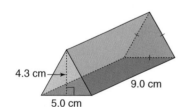

b)

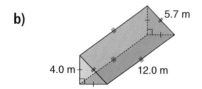

4. The base area and height for each triangular prism are given.
Find the volume of each prism.

a) 10 cm A = 8 cm²

b) 1.75 m

5. a) Use the variables below to sketch and label a triangular prism.
Each measurement has been rounded to the nearest whole number.
a, b, and c are the dimensions of a triangular face.
h is the height of the triangular face with b the base of the face.
l is the length of the prism.
$a = 14$ cm, $b = 34$ cm, $c = 22$ cm, $h = 6$ cm, $l = 24$ cm

b) Calculate the volume of this prism.

6. A glass crystal has the shape of a regular hexagonal prism. What volume of glass is needed to make the prism?

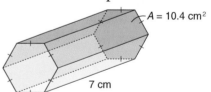

7. A triangular prism has volume 20 cm³.
Sketch a prism with this volume.
Label the dimensions you know.

8. The volume of a right isosceles triangular prism is 198 cm³.
The area of one triangular face is 18 cm². Find as many dimensions of the prism as you can.

1. When crossword puzzles were first designed, there was a rule that the black squares could not be more than $\frac{16}{100}$ of the grid.

    a) Look at the 3 grids below.
       Does each grid obey the rule?
       Explain.

    i)
    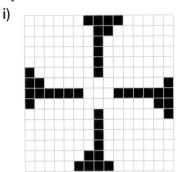

    ii)
    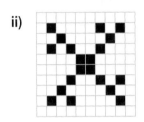

    iii)
    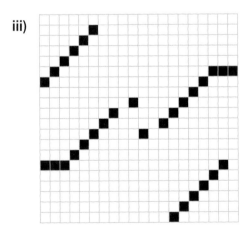

    b) Write the fraction of black squares on each grid. Order the fractions from least to greatest.

2. A fraction is written on each side of two counters.
   All the fractions are different.
   The counters are flipped and the fractions are added.
   Their possible sums are: 1, $1\frac{1}{4}$, $\frac{7}{12}$, $\frac{5}{6}$
   Which fractions are written on the counters? Explain how you know.

3. Add or subtract.
    a) $\frac{5}{3} + \frac{1}{6}$ b) $\frac{1}{4} + \frac{4}{3}$
    c) $\frac{5}{9} - \frac{1}{3}$ d) $\frac{3}{7} - \frac{1}{5}$
    e) $2\frac{2}{3} - 2\frac{1}{4}$ f) $4\frac{1}{4} + 2\frac{1}{2}$

4. Multiply.
    a) $\frac{5}{6} \times \frac{5}{6}$ b) $\frac{9}{4} \times \frac{8}{3}$
    c) $1\frac{1}{2} \times 2\frac{3}{4}$ d) $3\frac{1}{5} \times 4\frac{1}{8}$

5. Divide.
    a) $\frac{5}{3} \div \frac{3}{4}$ b) $\frac{5}{3} \div \frac{4}{3}$
    c) $1\frac{2}{3} \div 1\frac{1}{3}$ d) $2\frac{1}{4} \div 3\frac{1}{2}$

6. Simplify.
    a) $\frac{9}{5} - \frac{2}{3}$ b) $\frac{5}{2} + \frac{3}{4}$
    c) $\frac{9}{5} \times \frac{2}{3}$ d) $\frac{5}{2} \div \frac{3}{4}$

7. Write each decimal as a fraction.
    a) 0.75   b) 0.375
    c) 0.1875   d) 0.5625

8. Suppose you double the numerator of a fraction and halve the denominator. Is the new fraction greater than or less than the fraction you started with? Explain.

1. State if the data are collected from a census or a sample.
   Explain how you know.
   a) All government employees are asked to record their month of birth.
   b) Thirty of the 90 swimmers are asked to name their favourite make of goggles.

2. To find out if a new cover design should be developed for the school yearbook, the students on the yearbook committee ask their friends for their opinions.
   a) Is the sample biased or reliable? Explain.
   b) If the sampling method is biased, how can it be changed so the data collected better represent the population?

3. Marci recorded the times, in minutes, that Grade 8 students at her school spent travelling to school.
   Here are the results.
   12, 4, 10, 22, 53, 23, 34, 18, 15, 7, 16, 3, 19, 10, 45, 6, 28, 34, 47, 58, 6, 44, 1, 27, 30, 21, 2, 11, 41, 33, 5, 13, 18, 9, 23, 13, 24, 26, 8, 16, 20, 14, 31, 8, 10, 18, 14, 25, 3, 17, 26, 10
   a) Organize the data. Explain your method.
   b) Display the data using a suitable graph. Justify your choice.
   c) What information do you know from the graph that you do not know from the data?
   d) What can you infer from these data?

4. These data show Kelsi's and Courtney's heights, in centimetres, since birth.

Age	0	3	6	9	12
Kelsi's height (cm)	65	90	118	134	160
Courtney's height (cm)	55	85	112	130	153

   a) Graph the data.
   b) What trends do you see in the data? How does the graph show the trends?
   c) How could the graph be used to estimate each height?
      i) Kelsi's height at age 5 years
      ii) Courtney's height at age 15 years
      What assumptions do you make?
   d) Can the graph be used to predict each girl's height at 25 years of age? Explain.

5. This table shows the average number of births per day of the week in the United States in 2002.

Day of Week	Average Number of Births
Sunday	7 526
Monday	11 453
Tuesday	12 823
Wednesday	12 083
Thursday	12 365
Friday	12 285
Saturday	8 573

   a) Round the data to the nearest 100.
   b) Draw a circle graph to display the data.
   c) What percent of births occurred on:
      i) a Friday?   ii) the weekend?
   d) Why are there fewer births at the weekend?

Unit 6

1. Copy and complete this table.

	Radius	Diameter	Circumference
a)	6 cm		
b)		4.2 m	
c)			78.5 cm
d)	71.3 mm		

2. Write each measurement in question 1 in a different unit.

3. A designer has made a circular tablecloth with diameter 1 m. He wants to sew a fringe around the cloth. The fringe is sold by the tenth of a metre.
   a) What length of fringe is needed?
   b) One metre of fringe costs $4.70. How much will the fringe cost?

4. A circular pool has a circular concrete patio around it.

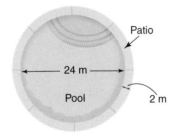

   a) What is the circumference of the pool?
   b) What is the combined radius of the pool and patio?
   c) What is the circumference of the patio?
   d) What is the surface area of the pool?
   e) What is the combined area of the pool and patio?
   f) What is the area of the patio?

5. A semicircular mat has a diameter of 60 cm.

60 cm

   What is the area of the mat?

6. A spinner has 8 congruent sectors: 6 yellow and 2 green. The radius of the spinner is 5 cm. What is the area of the yellow sectors?

7. A tunnel is cylindrical. The tunnel is 400 m long with radius 2.3 m. What is the capacity of the tunnel? Give the answer in litres.

8. A mug is cylindrical. Its diameter is 7 cm and its height 10 cm. The mug is half full of tea. How much tea is in the mug? Give the answer in millilitres.

9. Which cylinder do you think has the greater volume?
   • a cylinder with radius 1 m and height 2 m, or
   • a cylinder with radius 2 m and height 1 m

   How can you find out without using a calculator? Explain.

10. A hot water tank is cylindrical. Its interior is insulated to reduce heat loss. The interior has height 1.5 m and diameter 65 cm. What is the surface area of the interior of the tank? Give the answer in two different square units.

1. Name each angle described and its measure.
Justify your answers.

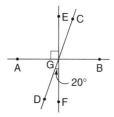

a) the angle opposite ∠EGC
b) the angle opposite ∠FGB
c) the supplement of ∠CGF
d) two angles complementary to ∠AGD
e) two angles supplementary to ∠DGF

2. Find the measures of ∠Q and ∠R.
Explain how you know.

3. Find the measure of each unmarked angle. Show your work.

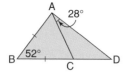

4. Find the measures of ∠SPT and ∠RPS.
Explain your reasoning.

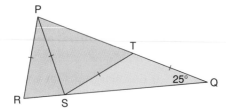

5. a) Can a supplementary angle also be a complementary angle? Explain.
b) Can a complementary angle also be a supplementary angle? Explain.

6. Look at this diagram.

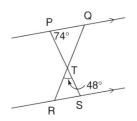

a) Name two parallel line segments.
b) Name two transversals.
c) Name two opposite angles.
d) Name two pairs of alternate angles.
e) Find the measures of ∠PTQ, ∠TSR, ∠TRS, and ∠PQT.

7. a) Draw line segment AB 10 cm long. Use a compass and ruler to draw the perpendicular bisector of AB.
b) Draw line segment AB again. Use a different method to draw the perpendicular bisector.
c) How can you check that you have drawn each bisector accurately?

8. a) Draw any obtuse angle. Label it ∠PQR. Use a compass and ruler to bisect ∠PQR.
b) Draw any acute angle. Label it ∠CDE. Use a different method to bisect ∠CDE.
c) How can you check that you have drawn each bisector correctly?

1. Copy each square on 1-cm grid paper. Find its area, then write the side length of the square.

   a)   b)

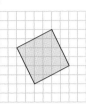 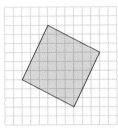

2. Use 1-cm grid paper.
   a) Draw a square with side length 1 cm.
      What is the area of the square?
      Draw a diagonal of the square.
      Find the length of the diagonal.
      Record the measurements in a table.
      Use these headings.

Side Length of Square (cm)	Area of Square (cm²)	Length of Diagonal (cm)

   b) Repeat part a for a square with each side length.
      i) 2 cm   ii) 3 cm   iii) 4 cm
   c) What patterns do you see in the table? Use these patterns to predict the length of a diagonal of a square with side length 7 cm. Show your work.

3. Estimate each square root. Explain how you estimated.
   a) $\sqrt{57}$ b) $\sqrt{157}$ c) $\sqrt{257}$

4. Use a calculator. Write each square root to 3 decimal places.
   a) $\sqrt{43}$ b) $\sqrt{1256}$ c) $\sqrt{2000}$

5. Look at the map below. The side length of each square is 10 km.
   How much farther is it to travel from Jonestown to Skene by car than by helicopter?

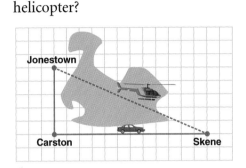

6. Find each length indicated. Sketch and label the triangle first.

   a)   b)

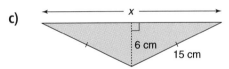

   c)

7. Part of a fence has equally-spaced boards that are 2 m high. Its diagonal board is 4 m long.

   About how long is the fence? Explain.

1. Write an addition expression for each scenario, then answer each question.
   a) At dawn, the temperature was $-6°$ C. It rose $4°C$.
      What was the final temperature?
   b) Surinder is overdrawn by $121. He pays $83 into his account.
      What is the result?
   c) A stock gains $15, then loses $23. What is the net result?

2. Evaluate.
   a) $(-4) + (-6)$
   b) $(-2) - (+2)$
   c) $(+8) + (-3)$
   d) $(-9) - (-8)$
   e) $(-3) + (+1)$
   f) $(-3) + (-5)$

3. Evaluate.
   a) $10 - 6$
   b) $-16 + 9$
   c) $(+7) - (+3) - (+4)$
   d) $(-13) + (-4) - (+7)$
   e) $(-44) - (-9) + (+91)$
   f) $(-78) - (-42) + (+12)$

4. The price of gold at the beginning of Week 1 was C$535 per ounce.
   Here are the weekly changes in the price of an ounce of gold.

Week 1	$+$5$
Week 2	$-$8$
Week 3	$-$3$
Week 4	$-$11$

   What was the price of an ounce of gold at the end of Week 4?
   Show your work.

5. In 2002, the mean summer temperature in Shingle Point, Yukon, was $+12°C$. The mean winter temperature was $-22°C$.
   What is the difference in these temperatures?

6. Which is greater? Can you tell without calculating? Explain.
   a) $(+8) + (-4)$ or $8 + 4$
   b) $(-10) + (-7)$ or $(-10) - (-7)$
   c) $(-2) + (-3)$ or $(-2)(+3)$

7. Evaluate.
   a) $(-56) \div (-4)$
   b) $(-4) \times (-15)$
   c) $(-18) \times (+2)$
   d) $\frac{-125}{+5}$

8. Evaluate.
   a) $(+3) - (-2)(+6) \div (-3)$
   b) $(+3)(-2) + (+6)(-3)$

9. Use these points:
   $A(-2, 3)$, $B(0, -1)$, and $C(4, -3)$
   a) Suppose the order of the coordinates is reversed. In which quadrant would each point be now?
   Draw $\triangle ABC$ on a coordinate grid.
   b) Which transformation changes the orientation of the triangle?
   Justify your answer with a diagram.
   c) For which transformation is the image of segment AC perpendicular to AC?
   Justify your answer with a diagram.

1. Simplify.
   a) $1 \times b$
   b) $a \times 0$
   c) $c + 0$
   d) $d \times 1$

2. Expand.
   a) $2(4x + 8y + 6)$
   b) $3(6c + 15d + 9)$
   c) $5(4x + y + 3)$
   d) $7(3a + 5b + 4)$

3. For each algebraic expression, substitute $n = 1, 2, 3, 4, 5, 6$ to generate a number pattern.
   Describe each pattern, then write a pattern rule.
   a) $2n$
   b) $2n + 1$
   c) $2n - 1$
   d) $2n + 3$

4. Here is a number pattern:
   $2, 5, 8, 11, 14, \ldots$
   a) Describe the pattern in words.
       Write a pattern rule.
   b) Make a table for the pattern.
       Graph the data in the table.
   c) Find the 15th term of the pattern.
   d) Write an expression for the nth term of the pattern.
   e) Use the expression in part d to find the 75th term in the pattern.

5. Use this information:
   Cost to rent bowling alley: $75
   Bowling shoe rental: $3.00 per pair
   a) Write a problem that can be solved using an equation.
   b) Write the equation, then solve the problem.

6. Solve each equation. Verify the solution.
   a) $4x + 3 = 23$
   b) $7x = 56$
   c) $17 = 4x + 1$
   d) $3x - 5 = x + 3$

7. Write, then solve, an equation to answer each question. Verify the solution.
   a) Seven more than four times a number is 63.
       What is the number?
   b) Six less than five times a number is 29.
       What is the number?
   c) Ten subtracted from one-third a number is 2.
       What is the number?
   d) One-third of a number is subtracted from 10. The answer is 1.
       What is the number?

8. The cost to park a car is $5 for the first hour, plus $3 for each additional half hour.
   It cost Tudor $14 to park his car.
   How long was he parked?
   a) Write an equation you can use to solve the problem.
   b) Solve your equation.
   c) Verify the solution.
   d) Is there more than one possible answer? Explain.

9. The nth term of a number pattern is $3n + 5$.
   Find the term number for each term value.
   a) 62
   b) 86
   c) 152

1. A number cube is labelled 1 to 6.
   It is rolled 90 times.
   Predict how many times:
   a) an odd number will show
   b) a number greater than 2
      will show
   c) a 1 will not show

2. Use a copy of the number line on
   page 457. Place the letter of each
   event below on the number line at the
   best estimate of its probability.
   a) A bucket contains 8 orange golf
      balls, 4 green golf balls, 2 white
      golf balls, and 5 blue golf balls.
      One ball is picked at random.
      It is white.
   b) There will be snow in February in
      Timmins.
   c) A card is randomly selected from a
      deck of cards. It is a spade or a club
      or a diamond.
   d) You read a novel in 2 days.
   e) A spinner has 16 congruent sectors
      labelled 1 to 16. The pointer is spun.
      It lands on 8.

3. a) List the events in question 2 for
      which you can calculate their
      probabilities.
   b) Explain why you can calculate each
      probability you listed in part a.
   c) Why can you not calculate the
      probabilities of the other events?

4. There is a multiple-choice test with
   5 questions. Each question has 3 choices.
   A student guesses the answer to each
   question.

   a) Describe a simulation you could do
      to estimate how many questions
      the student may get correct.
   b) Conduct the simulation.
      Record your results.
   c) Compare your results with those of
      another student. If the results are
      different, explain why.

5. The student council runs a coin toss
   game during School Spirit week.
   All the profits go to charity.
   Each player pays 50¢ to play.
   A player tosses two coins and wins a
   prize if the coins match.
   Each prize is $2.00.
   Can the student council expect to make
   a profit? Justify your answer.

6. The 4 queens are removed from a deck
   of cards. A number cube labelled 1 to 6
   is rolled and one of the queens is chosen
   at random.
   a) Use a tree diagram to list the possible
      outcomes.
   b) What is the probability of getting the
      queen of spades?
   c) What is the probability of rolling a 6
      and getting the queen of spades?
   d) What is the probability of rolling
      a 2 or a 3 and getting the queen
      of spades?

7. The coach of the basketball team said
   that the odds in favour of her team
   winning were 3 to 4. A team member
   said that this meant the probability that
   the team would lose is 25%. Was the
   team member correct? Explain.

1. Your sock drawer has 12 black socks, 18 blue socks, and 10 white socks. You reach into the drawer in the dark. How many socks do you need to pull out to be sure you have a matching pair?

2. Use eight 8s to equal 0, using only one mathematical operation.

3. Copy this diagram.
How can you remove 8 line segments to leave 3 squares?

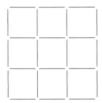

4. Use four 1s, four 2s, four 3s, and four 4s. Copy the grid below.
Arrange the numbers in the squares so the sum of each row, column, and main diagonal is 10.

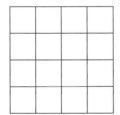

5. Find a 2-digit number that is 3 times the sum of its digits.

6. Use six 7s and multiplication. Write an expression whose product is greater than 30 and less than 300.

7. Some months, such as March, have 31 days. Other months, such as April, have 30 days. How many months have 28 days?

8. Copy this figure on grid paper. How many squares are there in the figure?

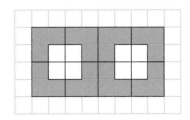

9. Why is it impossible to press a leaf between pages 121 and 122 of most atlases?

10. A bottle of men's aftershave costs $45. The aftershave costs $40 more than the empty bottle. How much does the bottle cost?

11. How many different ways can you arrange a red, a green, an orange, and a black marble in a row?

12. Divide 40 by $\frac{1}{2}$. Add 10 to the quotient. What is the answer?

13. The sum of the ages of Navid and her daughter Daria is 55. Navid's age is her daughter's age reversed. How old is Navid? How old is Daria?

14. Use any operations and brackets with the numbers 2, 5, 8, and 9. Use each number only once. Write an expression that simplifies to 19.

15. A number is perfect if the sum of its factors (excluding the number itself) is equal to the number. The first perfect number is 6 because its factors, 1, 2, and 3 add to 6. What is the next perfect number?

16. Fasil has three children. Their mean age is 11. Their median age is 10. The oldest child is 15. How old is the youngest child?

17. Bag A contains 3 red marbles and 2 yellow marbles. Bag B contains 2 red marbles and 1 yellow marble. To win a prize, you close your eyes and pick a red marble out of a bag. Which bag gives you the better chance of winning the prize? Explain.

18. How many different whole numbers can you make with the digits 3, 6, and 8? The whole number may have 1, 2, or 3 digits, but a digit may not be repeated within any one number.

19. Use the numbers 0 to 9. Arrange the numbers into 3 groups so that the sums of the numbers in each group are equal.

20. I looked out of my apartment window to the park below. I saw people and dogs. I counted 22 heads and 68 legs. How many dogs were in the park?

21. There are six people in a room. Each of them shakes hands with all the others. How many handshakes will there be?

22. Two girls had different numbers of baseball cards. Shauna said, "If you give me 5 cards, I'll have as many as you." Marissa said, "If you give me 5 cards, I'll have twice as many as you." How many cards did each girl have?

23. Here are 6 vertical line segments. Copy the segments. Draw five more line segments to make nine.

24. Find 4 consecutive odd numbers with a sum of 120.

25. Insert the digits 1, 2, 3, 4, 5, 6, 7, and 8 in the eight boxes, one digit to a box, so that there are no consecutive numbers next to each other, horizontally, vertically, or diagonally.

26. Find the next term in each number pattern. Write each pattern rule.
a) 92, 74, 46, 22, 18, …
b) 77, 49, 36, 18, …
c) 6, 9, 18, 21, 42, 45, …
Which patterns do not go on forever? Why not?

27. A taxi driver picks up a passenger at a hotel in downtown Toronto. The passenger wants to go to Pearson Airport. Due to traffic, the average speed was low and the trip took 80 min. At the airport, the taxi driver picks up another passenger who wants to go back to the same hotel downtown. The driver takes the same route, and drives at the same average speed. This time, the trip took 1 h 20 min. Why?

Illustrated Glossary

acute angle: an angle measuring less than 90°

acute triangle: a triangle with three acute angles

algebraic expression: a mathematical expression containing a variable: for example, $6x - 4$ is an algebraic expression

alternate angles: angles that are between two lines and are on opposite sides of the transversal that cuts the two lines

a and c are alternate angles.
d and b are alternate angles.

angle: the figure formed by two rays from the same endpoint

angle bisector: the line that divides an angle into two equal angles

approximate: a number close to the exact value of an expression; the symbol $\doteq$ means "is approximately equal to"

area: the number of square units needed to cover a region

array: an arrangement in rows and columns

assumption: something that is accepted as true, but has not been proved

average: a single number that represents a set of numbers; see *mean*, *median*, and *mode*

bar graph: a graph that uses horizontal or vertical bars to display data that can be counted (see page 195)

bar notation: the use of a horizontal bar over decimal digits to indicate that they repeat; for example, $1.\overline{34}$ means 1.343 434 …

base: the side of a polygon or the face of a solid from which the height is measured; the factor repeated in a power

bias: an emphasis on characteristics that are not typical of the entire population

bisector: a line that divides a line segment or an angle into two equal parts

capacity: the amount a container can hold

census: a method of data collection where all the people in a population are surveyed

circle graph: a graph in which a circle used to represent one whole is divided into sectors that represent parts of the whole

circumcentre: the point where the perpendicular bisectors of the sides of a triangle intersect; see *circumcircle*

circumcircle: a circle drawn through all vertices of a triangle and with its centre at the circumcentre of the triangle

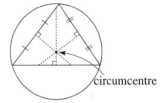
circumcentre

circumference: the distance around a circle, also known as the perimeter of the circle

commission: the percent paid to a salesperson based on the amount sold

common denominator: a number that is a multiple of each of the given denominators; for example, 12 is a common denominator for the fractions $\frac{1}{3}, \frac{5}{4}, \frac{7}{12}$

common factor: a number that is a factor of each of the given numbers; for example, 3 is a common factor of 15, 9, and 21

complementary angles: two angles whose sum is 90°

composite number: a number with three or more factors; for example, 8 is a composite number because its factors are 1, 2, 4, and 8

concave polygon: has at least one angle greater than 180°

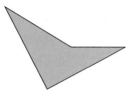

cone: a solid formed by a region and all line segments joining points on the boundary of the region to a point not in the region

congruent: figures that have the same size and shape, but not necessarily the same orientation

consecutive numbers: integers that come one after the other without any integers missing; for example, 34, 35, 36 are consecutive numbers, so are −2, −1, 0, and 1

convex polygon: has all angles less than 180°

coordinate axes: the horizontal and vertical axes on a grid

coordinate grid: a vertical number line and a horizontal number line that intersect at right angles at 0

coordinates: the numbers in an ordered pair that locate a point on the grid; see *ordered pair*

corresponding angles: angles that are on the same side of a transversal that cuts two lines, and on the same side of each line
Angles *f* and *b*,
a and *g*,
e and *c*,
d and *h*
are corresponding angles.

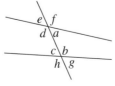

cube: a solid with six congruent square faces

cube number: a power with exponent 3; for example, 8 is a cube number because $8 = 2^3$

cubic units: units that measure volume

cylinder: a solid with two parallel, congruent, circular bases (see page 253)

data: facts or information

database: an organized collection of facts or information, often stored on a computer

denominator: the term below the line in a fraction

diagonal: a line segment that joins two vertices of a figure, but is not a side

diameter: the distance across a circle, measured through its centre

digit: any of the symbols used to write numerals; for example, in the base-ten system the digits are 0, 1, 2, 3, 4, 5, 6, 7, 8, and 9

dimensions: measurements, such as length, width, and height

discount: the amount by which a price is reduced

distributive property: a product can be written as a sum of two products; for example, $8(2 + 4) = 8 \times 2 + 8 \times 4$

double-bar graph: a bar graph that shows two sets of data

equation: a mathematical statement that two expressions are equal

equilateral triangle: a triangle with three equal sides

equivalent: having the same value; for example, $\frac{2}{3}$ and $\frac{6}{9}$ are equivalent fractions; 2:3 and 6:9 are equivalent ratios

estimate: a reasoned guess that is close to the actual value, without calculating it exactly

evaluate: to substitute a value for each variable in an expression, then simplify; to find the answer

even number: a number that has 2 as a factor; for example, 2, 4, 6

event: any set of outcomes of an experiment

expanded form of a number: a number written as the sum of its parts; for example,
$3297 = 3 \times 10^3 + 2 \times 10^2 + 9 \times 10 + 7$

experimental probability: the probability of an event calculated from experimental results; another name for the *relative frequency* of an outcome

exponent: a number, shown in a smaller size and raised, that tells how many times the base of the power is used as a factor; for example, 2 is the exponent in 6^2

expression: a mathematical phrase made up of numbers and/or variables connected by operations

factor: to factor means to write as a product; for example, $20 = 2 \times 2 \times 5$

formula: a rule that is expressed as an equation

fraction: an indicated quotient of two quantities

frequency: the number of times a particular number occurs in a set of data

greatest common factor (GCF): the greatest number that divides into each number in a set; for example, 5 is the greatest common factor of 10 and 15

hexagon: a six-sided polygon

histogram: a vertical bar graph where the height of each bar is proportional to the frequency, and the data are continuous (see page 218)

horizontal axis: the horizontal number line on a coordinate grid

hypotenuse: the side opposite the right angle in a right triangle

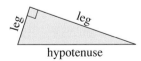

image: the figure that results from a transformation

improper fraction: a fraction with the numerator greater than the denominator; for example, $\frac{6}{5}$ and $\frac{5}{3}$

inequality: a statement that one quantity is greater than (>), greater than or equal to ($\geq$), less than (<), less than or equal to ($\leq$), or not equal to ($\neq$) another quantity

inference: a conclusion drawn from data

integers: the set of numbers
$\ldots, -3, -2, -1, 0, +1, +2, +3, \ldots$

interest: money paid for the use of money

interior angles: angles that are between two lines and are on the same side of the transversal that cuts the two lines

a and b are interior angles.
c and d are interior angles.

intersecting lines: lines that meet or cross; lines that have one point in common

inverse operation: an operation that reverses the result of another operation; for example, subtraction is the inverse of addition, and division is the inverse of multiplication

irrational number: a number that cannot be represented as a terminating or repeating decimal; for example, π

isometric diagram: a representation of an object as it would appear in three dimensions

isosceles acute triangle: a triangle with two equal sides, and all angles less than 90°

isosceles obtuse triangle: a triangle with two equal sides, and one angle greater than 90°

isosceles right triangle: a triangle with two equal sides, and a 90° angle

isosceles triangle: a triangle with two equal sides

kite: a quadrilateral with two pairs of equal adjacent sides

f a right triangle that form the
...potenuse

...that displays data that change
...g points joined by line segments
...05)

...e segment: the part of a line between two points on the line

line symmetry: a figure has line symmetry when it can be divided into 2 congruent parts, so that one part coincides with the other part when the figure is folded at the line of symmetry; for example, line *l* is the line of symmetry for figure ABCD

lowest common multiple (LCM): the lowest multiple that is the same for two numbers; for example, the lowest common multiple of 12 and 21 is 84

magic square: an array of numbers in which the sum of the numbers in any row, column, or main diagonal is always the same

mean: the sum of a set of numbers divided by the number of numbers in the set

median: the middle number when data are arranged in numerical order; if there is an even number of data, the median is the mean of the two middle numbers

midpoint: the point that divides a line segment into two equal parts

mixed number: a number consisting of a whole number and a fraction; for example, $1\frac{1}{18}$

mode: the number that occurs most often in a set of numbers

multiple: the product of a given number and a natural number; for example, some multiples of 8 are 8, 16, 24, …

natural numbers: the set of numbers 1, 2, 3, 4, 5, …

negative number: a number less than 0

net: a pattern that can be folded to make an object

numerator: the term above the line in a fraction

obtuse angle: an angle greater than 90° and less than 180°

obtuse triangle: a triangle with one angle greater than 90°

octagon: an eight-sided polygon

octahedron: a polyhedron with 8 faces

odd number: a number that does not have 2 as a factor; for example, 1, 3, 7

odds: the likelihood of the occurrence of one event rather than the occurrence of another event

operation: a mathematical process or action such as addition, subtraction, multiplication, or division

opposite angles: the equal angles that are formed by two intersecting lines

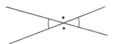

opposite integers: two integers with a sum of 0; for example, +3 and −3

ordered pair: two numbers in order, for example, (2, 4); on a coordinate grid, the first number is the horizontal coordinate of a point, and the second number is the vertical coordinate of the point

order of operations: the rules that are followed when simplifying or evaluating an expression

origin: the point where the *x*-axis and the *y*-axis intersect

outcome: a possible result of an experiment or a possible answer to a survey question

outlier: a number in a set of data that is significantly different from the other numbers

parallel lines: lines on the same flat surface that do not intersect

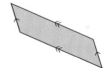

parallelogram: a quadrilateral with both pairs of opposite sides parallel

pentagon: a five-sided polygon

percent: the number of parts per 100; the numerator of a fraction with denominator 100

perfect square: a number that is the square of a whole number; for example, 16 is a perfect square because $16 = 4^2$

perimeter: the distance around a closed figure

perpendicular bisector: the line that is perpendicular to a line segment and divides the line segment into two equal parts

perpendicular lines: intersect at right angles

pictorial diagram: shows the shape of an object in two dimensions, but gives the impression of three dimensions

polygon: a closed figure that consists of line segments; for example, triangles and quadrilaterals

polyhedron (*plural,* **polyhedra**)**:** a solid with faces that are polygons

population: the set of all things or people being considered

positive number: a number greater than 0

power: an expression of a product of equal factors; for example, $4 \times 4 \times 4$ can be expressed as 4^3; 4 is the base and 3 is the exponent

prediction: a statement of what you think will happen

primary data: data collected by oneself; first-hand

prime factorization: a composite number written as a product of its prime factors; for example, $2 \times 2 \times 3 \times 3$ is the prime factorization of 36

prime number: a whole number with exactly two factors, itself and 1; for example, 2, 3, 5, 7, 11, 29

principal: the money borrowed, invested, or deposited

prism: a solid that has two congruent and parallel faces (the *bases*), and other faces that are rectangles; a prism is named by the shape of its bases

probability: the likelihood of a particular outcome

product: the result when two or more numbers are multiplied

proper fraction: a fraction with the numerator less than the denominator; for example, $\frac{5}{6}$

proportion: a statement that two ratios are equal; for example, $r{:}24 = 3{:}4$

pyramid: a solid that has one face that is a polygon (the *base*), and other faces that are triangles with a common vertex

Pythagorean Theorem: the rule that states that, for any right triangle, the area of the square on the hypotenuse is equal to the sum of the areas of the squares on the legs

Pythagorean triple: the three whole-number side lengths of a right triangle

quadrant: one of four regions into which coordinate axes divide a plane (see page 394)

quadrilateral: a four-sided polygon

quotient: the result when one number is divided by another

radius (*plural,* **radii**)**:** the distance from the centre of a circle to any point on the circle (see page 240)

range: the difference between the greatest and least numbers in a set of data

rate: a comparison of two quantities measured in different units

ratio: a comparison of two or more quantities with the same unit

reciprocals: two numbers whose product is 1; for example, $\frac{2}{3}$ and $\frac{3}{2}$

rectangle: a quadrilateral that has four right angles

rectangular prism: a prism that has rectangular faces

rectangular pyramid: a pyramid with a rectangular base

reflection: a transformation that is illustrated by a figure and its image in a mirror line

mirror line

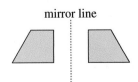

reflex angle: an angle between 180° and 360°

regular hexagon: a polygon that has six equal sides and six equal angles

regular octagon: a polygon that has eight equal sides and eight equal angles

regular octahedron: a regular polyhedron with 8 congruent faces; each face is an equilateral triangle

regular polygon: a polygon that has all sides equal and all angles equal

regular polyhedron: a solid with faces that are congruent regular polygons, with the same number of edges meeting at each vertex

relative frequency: the number of times a particular outcome occurred, written as a fraction of the total number of times the experiment was conducted

reliable survey: the results of a survey that can be duplicated in another survey

repeating decimal: a decimal with a repeating pattern in the digits that follow the decimal point; it is written with a bar above the repeating digits; for example, $\frac{1}{11} = 0.\overline{09}$

rhombus: a parallelogram with four equal sides

right angle: a 90° angle

right triangle: a triangle that has one right angle

rotation: a transformation in which a figure is turned about a fixed point

rotational symmetry: a figure has rotational symmetry when it coincides with itself in less than one full turn; for example, a square has rotational symmetry of order 4 about its centre O

sample/sampling: a representative portion of a population

scale: the ratio of the distance between two points on a map, model, or diagram to the distance between the actual locations; the numbers on the axes of a graph

scale drawing: a drawing in which the lengths are a reduction or an enlargement of actual lengths

scalene triangle: a triangle with all sides different

scatter plot: a graph that attempts to show a relationship between two variables by means of points plotted on a coordinate grid

scientific notation: a number written as the product of a number greater than or equal to 1, and less than 10, and a power of 10; for example, 4500 is written as 4.5×10^3

secondary data: data not collected by oneself, but by others; data found from the library, or the Internet; second-hand

similar figures: figures with the same shape, but not necessarily the same size

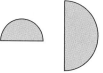

simple interest: interest that is calculated at the end of the period for which the money is borrowed or invested; the formula for calculating simple interest is: $I = Prt$, where P is the principal, r the annual interest rate as a decimal, and t the time in years

simplest form: a ratio with terms that have no common factors, other than 1; a fraction with numerator and denominator that have no common factors, other than 1

simulation: an experiment that is used to model a real-life situation to estimate the probability of an event

skeleton: an object shown with edges and vertices only

spreadsheet: a computer-generated arrangement of data in rows and columns, where a change in one value results in appropriate calculated changes in the other values

square: a rectangle with four equal sides

square number: the product of a number multiplied by itself; for example, 25 is the square of 5

square root: a number which, when multiplied by itself, results in a given number; for example, 5 is a square root of 25

statistics: the branch of mathematics that deals with the collection, organization, and interpretation of data

stem-and-leaf plot: an arrangement of data; for 3-digit numbers, the hundreds and tens digits are the "stems" and the ones digits are the "leaves" (see page 217)

straight angle: an angle measuring 180°

supplementary angles: two angles whose sum is 180°

surface area: the total area of the surface of an object

symmetrical: possessing symmetry; see *line symmetry* and *rotational symmetry*

terminating decimal: a decimal with a certain number of digits after the decimal point; for example, $\frac{1}{8}$ = 0.125

tessellation: a tiling pattern

tetrahedron: a solid with four triangular faces; a triangular pyramid (see page 464)

theoretical probability: the number of favourable outcomes written as a fraction of the total number of possible outcomes, when the outcomes are equally likely

three-dimensional: having length, width, and depth or height

transformation: a translation, rotation, or reflection

translation: a transformation that moves a point or a figure in a straight line to another position on the same flat surface

transversal: a line crossing two or more lines

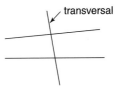

trapezoid: a quadrilateral that has at least one pair of parallel sides

tree diagram: a diagram that resembles the roots or branches of a tree, used to count outcomes (see page 462)

triangle: a three-sided polygon

triangular prism: a prism that has two congruent triangular faces and three rectangular faces (see page 112)

two-dimensional: having length and width, but no thickness, height, or depth

unit fraction: a fraction that has a numerator of 1

unit price: the price of one item, or the price of a particular mass or volume of an item

unit rate: the quantity associated with a single unit of another quantity; for example, 6 m in 1 s is a unit rate; it is written as 6 m/s

valid survey: the results of a survey that represent the population

variable: a letter or symbol representing a quantity that can vary

vertex (*plural,* vertices): the corner of a figure or a solid

vertical axis: the vertical number line on a coordinate grid

volume: the amount of space occupied by an object

whole numbers: the set of numbers 0, 1, 2, 3, …

***x*-axis:** the horizontal number line on a coordinate grid

***y*-axis:** the vertical number line on a coordinate grid

zero pair: two opposite numbers whose sum is 0

Index

Acknowledgments

The publisher wishes to thank the following sources for photographs, illustrations, and other materials used in this book. Care has been taken to determine and locate ownership of copyright material in this text. We will gladly receive information enabling us to rectify any errors or omissions in credits.

Photography
Cover: John Guistina/Imagestate/firstlight.ca
p. 3 Ray Boudreau; p. 4 (top) Royalty-Free/CORBIS; p. 4 (bottom) Emmanuel Faure/Taxi/Getty Images; p. 5 (top) Bernd Fuchs/firstlight.ca; p. 5 (bottom) Perry Mastrovito/firstlight.ca; p. 9 Ray Boudreau; p. 12 Omni-Photo Communications, Inc.; p. 13 Ray Boudreau; p. 19 Creatas/firstlight.ca; p. 20 20th Century Fox/Fotos International/Getty Images; p. 23 Canadian Press/Tom Hanson; p. 25 Ray Boudreau; p. 27 Canadian Press/Mike Ridewood; p. 29 Ray Boudreau; p. 33 (top right) Image courtesy of Computer History Museum; p. 33 (bottom left) Centrum voor Wiskunde en Informatica (CWI), The Netherlands, thanks to Herman te Riele; p. 34 Ray Boudreau; p. 41 Ray Boudreau; p. 46 Greg Griffith/firstlight.ca; p. 47 Ray Boudreau; p. 48 (top) Photodisc Collection/Getty Images; p. 48 (bottom) Chris Cheadle/firstlight.ca; p. 49 (top) B & C Alexander/firstlight; p. 49 (bottom) Chris Cheadle/firstlight.ca; p. 50 Canadian Press/Adrian Wyld; p. 55 Photodisc Collection/ Getty Images; p. 57 Ray Boudreau; p. 60 (top) Dorling Kindersley; p. 60 (bottom) Nicolas Russell/Photographer's Choice/Getty Images; p. 65 Ray Boudreau; p. 78 Owen Franken/ CORBIS; p. 80 Jack Star/PhotoLink/Getty Images; p. 87 Ray Boudreau; p. 92 (left) Mark Scott/Taxi/Getty Images; p. 92 (right) Digital Vision; p. 93 (left) Nick Daly/Digital Vision; p. 93 (top) Dennis MacDonald/Photo Edit, Inc.; p. 93 (bottom) Richard Lam/Canadian Press; p. 94 David Young-Wolff-Photo Edit, Inc.; p. 95 Phil Schemeister/CORBIS; p. 96 (top and bottom) Ray Boudreau; p. 102 (top and bottom) Ray Boudreau; p. 109 Ray Boudreau; p. 120 Prentice Hall, Inc.; p. 128 Ray Boudreau; p. 129 Canadian Press/Paul Chiasson; p. 131 Geray Sweeney/CORBIS; p. 135 Ray Boudreau; p. 143 Ray Boudreau; p. 154 Ray Boudreau; p. 169 Ray Boudreau; p. 172 Ray Boudreau; p. 179 Ray Boudreau; p. 182 Abaca Press (2004) All Rights Reserved; p. 183 (top) Flying Colours Ltd./Digital Vision; p. 183 (bottom left) Martial Colomb/Photodisc Collection/Getty Images; p. 183 (bottom right) Canadian Press/Don Denton; p. 187 Ray Boudreau; p. 188 Canadian Press/Miles Kennedy; p. 190 Rick Madonik/Toronto Star; p. 195 Royalty-Free/CORBIS; p. 199 Corel Collections, *Food*; p. 224 Ray Boudreau; p. 227 Canadian Press/Corey Larocque; p. 235 Nancy Ney/Digital Vision; p. 236 (top) Photodisc Collection/Getty Images; p. 236 (bottom) Ryan McVay/ Photodisc Collection/Getty Images; p. 237 (top left) Corel Collection, *Food*; p. 237 (top right) Royalty-Free/CORBIS; p. 237 (centre) Digital Vision; p. 237 (bottom) Photodisc Collection/Getty Images; p. 239 (top left) Andrew Ward/Life File/Photodisc Collection/Getty Images; p. 239 (top left centre) Janis Christie/Photodisc Collection/Getty Images; p. 239 (top right centre) Steve Cole/Photodisc Collection/Getty Images; p. 239 (top right) Ryan McVay/Photodisc Collection/Getty Images; p. 239 (bottom) Ray Boudreau; p. 241 LessLIE, Coast Salish artist; p. 242 Ray Boudreau; p. 247 Ray Boudreau; p. 250 Canadian Press/Jacques Boissinot; p. 256 Ray Boudreau; p. 258 Ray Boudreau; p. 260 Katherine Fawssett/The Image Bank/ Getty Images; p. 265 Ray Boudreau; p. 267 (bottom) Photodisc Collection/Getty Images; p. 271 Ray Boudreau; p. 278 Ray Boudreau; p. 284 Ray Boudreau; p. 289 Spike Mafford/ Photodisc Collection/Getty Images; p. 293 Ray Boudreau; p. 295 Ray Boudreau; p. 296 Ray Boudreau; p. 299 Ray Boudreau; p. 302 Skip Nall/Photodisc Collection/Getty Images; p. 306 Guy Grenier/Masterfile Corporation; p. 308 Ray Boudreau; p. 310 NASA/Science Photo Library; p. 316 Ray Boudreau; p. 317 AP Photo/Lawrence Jackson; p. 319 Ray Boudreau; p. 320 (top) Arthur S. Aubry/Photodisc Collection/ Getty Images; p. 320 (centre) bridge Corel Collection Southwestern U.S.; p. 321 (top left) Royalty-Free/CORBIS; p. 321 (top right) Martin Bond/Photo Researchers, Inc.; p. 321 (bottom) Vision/Cordelli/Digital Vision; p. 325 Ray Boudreau; p. 328 Royalty-Free/CORBIS; p. 329 Ray Boudreau; p. 333 Ray Boudreau; p. 338 SEF/Art Resource N.Y.; p. 349 Burke/Triolo/ Brand X/Getty Images; p. 358 (top) Adam Crowley/Photodisc Collection/Getty Images; p. 358 (bottom) Ray Boudreau; p. 359 (top) Nicholas Pitt/Digital Vision; p. 359 (bottom) Yale Babylonian Collection; pp. 362–363 (top) Corel Collections, *Lakes and Rivers*; p. 362 (top) imagesource/firstlight.ca; p. 363 (middle left) Canadian Press/Elise Amendola; p. 363 (middle right) Canadian Press/Jonathan Hayward; p. 363 (bottom) Lawson Wood/CORBIS; p. 369 Ray Boudreau; p. 371 Chris Cheadle/firstlight.ca; p. 375 (top) Dave Reede/firstlight.ca; p. 375 (bottom) Dave Reede/firstlight.ca; p. 393 Ray Boudreau; p. 408 Ray Boudreau; p. 414 Ray Boudreau; p. 416 (top) Rubberball Images; p. 416 (bottom) Photodisc Collection/Getty Images; p. 417 (top) Photodisc Collection/ Getty Images; p. 417 (middle) Royalty-Free/CORBIS; p. 417 (bottom) Creatas/firstlight.ca; p. 428 Ray Boudreau; p. 443 Blend Images Royalty-Free; p. 444 Ray Boudreau; p. 446 Stockbyte Images; p. 450 Chad Baker/Ryan McVay/ Photodisc Collection/Getty Images; p. 451 Ray Boudreau; pp. 452–453 Royalty-Free/ CORBIS; p. 461 Ray Boudreau; p. 464 Canadian Press/Joerg Sarbach; p. 469 Medioimages/Getty Images; p. 471 Ray Boudreau; p. 474 Ray Boudreau; p. 480 J.A. Giordano/CORBIS; p. 481 (top) David Young-Wolff/Photo Edit Inc.; p. 481 (bottom) Patrick Giardino/CORBIS; p. 483 Ray Boudreau

Illustrations
Steve Attoe, Philippe Germain, Stephen MacEachern, Dave Mazierski, Paul McCusker, Allan Moon, NSV Productions/ Neil Stewart, Dusan Petricic, Pronk&Associates, Michel Rabagliati, Carl Wiens

AppleWorks is a trade-mark of Apple Computer, Inc. registered in the U.S. and other countries.

Graphers—Copyright 1996 Sunburst Technology Corporation. All rights reserved.

Fathom Dynamic Statistics, Key Curriculum Press, 1150 65th St., Emeryville, CA 94608, 1-800-995-MATH, www.keypress.com/fathom

The Geometer's Sketchpad, Key Curriculum Press, 1150 65th St., Emeryville, CA 94608, 1-800-995-MATH, www.keypress.com/sketchpad